T0191708

Numerical Computations with GPUs

Volodymyr Kindratenko
Editor

Numerical Computations with GPUs

 Springer

Editor
Volodymyr Kindratenko
National Center for Supercomputing
  Applications
University of Illinois
Urbana, IL, USA

ISBN 978-3-319-37994-4   ISBN 978-3-319-06548-9 (eBook)
DOI 10.1007/978-3-319-06548-9
Springer Cham Heidelberg New York Dordrecht London

Printed on acid-free paper

Springer is part of Springer Science+Business Media (www.springer.com)

Preface

This book is intended to serve as a practical guide for the development and implementation of numerical algorithms on Graphics Processing Units (GPUs). The book assumes that the reader is familiar with the mathematical context and has a good working knowledge of GPU architecture and its programming sufficient to translate specialized mathematical algorithms and pseudo-codes presented in the book into a fully functional CUDA or OpenCL software. In case the reader is not familiar with the GPU programming, the reader is directed to other sources, such as NVIDIA's *CUDA Parallel Computing Platform* website, for low-level programming details, tools, and techniques prior to reading this book.

Book Focus

The main focus of this book is on the efficient implementation of numerical methods on GPUs. The book chapters are written by the leaders in the field working for many years on the development and implementation of computationally intensive numerical algorithms for solving scientific computing and engineering problems.

It is widely understood and accepted that modern scientific discovery in all of the disciplines requires extensive computations. It is also the case that modern engineering heavily utilizes advanced computational models and tools. At the heart of many such computations are libraries of mathematical codes for solving systems of linear equations, computing solutions of differential equations, finding integrals and function values, transforming time series, etc. These libraries have been developed over several decades and have been constantly updated to track the ever changing architecture and capabilities of computing hardware. With the introduction of GPUs, many of the existing numerical libraries are currently undergoing another phase of transformation in order to continue serving the computational science and engineering community by providing the required level of performance. Simultaneously, new numerical methods are under development to take advantage of the revolutionary architecture of GPUs. In either case, the developers of such numerical codes face the challenge of extracting parallelism present in numerical methods and expressing it in the form that can be successfully utilized by the massively parallel GPU architecture. This frequently requires reformulating the original algorithmic structure of the code, tuning its performance, and developing and validating entirely new algorithms that can take advantage of the new hardware. It is my hope that this book will serve as a reference implementation and will provide the guidance for the developers of such codes by presenting a collective experience from many recent successful efforts.

Audience and Organization

This book targets practitioners working on the implementation of numerical codes on GPUs, researchers and software developers attempting to extend existing numerical libraries to GPUs, and readers interested in all aspects of GPU programming. It especially targets community of computational scientists from disciplines known to make use of linear algebra, differential equations, Monte Carlo methods, and Fourier transform.

The book is organized in four parts, each covering a particular set of numerical methods. First part is dedicated to the solution of linear algebra problems, ranging from the matrix–matrix multiplication, to the solution of systems of linear equations, to the computation of eigenvalues. Several chapters in this part address the problem of computing on a very large number of small matrixes. The final chapter also addresses the sparse matrix–vector product problem.

Second part is dedicated to the solution of differential equations and problems based on the space discretization of differential equations. Methods such as finite elements, finite difference, and successive over-relaxation with the applications to problem domains such as flow and wave propagation and solution of Maxwell's equations are presented. One chapter also addresses the challenge of integrating a large number of independent ordinary differential equations.

Third part is dedicated to the use of Monte Carlo methods for numerical integration. Monte Carlo techniques are well suited for GPU implementation and their use is widening. The part also includes chapters about random number generation on GPUs as a necessary first step in Monte Carlo methods.

The final part consists of two chapters dedicated to the efficient implementation of Fourier transform and one chapter discussing N-body simulations.

Acknowledgments

This book consists of contributed chapters provided by the experts in various fields involved with numerical computations on GPUs. I would like to thank all of the contributing authors whose work appears in this edition. I am also thankful to the Directors of the National Center for Supercomputing Applications at the University of Illinois at Urbana-Champaign for the support and encouragement.

Urbana, IL, USA   Volodymyr Kindratenko

Contents

Part I
Linear Algebra

Chapter 1
Accelerating Numerical Dense Linear Algebra Calculations with GPUs

Jack Dongarra, Mark Gates, Azzam Haidar, Jakub Kurzak, Piotr Luszczek, Stanimire Tomov, and Ichitaro Yamazaki

1.1 Introduction

Enabling large scale use of GPU-based architectures for high performance computational science depends on the successful development of fundamental numerical libraries for GPUs. Of particular interest are libraries in the area of dense linear algebra (DLA), as many science and engineering applications depend on them; these applications will not perform well unless the linear algebra libraries perform well.

Drivers for DLA developments have been significant hardware changes. In particular, the development of LAPACK [1]—the contemporary library for DLA computations—was motivated by the hardware changes in the late 1980s when its predecessors (EISPACK and LINPACK) needed to be redesigned to run efficiently on shared-memory vector and parallel processors with multilayered memory hierarchies. Memory hierarchies enable the caching of data for its reuse in computations, while reducing its movement. To account for this, the main DLA algorithms were reorganized to use block matrix operations, such as matrix multiplication, in their innermost loops. These block operations can be optimized for various architectures to account for memory hierarchy, and so provide a way to achieve high-efficiency on diverse architectures.

J. Dongarra
University of Tennessee Knoxville, Knoxville, TN 37996-3450, USA

Oak Ridge National Laboratory, Oak Ridge, TN 37830, USA

University of Manchester, Manchester M13 9PL, UK
e-mail: dongarra@eecs.utk.edu

M. Gates • A. Haidar • J. Kurzak • P. Luszczek • S. Tomov (✉) • I. Yamazaki
University of Tennessee Knoxville, Knoxville, TN 37996-3450, USA
e-mail: mgates3@eecs.utk.edu; haidar@eecs.utk.edu; kurzak@eecs.utk.edu;
luszczek@eecs.utk.edu; tomov@eecs.utk.edu; iyamazak@eecs.utk.edu

V. Kindratenko (ed.), *Numerical Computations with GPUs*,
DOI 10.1007/978-3-319-06548-9__1, © Springer International Publishing Switzerland 2014

Challenges for DLA on GPUs stem from present-day hardware changes that require yet another major redesign of DLA algorithms and software in order to be efficient on modern architectures. This is provided through the MAGMA library [12], a redesign for GPUs of the popular LAPACK.

There are two main hardware trends that challenge and motivate the development of new algorithms and programming models, namely:

The explosion of parallelism where a single GPU can have thousands of cores (e.g., there are 2,880 CUDA cores in a K40), and algorithms must account for this level of parallelism in order to use the GPUs efficiently;

The growing gap of compute vs. data-movement capabilities that has been increasing exponentially over the years. To use modern architectures efficiently new algorithms must be designed to reduce their data movements. Current discrepancies between the compute- vs. memory-bound computations can be orders of magnitude, e.g., a K40 achieves about 1,240 Gflop/s on dgemm but only about 46 Gflop/s on dgemv.

This chapter presents the current best design and implementation practices that tackle the above mentioned challenges in the area of DLA. Examples are given with fundamental algorithms—from the matrix–matrix multiplication kernel written in CUDA (in Sect. 1.2) to the higher level algorithms for solving linear systems (Sects. 1.3 and 1.4), to eigenvalue and SVD problems (Sect. 1.5).

The complete implementations and more are available through the MAGMA library.[1] Similar to LAPACK, MAGMA is an open source library and incorporates the newest algorithmic developments from the linear algebra community.

1.2 BLAS

The *Basic Linear Algebra Subroutines* (BLAS) are the main building blocks for dense matrix software packages. The matrix multiplication routine is the most common and most performance-critical BLAS routine. This section presents the process of building a fast matrix multiplication GPU kernel in double precision, real arithmetic (dgemm), using the process of autotuning. The target is the Nvidia K40c card.

In the canonical form, matrix multiplication is represented by three nested loops (Fig. 1.1). The primary tool in optimizing matrix multiplication is the technique of loop tiling. Tiling replaces one loop with two loops: the inner loop incrementing the loop counter by one, and the outer loop incrementing the loop counter by the tiling factor. In the case of matrix multiplication, tiling replaces the three loops of Fig. 1.1 with the six loops of Fig. 1.2. Tiling of matrix multiplication exploits the *surface to volume effect*, i.e., execution of $O(n^3)$ floating-point operations over $O(n^2)$ data.

---

[1]http://icl.cs.utk.edu/magma/.

Fig. 1.1 Canonical form of matrix multiplication

```
1    for (m = 0; m< M; m++)
2      for (n = 0; n < N; n++)
3        for (k = 0; k< K; k++)
4          C[n][m] += A[k][m]*B[n][k];
```

```
1    for (m_ = 0; m_ < M; m_+=tileM)
2    for (n_ = 0; n_ < N; n_+=tileN)
3    for (k_ = 0; k_ < K; k_+=tileK)
4      for (m = 0; m< tileM; m++)
5       for (n = 0; n< tileN; n++)
6        for (k = 0; k< tileK; k++)
7         C[n_+n][m_+n] +=
8         A[k_+k][m_+m]*
9         B[n_+n][k_+k];
```

Fig. 1.2 Matrix multiplication with loop tiling

```
1    for (m_ = 0; m_ < M; m_+=tileM)
2    for (n_ = 0; n_ < N; n_+=tileN)
3    for (k_ = 0; k_ < K; k_+=tileK)
4    {
5      instruction
6      instruction
7      instruction
8      ...
9    }
```

Fig. 1.3 Matrix multiplication with complete unrolling of tile operations

Next, the technique of loop unrolling is applied, which replaces the three innermost loops with a single block of straight-line code (a single *basic block*), as shown in Fig. 1.3. The purpose of unrolling is twofold: to reduce the penalty of looping (the overhead of incrementing loop counters, advancing data pointers and branching), and to increase instruction-level parallelism by creating sequences of independent instructions, which can fill out the processor's pipeline.

This optimization sequence is universal for almost any computer architecture, including "standard" superscalar processors with cache memories, as well as GPU accelerators and other less conventional architectures. Tiling, also referred to as blocking, is often applied at multiple levels, e.g., L2 cache, L1 cache, registers file, etc.

In the case of a GPU, the C matrix is overlaid with a 2D grid of thread blocks, each one responsible for computing a single tile of C. Since the code of a GPU kernel spells out the operation of a single thread block, the two outer loops disappear, and only one loop remains—the loop advancing along the k dimension, tile by tile.

Figure 1.4 shows the GPU implementation of matrix multiplication at the device level. Each thread block computes a tile of C (dark gray) by passing through a stripe of A and a stripe of B (light gray). The code iterates over A and B in chunks of K_{blk} (dark gray). The thread block follows the cycle of:

- making texture reads of the small, dark gray, stripes of A and B and storing them in shared memory,
- synchronizing threads with the __syncthreads() call,
- loading A and B from shared memory to registers and computing the product,
- synchronizing threads with the __syncthreads() call.

After the light gray stripes of A and B are completely swept, the tile of C is read, updated and stored back to device memory. Figure 1.5 shows closer what happens in the inner loop. The light gray area shows the shape of the thread block. The dark gray regions show how a single thread iterates over the tile.

Fig. 1.4 gemm at the device level

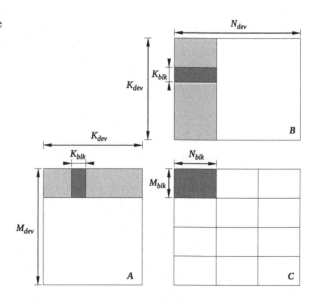

Figure 1.6 shows the complete kernel implementation in CUDA. Tiling is defined by BLK_M, BLK_N, and BLK_K. DIM_X and DIM_Y define how the thread block covers the tile of C, DIM_XA and DIM_YA define how the thread block covers a stripe of A, and DIM_XB and DIM_YB define how the thread block covers a stripe of B.

In lines 24–28 the values of C are set to zero. In lines 32–38 a stripe of A is read (texture reads) and stored in shared memory. In lines 40–46 a stripe of B is read (texture reads) and stored in shared memory. The __syncthreads() call in line

Fig. 1.5 gemm at the block level

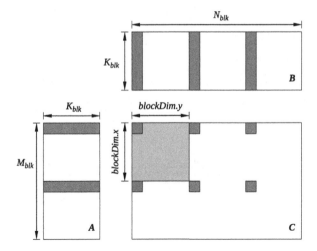

48 ensures that reading of A and B, and storing in shared memory, is finished before operation continues. In lines 50–56 the product is computed, using the values from shared memory. The __syncthreads() call in line 58 ensures that computing the product is finished and the shared memory can be overwritten with new stripes of A and B. In lines 60 and 61 the pointers are advanced to the location of new stripes. When the main loop completes, C is read from device memory, modified with the accumulated product, and written back, in lines 64–77. The use of texture reads with clamping eliminates the need for *cleanup code* to handle matrix sizes not exactly divisible by the tiling factors.

With the parametrized code in place, what remains is the actual autotuning part, i.e., finding good values for the nine tuning parameters. Here the process used in the BEAST project (*Bench-testing Environment for Automated Software Tuning*) is described. It relies on three components: (1) defining the search space, (2) pruning the search space by applying filtering constraints, (3) benchmarking the remaining configurations and selecting the best performer. The important point in the BEAST project is to not introduce artificial, arbitrary limitations to the search process.

The loops of Fig. 1.7 define the search space for the autotuning of the matrix multiplication of Fig. 1.6. The two outer loops sweep through all possible 2D shapes of the thread block, up to the device limit in each dimension. The three inner loops sweep through all possible tiling sizes, up to arbitrarily high values, represented by the INF symbol. In practice, the actual values to substitute the INF symbols can be found by choosing a small starting point, e.g., (64, 64, 8), and moving up until further increase has no effect on the number of kernels that pass the selection.

The list of pruning constraints consists of nine simple checks that eliminate kernels deemed inadequate for one of several reasons:

- The kernel would not compile due to exceeding a hardware limit.
- The kernel would compile but fail to launch due to exceeding a hardware limit.

```
1    extern "C" __global__
2    void beast_gemm_kernel(
3       int M, int N, int K,
4       double alpha, double *A, int lda,
5                     double *B, int ldb,
6       double beta, double *C, int ldc )
7    {
8       int blx = blockIdx.x;         // block's m position
9       int bly = blockIdx.y;         // block's n position
10      int idx = threadIdx.x;        // thread's m position in C
11      int idy = threadIdx.y;        // thread's n position in C
12      int idt = DIM_X*idy+idx;      // thread's number
13
14      int idxA = idt % DIM_XA;      // thread's m position for loading A
15      int idyA = idt / DIM_XA;      // thread's n position for loading A
16      int idxB = idt % DIM_XB;      // thread's m position for loding B
17      int idyB = idt / DIM_XB;      // thread's n position for loading B
18
19      __shared__ double sA[BLK_K][BLK_M+1];    // shared memory buffer for A
20      __shared__ double sB[BLK_N][BLK_K+1];    // shared memory buffer for B
21      double rC[BLK_N/DIM_Y][BLK_M/DIM_X];     // registers for C
22
23      int coord_A = blx*BLK_M   + idyA*lda+idxA;    // A stripe's initial location
24      int coord_B = bly*BLK_N*ldb + idyB*ldb+idxB;  // B stripe's initial location
25      int m, n, k, kk;                              // loop counters
26
27      #pragma unroll
28      for (n = 0; n < BLK_N/DIM_Y; n++)
29        #pragma unroll
30        for (m = 0; m < BLK_M/DIM_X; m++)
31          rC[n][m] = 0.0;
32
33      for (kk = 0; kk < K; kk += BLK_K)
34      {
35        #pragma unroll
36        for (n = 0; n < BLK_K; n += DIM_YA)
37          #pragma unroll
38          for (m = 0; m < BLK_M; m += DIM_XA) {
39            int2 v = tex1Dfetch(tex_ref_A, coord_A + n*lda+m);
40            sA[n+idyA][m+idxA] = __hiloint2double(v.y, v.x);
41          }
42
43        #pragma unroll
44        for (n = 0; n < BLK_N; n += DIM_YB)
45          #pragma unroll
46          for (m = 0; m < BLK_K; m += DIM_XB) {
47            int2 v = tex1Dfetch(tex_ref_B, coord_B + n*ldb+m);
48            sB[n+idyB][m+idxB] = __hiloint2double(v.y, v.x);
49          }
50
51        __syncthreads();
52
53        #pragma unroll
54        for (k = 0; k < BLK_K; k++)
55          #pragma unroll
56          for (n = 0; n < BLK_N/DIM_Y; n++)
57            #pragma unroll
58            for (m = 0; m < BLK_M/DIM_X; m++)
59              rC[n][m] += sA[k][m*DIM_X+idx] * sB[n*DIM_Y+idy][k];
60
61        __syncthreads();
62
63        coord_A += BLK_K*lda;
64        coord_B += BLK_K;
65      }
66
67      #pragma unroll
68      for (n = 0; n < BLK_N/DIM_Y; n++) {
69        int coord_dCn = bly*BLK_N + n*DIM_Y+idy;
70        #pragma unroll
71        for (m = 0; m < BLK_M/DIM_X; m++) {
72          int coord_dCm = blx*BLK_M + m*DIM_X+idx;
73          if (coord_dCm < M && coord_dCn < N) {
74            int offsC = coord_dCn*ldc + coord_dCm;
75            double &regC = rC[n][m];
76            double &memC = C[offsC];
77            memC = alpha*regC + beta*memC;
78          }
79        }
80      }
81    }
```

Fig. 1.6 Complete dgemm (C = alpha A B + beta C) implementation in CUDA

```
 1    // Sweep thread block dimensions.
 2    for (dim_m = 1; dim_m <=MAX_THREADS_DIM_X; dim_m++)
 3     for (dim_n = 1; dim_n <=MAX_THREADS_DIM_Y; dim_n++)
 4      // Sweep tiling sizes.
 5      for (blk_m = dim_m; blk_m < INF; blk_m += dim_m)
 6       for (blk_n = dim_n; blk_n < INF; blk_n += dim_n)
 7        for (blk_k = 1; blk_k < INF; blk_k++)
 8        {
 9          // Apply pruning constraints.
10        }
```

Fig. 1.7 The parameter search space for the autotuning of matrix multiplication

- The kernel would compile and launch, but produce invalid results due to the limitations of the implementation, e.g., unimplemented corner case.
- The kernel would compile, launch and produce correct results, but have no chance of running fast, due to an obvious performance shortcoming, such as very low occupancy.

The nine checks rely on basic hardware parameters, which can be obtained by querying the card with the CUDA API, and include:

1. The number of threads in the block is not divisible by the warp size.
2. The number of threads in the block exceeds the hardware maximum.
3. The number of registers per thread, to store C, exceeds the hardware maximum.
4. The number of registers per block, to store C, exceeds the hardware maximum.
5. The shared memory per block, to store A and B, exceeds the hardware maximum.
6. The thread block cannot be shaped to read A and B without cleanup code.
7. The number of load instructions, from shared memory to registers, in the innermost loop, in the PTX code, exceeds the number of *Fused Multiply-Adds* (FMAs).
8. Low occupancy due to high number of registers per block to store C.
9. Low occupancy due to the amount of shared memory per block to read A and B.

In order to check the last two conditions, the number of registers per block, and the amount of shared memory per block are computed. Then the maximum number of possible blocks per multiprocessor is found, which gives the maximum possible number of threads per multiprocessor. If that number is lower than the minimum occupancy requirement, the kernel is discarded. Here the threshold is set to a fairly low number of 256 threads, which translates to minimum occupancy of 0.125 on the Nvidia K40 card, with the maximum number of 2,048 threads per multiprocessor.

This process produces 14,767 kernels, which can be benchmarked in roughly 1 day. Three thousand two hundred and fifty six kernels fail to launch due to excessive number of registers per block. The reason is that the pruning process uses a lower estimate on the number of registers, and the compiler actually produces code requiring more registers. We could detect it in compilation and skip benchmarking of such kernels or we can run them and let them fail. For simplicity we chose the latter. We could also cap the register usage to prevent the failure to launch. However, capping register usage usually produces code of inferior performance.

Eventually, 11,511 kernels run successfully and pass correctness checks. Figure 1.8 shows the performance distribution of these kernels. The fastest kernel achieves 900 Gflop/s with tiling of $96 \times 64 \times 12$, with 128 threads (16×8 to compute C, 32×4 to read A, and 4×32 to read B). The achieved occupancy number of 0.1875 indicates that, most of the time, each multiprocessor executes 384 threads (three blocks).

Fig. 1.8 Distribution of the dgemm kernels

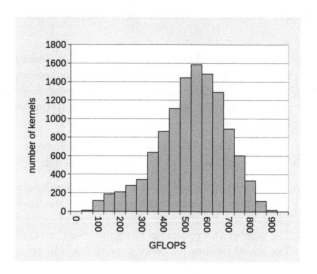

In comparison, CUBLAS achieves the performance of 1,225 Gflop/s using 256 threads per multiprocessor. Although CUBLAS achieves a higher number, this example shows the effectiveness of the autotuning process in quickly creating well performing kernels from high level language source codes. This technique can be used to build kernels for routines not provided in vendor libraries, such as extended precision BLAS (double–double and triple-float), BLAS for misshaped matrices (tall and skinny), etc. Even more importantly, this technique can be used to build domain specific kernels for many application areas.

As the last interesting observation, we offer a look at the PTX code produced by the nvcc compiler (Fig. 1.9). We can see that the compiler does exactly what is expected, which is completely unrolling the loops in lines 50–56 of the C code in Fig. 1.6, into a stream of loads from shared memory to registers and FMA instructions, with substantially more FMAs than loads.

```
 1     ld.shared.f64    %fd258, [%rd3];
 2     ld.shared.f64    %fd259, [%rd4];
 3     fma.rn.f64   %fd260, %fd258, %fd259, %fd1145;
 4     ld.shared.f64    %fd261, [%rd3+128];
 5     fma.rn.f64   %fd262, %fd261, %fd259, %fd1144;
 6     ld.shared.f64    %fd263, [%rd3+256];
 7     fma.rn.f64   %fd264, %fd263, %fd259, %fd1143;
 8     ld.shared.f64    %fd265, [%rd3+384];
 9     fma.rn.f64   %fd266, %fd265, %fd259, %fd1142;
10     ld.shared.f64    %fd267, [%rd3+512];
11     fma.rn.f64   %fd268, %fd267, %fd259, %fd1141;
12     ld.shared.f64    %fd269, [%rd3+640];
13     fma.rn.f64   %fd270, %fd269, %fd259, %fd1140;
14     ld.shared.f64    %fd271, [%rd4+832];
15     fma.rn.f64   %fd272, %fd258, %fd271, %fd1139;
16     fma.rn.f64   %fd273, %fd261, %fd271, %fd1138;
17     fma.rn.f64   %fd274, %fd263, %fd271, %fd1137;
18     fma.rn.f64   %fd275, %fd265, %fd271, %fd1136;
19     fma.rn.f64   %fd276, %fd267, %fd271, %fd1135;
20     fma.rn.f64   %fd277, %fd269, %fd271, %fd1134;
21     ld.shared.f64    %fd278, [%rd4+1664];
22     fma.rn.f64   %fd279, %fd258, %fd278, %fd1133;
23     fma.rn.f64   %fd280, %fd261, %fd278, %fd1132;
24     fma.rn.f64   %fd281, %fd263, %fd278, %fd1131;
25     fma.rn.f64   %fd282, %fd265, %fd278, %fd1130;
26     fma.rn.f64   %fd283, %fd267, %fd278, %fd1129;
27     fma.rn.f64   %fd284, %fd269, %fd278, %fd1128;
28     ld.shared.f64    %fd285, [%rd4+2496];
29     fma.rn.f64   %fd286, %fd258, %fd285, %fd1127;
30     fma.rn.f64   %fd287, %fd261, %fd285, %fd1126;
31     fma.rn.f64   %fd288, %fd263, %fd285, %fd1125;
32     fma.rn.f64   %fd289, %fd265, %fd285, %fd1124;
33     fma.rn.f64   %fd290, %fd267, %fd285, %fd1123;
34     fma.rn.f64   %fd291, %fd269, %fd285, %fd1122;
35     ld.shared.f64    %fd292, [%rd4+3328];
36     fma.rn.f64   %fd293, %fd258, %fd292, %fd1121;
37     fma.rn.f64   %fd294, %fd261, %fd292, %fd1120;
38     fma.rn.f64   %fd295, %fd263, %fd292, %fd1119;
39     fma.rn.f64   %fd296, %fd265, %fd292, %fd1118;
40     fma.rn.f64   %fd297, %fd267, %fd292, %fd1117;
41     fma.rn.f64   %fd298, %fd269, %fd292, %fd1116;
42     ld.shared.f64    %fd299, [%rd4+4160];
43     fma.rn.f64   %fd300, %fd258, %fd299, %fd1115;
44     fma.rn.f64   %fd301, %fd261, %fd299, %fd1114;
45     fma.rn.f64   %fd302, %fd263, %fd299, %fd1113;
46     fma.rn.f64   %fd303, %fd265, %fd299, %fd1112;
47     fma.rn.f64   %fd304, %fd267, %fd299, %fd1111;
48     fma.rn.f64   %fd305, %fd269, %fd299, %fd1110;
49     ld.shared.f64    %fd306, [%rd4+4992];
50     fma.rn.f64   %fd307, %fd258, %fd306, %fd1109;
51     fma.rn.f64   %fd308, %fd261, %fd306, %fd1108;
52     fma.rn.f64   %fd309, %fd263, %fd306, %fd1107;
53     fma.rn.f64   %fd310, %fd265, %fd306, %fd1106;
54     fma.rn.f64   %fd311, %fd267, %fd306, %fd1105;
55     fma.rn.f64   %fd312, %fd269, %fd306, %fd1104;
56     ld.shared.f64    %fd313, [%rd4+5824];
57     fma.rn.f64   %fd314, %fd258, %fd313, %fd1103;
58     fma.rn.f64   %fd315, %fd261, %fd313, %fd1102;
59     fma.rn.f64   %fd316, %fd263, %fd313, %fd1101;
60     fma.rn.f64   %fd317, %fd265, %fd313, %fd1100;
61     fma.rn.f64   %fd318, %fd267, %fd313, %fd1099;
62     fma.rn.f64   %fd319, %fd269, %fd313, %fd1098;
63     ld.shared.f64    %fd320, [%rd3+776];
64     ld.shared.f64    %fd321, [%rd4+8];
65     fma.rn.f64   %fd322, %fd320, %fd321, %fd260;
66     ld.shared.f64    %fd323, [%rd3+904];
67     fma.rn.f64   %fd324, %fd323, %fd321, %fd262;
68     ld.shared.f64    %fd325, [%rd3+1032];
69     fma.rn.f64   %fd326, %fd325, %fd321, %fd264;
70     ld.shared.f64    %fd327, [%rd3+1160];
71     fma.rn.f64   %fd328, %fd327, %fd321, %fd266;
72     ld.shared.f64    %fd329, [%rd3+1288];
73     fma.rn.f64   %fd330, %fd329, %fd321, %fd268;
74     ld.shared.f64    %fd331, [%rd3+1416];
75     fma.rn.f64   %fd332, %fd331, %fd321, %fd270;
76     ld.shared.f64    %fd333, [%rd4+840];
77     fma.rn.f64   %fd334, %fd320, %fd333, %fd272;
78     fma.rn.f64   %fd335, %fd323, %fd333, %fd273;
79     fma.rn.f64   %fd336, %fd325, %fd333, %fd274;
80     fma.rn.f64   %fd337, %fd327, %fd333, %fd275;
81     fma.rn.f64   %fd338, %fd329, %fd333, %fd276;
82     fma.rn.f64   %fd339, %fd331, %fd333, %fd277;
83     ld.shared.f64    %fd340, [%rd4+1672];
84     fma.rn.f64   %fd341, %fd320, %fd340, %fd279;
85     fma.rn.f64   %fd342, %fd323, %fd340, %fd280;
86     fma.rn.f64   %fd343, %fd325, %fd340, %fd281;
87     fma.rn.f64   %fd344, %fd327, %fd340, %fd282;
88     fma.rn.f64   %fd345, %fd329, %fd340, %fd283;
89     fma.rn.f64   %fd346, %fd331, %fd340, %fd284;
```

Fig. 1.9 A portion of the PTX for the innermost loop of the fastest dgemm kernel

1.3 Solving Linear Systems

Solving dense linear systems of equations is a fundamental problem in scientific computing. Numerical simulations involving complex systems represented in terms of unknown variables and relations between them often lead to linear systems of equations that must be solved as fast as possible. This section presents a methodology for developing these solvers. The technique is illustrated using the Cholesky factorization.

1.3.1 Cholesky Factorization

The Cholesky factorization (or Cholesky decomposition) of an $n \times n$ real symmetric positive definite matrix A has the form $A = LL^T$, where L is an $n \times n$ real lower triangular matrix with positive diagonal elements [5]. This factorization is mainly used as a first step for the numerical solution of linear equations $Ax = b$, where A is a symmetric positive definite matrix. Such systems arise often in physics applications, where A is positive definite due to the nature of the modeled physical phenomenon. The reference implementation of the Cholesky factorization for machines with hierarchical levels of memory is part of the LAPACK library. It consists of a succession of panel (or block column) factorizations followed by updates of the trailing submatrix.

1.3.2 Hybrid Algorithms

The Cholesky factorization algorithm can easily be parallelized using a fork-join approach since each update—consisting of a matrix–matrix multiplication—can be performed in parallel (fork) but that a synchronization is needed before performing the next panel factorization (join). The number of synchronizations of this algorithm and the synchronous nature of the panel factorization would be prohibitive bottlenecks for performance on highly parallel devices such as GPUs.

Instead, the panel factorization and the update of the trailing submatrix are broken into tasks, where the less parallel panel tasks are scheduled for execution on multicore CPUs, and the parallel updates mainly on GPUs. Figure 1.10 illustrates this concept of developing hybrid algorithms by splitting the computation into tasks, data dependencies, and consequently scheduling the execution over GPUs and multicore CPUs. The scheduling can be static (described next), or dynamic (see Sect. 1.4). In either case, the small and not easy to parallelize tasks from the critical path (e.g., panel factorizations) are executed on CPUs, and the large and highly parallel task (like the matrix updates) are executed mostly on the GPUs.

Fig. 1.10 Algorithms as a collection of tasks and dependencies among them for hybrid GPU-CPU computing

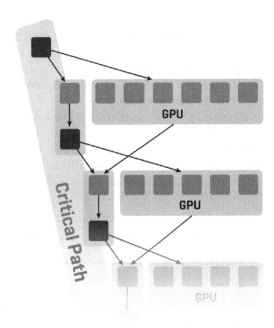

1.3.3 Hybrid Cholesky Factorization for a Single GPU

Figure 1.11 gives the hybrid Cholesky factorization implementation for a single GPU. Here da points to the input matrix that is in the GPU memory, work is a work-space array in the CPU memory, and nb is the blocking size. This algorithm assumes the input matrix is stored in the leading *n*-by-*n* lower triangular part of da, which is overwritten on exit by the result. The rest of the matrix is not referenced. Compared to the LAPACK reference algorithm, the only difference is that the hybrid

```
1   for (j = 0;  j < *n;  j += nb) {
2     jb = min(nb, *n-j);
3     cublasDsyrk('l','n', jb, j,-1, da(j,0),*lda, 1, da(j,j),*lda);
4     cudaMemcpy2DAsync (work, jb*sizeof(double), da(j,j), *lda*sizeof(double),
5                        sizeof(double)*jb, jb, cudaMemcpyDeviceToHost, stream[1]);
6     if (j + jb < *n)
7       cublasDgemm('n','t', *n-j-jb, jb, j, -1, da(j+jb,0),*lda, da(j,0),
8                   *lda, 1, da(j+jb,j),*lda);
9     cudaStreamSynchronize (stream[1]);
10    dpotrf ("Lower", &jb, work, &jb, info);
11    if (*info != 0)
12      *info = *info + j, break;
13    cudaMemcpy2DAsync (da(j,j), *lda*sizeof(double), work, jb*sizeof(double),
14                       sizeof(double)*jb, jb, cudaMemcpyHostToDevice, stream[0]);
15    if (j + jb < *n)
16      cublasDtrsm('r','l','t','n', *n-j-jb, jb, 1, da(j,j), *lda,
17                  da(j+jb,j),*lda);
18  }
```

Fig. 1.11 Hybrid Cholesky factorization for single CPU-GPU pair (*dpotrf*)

one has three extra lines—4, 9, and 13. These extra lines implement our intent in the hybrid code to have the jb-by-jb diagonal block starting at da(j,j) factored on the CPU, instead of on the GPU. Therefore, at line 4 we send the block to the CPU, at line 9 we synchronize to ensure that the data has arrived, then factor it on the CPU using a call to LAPACK at line 10, and send the result back to the GPU at line 13. Note that the computation at line 7 is independent of the factorization of the diagonal block, allowing us to do these two tasks in parallel on the CPU and on the GPU. This is implemented by statically scheduling first the *dgemm* (line 7) on the GPU; this is an asynchronous call, hence the CPU continues immediately with the *dpotrf* (line 10) while the GPU is running the *dgemm*.

The hybrid algorithm is given an LAPACK interface to simplify its use and adoption. Thus, codes that use LAPACK can be seamlessly accelerated multiple times with GPUs.

To summarize, the following is achieved with this algorithm:

- The LAPACK Cholesky factorization is split into tasks;
- Large, highly data parallel tasks, suitable for efficient GPU computing, are statically assigned for execution on the GPU;
- Small, inherently sequential *dpotrf* tasks (line 10), not suitable for efficient GPU computing, are executed on the CPU using LAPACK;
- Small CPU tasks (line 10) are overlapped by large GPU tasks (line 7);
- Communications are asynchronous to overlap them with computation;
- Communications are in a surface-to-volume ratio with computations: sending nb^2 elements at iteration j is tied to $O(nb \times j^2)$ flops, $j \geq nb$.

1.4   The Case for Dynamic Scheduling

In this section, we present the linear algebra aspects of our generic solution for development of either Cholesky, Gaussian, and Householder factorizations based on block outer-product updates of the trailing matrix.

Conceptually, one-sided factorization $\mathscr{F}$ maps a matrix A into a product of two matrices X and Y:

$$\mathscr{F} : \begin{bmatrix} A_{11} & A_{12} \\ A_{21} & A_{22} \end{bmatrix} \mapsto \begin{bmatrix} X_{11} & X_{12} \\ X_{21} & X_{22} \end{bmatrix} \times \begin{bmatrix} Y_{11} & Y_{12} \\ Y_{21} & Y_{22} \end{bmatrix}$$

Algorithmically, this corresponds to a sequence of in-place transformations of A, whose storage is overwritten with the entries of matrices X and Y (P_{ij} indicates the currently factorized panels):

$$\begin{bmatrix} A_{11}^{(0)} & A_{12}^{(0)} & A_{13}^{(0)} \\ A_{21}^{(0)} & A_{22}^{(0)} & A_{23}^{(0)} \\ A_{31}^{(0)} & A_{32}^{(0)} & A_{33}^{(0)} \end{bmatrix} \rightarrow \begin{bmatrix} P_{11} & A_{12}^{(0)} & A_{13}^{(0)} \\ P_{21} & A_{22}^{(0)} & A_{23}^{(0)} \\ P_{31} & A_{32}^{(0)} & A_{33}^{(0)} \end{bmatrix} \rightarrow \begin{bmatrix} XY_{11} & Y_{12} & Y_{13} \\ X_{21} & A_{22}^{(1)} & A_{23}^{(1)} \\ X_{31} & A_{32}^{(1)} & A_{33}^{(1)} \end{bmatrix} \rightarrow \begin{bmatrix} XY_{11} & Y_{12} & Y_{13} \\ X_{21} & P_{22} & A_{23}^{(1)} \\ X_{31} & P_{32} & A_{33}^{(1)} \end{bmatrix} \rightarrow$$

Algorithm 1 Two-phase implementation of a one-sided factorization

---

// iterate over all matrix panels
for $P_i \in \{P_1, P_2, \ldots, P_n\}$
    FactorizePanel(P_i)
    UpdateTrailingMatrix($A^{(i)}$)
end

Table 1.1 Routines for panel factorization and the trailing matrix update

	Cholesky	Householder	Gauss
FactorizePanel	dpotf2	dgeqf2	dgetf2
	dtrsm		
	dsyrk	dlarfb	dlaswp
UpdateTrailingMatrix	dgemm		dtrsm
			dgemm

Algorithm 2 Two-phase implementation with the update split between Fermi and Kepler GPUs

---

// iterate over all matrix panels
for $P_i \in \{P_1, P_2, \ldots\}$
    FactorizePanel(P_i)
    UpdateTrailingMatrix$_{\text{Kepler}}$($A^{(i)}$)
    UpdateTrailingMatrix$_{\text{Fermi}}$($A^{(i)}$)
end

$$\rightarrow \begin{bmatrix} XY_{11} & Y_{12} & Y_{13} \\ X_{21} & XY_{22} & Y_{23} \\ X_{31} & X_{32} & A_{33}^{(2)} \end{bmatrix} \rightarrow \begin{bmatrix} XY_{11} & Y_{12} & Y_{13} \\ X_{21} & X_{22} & Y_{23} \\ X_{31} & X_{32} & P_{33} \end{bmatrix} \rightarrow \begin{bmatrix} XY_{11} & Y_{12} & Y_{13} \\ X_{21} & XY_{22} & Y_{23} \\ X_{31} & X_{32} & XY_{33} \end{bmatrix} \rightarrow [XY],$$

where XY_{ij} is a compact representation of both X_{ij} and Y_{ij} in the space originally occupied by A_{ij}.

Observe two distinct phases in each step of the transformation from $[A]$ to $[XY]$: *panel factorization* (P) and trailing matrix update: $A^{(i)} \rightarrow A^{(i+1)}$. Implementation of these two phases leads to a straightforward iterative scheme shown in Algorithm 1. Table 1.1 shows BLAS and LAPACK routines that should be substituted for the generic routines named in the algorithm.

The use of multiple accelerators complicates the simple loop from Algorithm 1: we must split the update operation into multiple instances for each of the accelerators. This was done in Algorithm 2. Notice that FactorizePanel() is not split for execution on accelerators because it exhibits properties of latency-bound workloads, which face a number of inefficiencies on throughput-oriented GPU devices. Due to their high performance rate exhibited on the update operation, and the fact that the update requires the majority of floating-point operations, it is the trailing matrix update that is a good target for off-load. The problem of keeping track of the computational activities is exacerbated by the separation between the address spaces of main memory of the CPU and the GPUs. This requires synchronization between memory buffers and is included in the implementation shown in Algorithm 3.

Algorithm 3 Two-phase implementation with a split update and explicit communication

// *iterate over all matrix panels*
for $P_i \in \{P_1, P_2, \ldots\}$
    FactorizePanel(P_i)
    SendPanel$_{\text{Kepler}}$(P_i)
    UpdateTrailingMatrix$_{\text{Kepler}}$($A^{(i)}$)
    SendPanel$_{\text{Fermi}}$(P_i)
    UpdateTrailingMatrix$_{\text{Fermi}}$($A^{(i)}$)
end

Algorithm 4 Lookahead of depth 1 for the two-phase factorization

FactorizePanel(P_1)
SendPanel(P_1)
UpdateTrailingMatrix$_{\{\text{Kepler, Fermi}\}}$($P_1$)
PanelStartReceiving(P_2)
UpdateTrailingMatrix$_{\{\text{Kepler, Fermi}\}}$($R^{(1)}$)
// *iterate over remaining matrix panels*
for $P_i \in \{P_2, P_3, \ldots\}$
    PanelReceive(P_i)
    PanelFactor(P_i)
    SendPanel(P_i)
    UpdateTrailingMatrix$_{\{\text{Kepler, Fermi}\}}$($P_i$)
    PanelStartReceiving(P_i)
    UpdateTrailingMatrix$_{\{\text{Kepler, Fermi}\}}$($R^{(i)}$)
end
PanelReceive(P_n)
PanelFactor(P_n)

The complexity increases further as the code must be modified further to achieve close to peak performance. In fact, the bandwidth between the CPU and the GPUs is orders of magnitude too slow to sustain computational rates of GPUs.[2] The common technique to alleviate this imbalance is to use *lookahead* [14, 15].

Algorithm 4 shows a very simple case of a lookahead of depth 1. The update operation is split into an update of the next panel, the start of the receiving of the next panel that just got updated, and an update of the rest of the trailing matrix R. The splitting is done to overlap the communication of the panel and the update operation. The complication of this approach comes from the fact that depending on the communication bandwidth and the accelerator speed, a different lookahead depth might be required for optimal overlap. In fact, the adjustment of the depth is often required throughout the factorization's runtime to yield good performance: the updates consume progressively less time when compared to the time spent in the panel factorization.

---

[2]The bandwidth for the current generation PCI Express is at most 16 GB/s while the devices achieve over 1,000 Gflop/s performance.

Since the management of adaptive lookahead is tedious, it is desirable to use a dynamic scheduler to keep track of data dependences and communication events. The only issue is the homogeneity inherent in most of the schedulers which is violated here due to the use of three different computing devices that we used. Also, common scheduling techniques, such as task stealing, are not applicable here due to the disjoint address spaces and the associated large overheads. These caveats are dealt with comprehensively in the remainder of the chapter.

1.5 Eigenvalue and Singular Value Problems

Eigenvalue and singular value decomposition (SVD) problems are fundamental for many engineering and physics applications. For example, image processing, compression, facial recognition, vibrational analysis of mechanical structures, and computing energy levels of electrons in nanostructure materials can all be expressed as eigenvalue problems. Also, the SVD plays a very important role in statistics where it is directly related to the principal component analysis method, in signal processing and pattern recognition as an essential filtering tool, and in analysis of control systems. It has applications in such areas as least squares problems, computing the pseudoinverse, and computing the Jordan canonical form. In addition, the SVD is used in solving integral equations, digital image processing, information retrieval, seismic reflection tomography, and optimization.

1.5.1 Background

The eigenvalue problem is to find an eigenvector x and eigenvalue λ that satisfy

$$Ax = \lambda x,$$

where A is a symmetric or nonsymmetric $n \times n$ matrix. When the entire eigenvalue decomposition is computed we have $A = X \Lambda X^{-1}$, where Λ is a diagonal matrix of eigenvalues and X is a matrix of eigenvectors. The SVD finds orthogonal matrices U, V, and a diagonal matrix Σ with nonnegative elements, such that $A = U \Sigma V^T$, where A is an $m \times n$ matrix. The diagonal elements of Σ are singular values of A, the columns of U are called its left singular vectors, and the columns of V are called its right singular vectors.

All of these problems are solved by a similar three-phase process:

1. **Reduction phase**: orthogonal matrices Q (Q and P for singular value decomposition) are applied on both the left and the right side of A to reduce it to a condensed form matrix—hence these are called "two-sided factorizations." Note that the use of two-sided orthogonal transformations guarantees that A has the

same eigen/singular-values as the reduced matrix, and the eigen/singular-vectors of A can be easily derived from those of the reduced matrix (step 3);

2. **Solution phase**: an eigenvalue (respectively, singular value) solver further computes the eigenpairs Λ and Z (respectively, singular values Σ and the left and right vectors $\tilde{U}$ and $\tilde{V}^T$) of the condensed form matrix;

3. **Back transformation phase**: if required, the eigenvectors (respectively, left and right singular vectors) of A are computed by multiplying Z (respectively, $\tilde{U}$ and $\tilde{V}^T$) by the orthogonal matrices used in the reduction phase.

For the nonsymmetric eigenvalue problem, the reduction phase is to upper Hessenberg form, $H = Q^T A Q$. For the second phase, QR iteration is used to find the eigenpairs of the reduced Hessenberg matrix H by further reducing it to (quasi) upper triangular Schur form, $S = E^T H E$. Since S is in a (quasi) upper triangular form, its eigenvalues are on its diagonal and its eigenvectors Z can be easily derived. Thus, A can be expressed as:

$$A = QHQ^T = Q\, E\, S\, E^T\, Q^T,$$

which reveals that the eigenvalues of A are those of S, and the eigenvectors Z of S can be back-transformed to eigenvectors of A as $X = Q\, E\, Z$.

When A is symmetric (or Hermitian in the complex case), the reduction phase is to symmetric tridiagonal $T = Q^T A Q$, instead of upper Hessenberg form. Since T is tridiagonal, computations with T are very efficient. Several eigensolvers are applicable to the symmetric case, such as the divide and conquer (D&C), the multiple relatively robust representations (MRRR), the bisection algorithm, and the QR iteration method. These solvers compute the eigenvalues and eigenvectors of $T = Z\Lambda Z^T$, yielding Λ to be the eigenvalues of A. Finally, if eigenvectors are desired, the eigenvectors Z of T are back-transformed to eigenvectors of A as $X = Q\, Z$.

For the singular value decomposition (SVD), two orthogonal matrices Q and P are applied on the left and on the right, respectively, to reduce A to bidiagonal form, $B = Q^T A P$. Divide and conquer or QR iteration is then used as a solver to find both the singular values and the left and the right singular vectors of B as $B = \tilde{U}\Sigma\tilde{V}^T$, yielding the singular values of A. If desired, singular vectors of B are back-transformed to singular vectors of A as $U = Q\,\tilde{U}$ and $V^T = P^T \tilde{V}^T$.

There are many ways to formulate mathematically and solve these problems numerically, but in all cases, designing an efficient computation is challenging because of the nature of the algorithms. In particular, the orthogonal transformations applied to the matrix are two-sided, i.e., transformations are applied on both the left and right side of the matrix. This creates data dependencies that prevent the use of standard techniques to increase the computational intensity of the computation, such as blocking and look-ahead, which are used extensively in the one-sided LU, QR, and Cholesky factorizations. Thus, the reduction phase can take a large portion of the overall time. Recent research has been into two-stage algorithms [2, 6, 7, 10, 11], where the first stage uses Level 3 BLAS operations to reduce A

to band form, followed by a second stage to reduce it to the final condensed form. Because it is the most time consuming phase, it is very important to identify the bottlenecks of the reduction phase, as implemented in the classical approaches [1]. The classical approach is discussed in the next section, while Sect. 1.5.4 covers two-stage algorithms.

The initial reduction to condensed form (Hessenberg, tridiagonal, or bidiagonal) and the final back-transformation are particularly amenable to GPU computation. The eigenvalue solver itself (QR iteration or divide and conquer) has significant control flow and limited parallelism, making it less suited for GPU computation.

1.5.2 Classical Reduction to Hessenberg, Tridiagonal, or Bidiagonal Condensed Form

The classical approach (*"LAPACK algorithms"*) to reduce a matrix to condensed form is to use one-stage algorithms [5]. Similar to the one-sided factorizations (LU, Cholesky, QR), the two-sided factorizations are split into a *panel factorization* and a *trailing matrix update*. Pseudocode for the Hessenberg factorization is given in Algorithm 5 and shown schematically in Fig. 1.12; the tridiagonal and bidiagonal factorizations follow a similar form, though the details differ [17]. Unlike the one-sided factorizations, the panel factorization requires computing Level 2 BLAS matrix-vector products with the entire trailing matrix. This requires loading the entire trailing matrix into memory, incurring a significant amount of memory bound operations. It also produces synchronization points between the panel factorization and the trailing submatrix update steps. As a result, the algorithm follows the expensive fork-and-join model, preventing overlap between the CPU computation and the GPU computation. Also it prevents having a look-ahead panel and hiding communication costs by overlapping with computation. For instance, in the Hessenberg factorization, these Level 2 BLAS operations account for about 20 % of the floating point operations, but can take 70 % of the time in a CPU implementation [16]. Note that the computational complexity of the reduction phase is about $\frac{10}{3}n^3$, $\frac{8}{3}n^3$, and $\frac{4}{3}n^3$ for the reduction to Hessenberg, bidiagonal, and tridiagonal form respectively.

In the panel factorization, each column is factored by introducing zeros below the subdiagonal using an orthogonal Householder reflector, $H_j = I - \tau v_j v_j^T$. The matrix Q is represented as a product of $n - 1$ of these reflectors,

$$Q = H_1 H_2 \ldots H_{n-1}.$$

Before the next column can be factored, it must be updated as if H_j were applied on both sides of A, though we delay actually updating the trailing matrix. For each column, performing this update requires computing $y_j = A v_j$. For a GPU implementation, we compute these matrix-vector products on the GPU, using cublasDgemv for the Hessenberg and bidiagonal, and cublasDsymv for the tridiagonal factorization. Optimized versions of symv and hemv also exist in

Algorithm 5 Hessenberg reduction, magma_*gehrd

for $i = 1, \ldots, n$ by nb
   // panel factorization, in magma_*lahr2.
   get panel $A_{i:n,i:i+nb-1}$ from GPU
   for $j = i, \ldots, i + nb$
     $(v_j, \tau_j) = \text{householder}(a_j)$
     send v_j to GPU
     $y_j = A_{i+1:n,j:n} v_j$ on GPU
     get y_j from GPU

     compute $T_{(j)} = \begin{bmatrix} T_{(j-1)} & -\tau_j T_{(j-1)} V_{(j-1)}^T v_j \\ 0 & \tau_j \end{bmatrix}$

     update column $a_{j+1} = (I - VT^T V^T)(a_{j+1} - YT\{V^T\}_{j+1})$
   end

   // trailing matrix update, in magma_*lahru.
   $Y_{1:i,1:nb} = A_{1:i,i:n} V$ on GPU
   $A = (I - VT^T V^T)(A - YTV^T)$ on GPU
end

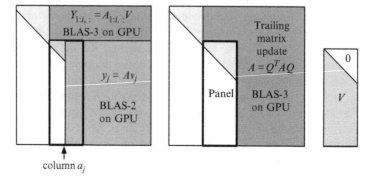

Fig. 1.12 Hessenberg panel factorization, trailing matrix update, and V matrix on GPU with upper triangle set to zero

MAGMA [13], which achieve higher performance by reading A only once and using extra workspace to store intermediate results. While these are memory-bound Level 2 BLAS operations, computing them on the GPU leverages the GPU's high memory bandwidth.

After factoring each panel of nb columns, the trailing matrix must be updated. Instead of applying each H_j individually to the entire trailing matrix, they are blocked together into a block Hessenberg update,

$$Q_i = H_1 H_2 \ldots H_{nb} = I - V_i T_i V_i^T.$$

The trailing matrix is then updated as

$$\hat{A} = Q_i^T A Q_i = (I - V_i T_i^T V_i^T)(A - Y_i T_i V_i^T) \tag{1.1}$$

for the nonsymmetric case, or using the alternate representation

$$\hat{A} = A - W_i V_i^T - V_i W_i^T \tag{1.2}$$

for the symmetric case. In all cases, the update is a series of efficient Level 3 BLAS operations executed on the GPU, either general matrix–matrix multiplies (dgemm) for the Hessenberg and bidiagonal factorizations, or a symmetric rank-$2k$ update (dsyr2k) for the symmetric tridiagonal factorization.

Several additional considerations are made for an efficient GPU implementation. In the LAPACK CPU implementation, the matrix V of Householder vectors is stored below the subdiagonal of A. This requires multiplies to be split into two operations, a triangular multiply (dtrmm) for the top triangular portion, and a dgemm for the bottom portion. On the GPU, we explicitly set the upper triangle of V to zero, as shown in Fig. 1.12, so the entire product can be computed using a single dgemm. Second, it is beneficial to store the small $nb \times nb$ T_i matrices used in the reduction, for later use in the back-transformation, whereas LAPACK recomputes them later from V_i.

1.5.3   Back-Transform Eigenvectors

For eigenvalue problems, after the reduction to condensed form, the eigensolver finds the eigenvalues Λ and eigenvectors Z of H or T. For the SVD, it finds the singular values Σ and singular vectors $\tilde{U}$ and $\tilde{V}$ of B. The eigenvalues and singular values are the same as for the original matrix A. To find the eigenvectors or singular vectors of the original matrix A, the vectors need to be back-transformed by multiplying by the same orthogonal matrix Q (and P, for the SVD) used in the reduction to condensed form. As in the reduction, the block Householder transformation $Q_i = I - V_i T_i V_i^T$ is used. From this representation, either Q can be formed explicitly using dorghr, dorgtr, or dorgbr; or we can multiply by the implicitly represented Q using dormhr, dormtr, or dormbr. In either case, applying it becomes a series of dgemm operations executed on the GPU.

The entire procedure is implemented in the MAGMA library: magma_dgeev for nonsymmetric eigenvalues, magma_dsyevd for real symmetric, and magma_dgesvd for the singular value decomposition.

1.5.4   Two Stage Reduction

Because of the expense of the reduction step, renewed research has focused on improving this step, resulting in a novel technique based on a two-stage reduction [6, 9]. The two-stage reduction is designed to increase the utilization of compute-intensive operations. Many algorithms have been investigated using this two-stage approach. The idea is to split the original one-stage approach into a compute-intensive phase (first stage) and a memory-bound phase (second or "bulge chasing" stage). In this section we will cover the description for the symmetric case. The first stage reduces the original symmetric dense matrix to a symmetric band form, while the second stage reduces from band to tridiagonal form, as depicted in Fig. 1.13.

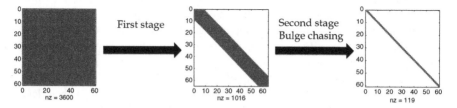

Fig. 1.13 Two stage technique for the reduction phase

1.5.4.1   First Stage: Hybrid CPU-GPU Band Reduction

The first stage applies a sequence of block Householder transformations to reduce a symmetric dense matrix to a symmetric band matrix. This stage uses compute-intensive matrix-multiply kernels, eliminating the memory-bound matrix-vector product in the one-stage panel factorization, and has been shown to have a good data access pattern and large portion of Level 3 BLAS operations [3, 4, 8]. It also enables the efficient use of GPUs by minimizing communication and allowing overlap of computation and communication. Given a dense $n \times n$ symmetric matrix A, the matrix is divided into $nt = n/b$ block-columns of size nb. The algorithm proceeds panel by panel, performing a QR decomposition for each panel to generate the Householder reflectors V (i.e., the orthogonal transformations) required to zero out elements below the bandwidth nb. Then the generated block Householder reflectors are applied from the left and the right to the trailing symmetric matrix, according to

$$A = A - WV^T - VW^T, \tag{1.3}$$

where V and T define the block of Householder reflectors and W is computed as

$$W = X - \tfrac{1}{2}VT^TV^TX, \text{ where} \tag{1.4}$$
$$X = AVT.$$

Since the panel factorization consists of a QR factorization performed on a panel of size $l \times b$ shifted by nb rows below the diagonal, this will remove both the synchronization and the data dependency constraints seen using the classical one stage technique. In contrast to the classical approach, the panel factorization by itself does not require any operation on the data of the trailing matrix, making it an independent task. Moreover, we can factorize the next panel once we have finished its update, without waiting for the total trailing matrix update. Thus this kind of technique removes the bottlenecks of the classical approach: there are no BLAS-2 operations concerning the trailing matrix and also there is no need to wait for the update of the trailing matrix in order to start the next panel. However, the resulting matrix is banded, instead of tridiagonal. The hybrid CPU-GPU algorithm is illustrated in Fig. 1.14. We first run the QR decomposition (dgeqrf panel on step i of Fig. 1.14) of a panel on the CPUs. Once the panel factorization of step i is finished, then we compute W on the GPU, as defined by Eq. (1.4). In particular, it involves a dgemm to compute VT, then a dsymm to compute $X = AVT$, which is the dominant cost of computing W, consisting of 95 % of the time spent in computing W, and finally another inexpensive dgemm. Once W is computed, the trailing matrix update (applying transformations on the left and right) defined by Eq. (1.3) can be performed using a rank-$2k$ update.

However, to allow overlap of CPU and GPU computation, the trailing submatrix update is split into two pieces. First, the next panel for step $i + 1$ (medium gray panel of Fig. 1.14) is updated using two dgemm's on the GPU. Next, the remainder of the trailing submatrix (dark gray triangle of Fig. 1.14) is updated using a dsyr2k. While the dsyr2k is executing, the CPUs receive the panel for step $i + 1$, perform the next panel factorization (dgeqrf), and send the resulting V_{i+1} back to the GPU. In this way, the factorization of panels $i = 2, \ldots, nt$ and the associated communication are hidden by overlapping with GPU computation, as demonstrated in Fig. 1.15. This is similar to the look-ahead technique typically used in the one-sided dense

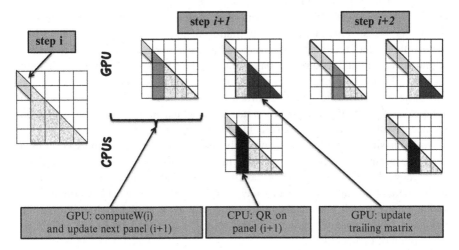

Fig. 1.14 Description of the reduction to band form, stage 1

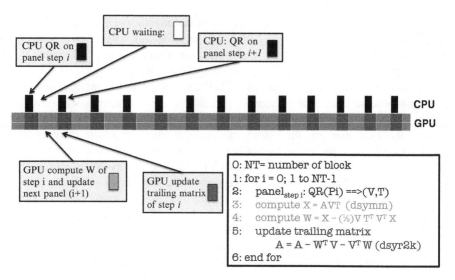

Fig. 1.15 Execution trace of reduction to band form

matrix factorizations. Figure 1.15 shows a snapshot of the execution trace of the reduction to band form, where we can easily identify the overlap between CPU and GPU computation. Note that the high-performance GPU is continuously busy, either computing W or updating the trailing matrix, while the lower performance CPUs wait for the GPU as necessary.

1.5.4.2 Second Stage: Cache-Friendly Computational Kernels

The band form is further reduced to the final condensed form using the bulge chasing technique. This procedure annihilates the extra off-diagonal elements by chasing the created fill-in elements down to the bottom right side of the matrix using successive orthogonal transformations. Each annihilation of the nb non-zero element below the off-diagonal of the band matrix is called a sweep. This stage involves memory-bound operations and requires the band matrix to be accessed from multiple disjoint locations. In other words, there is an accumulation of substantial latency overhead each time different portions of the matrix are loaded into cache memory, which is not compensated for by the low execution rate of the actual computations (the so-called surface-to-volume effect). To overcome these critical limitations, we developed a bulge chasing algorithm, to extensively use cache friendly kernels combined with fine grained, memory aware tasks in an out-of-order scheduling technique which considerably enhances data locality. This reduction has been designed for multicore architectures, and results have shown its efficiency. This step has been well optimized such that it takes between 5 and 10 % of the global time of the reduction from dense to tridiagonal. We refer the reader to [6, 8] for a detailed description of the technique.

We decide to develop a hybrid CPU-GPU implementation of only the first stage of the two stage algorithm, and leave the second stage executed entirely on the CPU. The main motivation is that the first stage is the most expensive computational phase of the reduction. Results show that 90 % of the time is spent in the first stage reduction. Another motivation for this direction is that accelerators perform poorly when dealing with memory-bound fine-grained computational tasks (such as bulge chasing), limiting the potential benefit of a GPU implementation of the second stage. Experiments showed that the two-stage algorithm can be up to six times faster than the standard one-stage approach.

1.5.5   Back Transform the Eigenvectors of the Two Stage Technique

The standard one-stage approach reduces the dense matrix A to condensed form (e.g., tridiagonal T in the case of symmetric matrix), computes its eigenvalues/eigenvectors (Λ, Z) and back transform its eigenvectors Z to computes the eigenvectors $X = Q\,Z$ of the original matrix A as mentioned earlier in Sect. 1.5.3. In the case of the two-stage approach, the first stage reduces the original dense matrix A to a band matrix by applying a two-sided transformations to A such that $Q_1^T A Q_1 = B$. Similarly, the second, bulge-chasing stage reduces the band matrix B to the condensed form (e.g, tridiagonal T) by applying two-sided transformations to B such that $Q_2^T B Q_2 = T$. Thus, when the eigenvectors matrix X of A are requested, the eigenvectors matrix Z resulting from the eigensolver needs to be back transformed by the Householder reflectors generated during the reduction phase, according to

$$X = Q_1 Q_2 Z = (I - V_1 t_1 V_1^T)\,(I - V_2 t_2 V_2^T)\,Z, \tag{1.5}$$

where (V_1, t_1) and (V_2, t_2) represent the Householder reflectors generated during the reduction stages one and two, respectively. Note that when the eigenvectors are requested, the two stage approach has the extra cost of the back transformation of Q_2. However, experiments show that even with this extra cost the overall performance of the eigen/singular-solvers using the two stage approach can be several times faster than solvers using the one stage approach.

From the practical standpoint, the back transformation Q_2 is not as straight-forward as the one of Q_1, which is similar to the classical back transformation described in Sect. 1.5.3. In particular, because of complications of the bulge-chasing mechanism, the order and the overlap of the Householder reflector generated during this stage is intricate. Let us first begin by describing the complexity and the design of the algorithm for applying Q_2. We present the structure of V_2 (the Householder reflectors that form the orthogonal matrix Q_2) in Fig. 1.16a. Note that these reflectors represent the annihilation of the band matrix, and thus each is of length nb—the bandwidth size. A naïve implementation would take each reflector and apply it in isolation to the matrix Z. Such an implementation is memory-bound and relies on Level 2 BLAS operations. A better procedure is to apply with calls to Level 3 BLAS, which achieves both very good scalability and performance. The priority is to create compute intensive operations to take advantage of the efficiency of Level 3 BLAS. We proposed and implemented accumulation and combination of the Householder reflectors. This is not always easy, and to achieve this goal we must pay attention to the overlap between the data they access as well as the fact that their application must follow the specific dependency order of the bulge chasing procedure in which they have been created. To stress these issues, we will clarify it by giving an example. For sweep i (e.g., the column at position B(i,i):B(i+nb,i)), its annihilation generates a set of k Householder reflectors (v_i^k), each of length nb, the v_i^k are represented in column i of the matrix V_2 depicted in Fig. 1.16a. Likewise, the ones related to the annihilation of sweep $i + 1$, are those presented in column $i + 1$, where they are shifted one element down compared to those of sweep i. It is possible to combine the reflectors $v_i^{(k)}$ from sweep i with those from sweep $i + 1, i + 2,\ldots, i + \ell$ and to apply them together in blocked fashion. This grouping is represented by the diamond-shaped region in Fig. 1.16a. While each of those diamonds is considered as one block, their back transformation (application to the matrix Z) needs to follow the dependency order. For example, applying block 4 and block 5 of the V_2's in Fig. 1.16a modifies block row 4 and block row 5, respectively, of the eigenvector matrix Z drawn in Fig. 1.16b where one can easily observe the overlapped region. The order dictates that block 4 needs to be applied before block 5. It is possible to compute this phase efficiently by splitting Z by blocks of columns over both the CPUs and the GPU as shown in Fig. 1.16b, where we can apply each diamond independently to each portion of E. Moreover, this method does not require any data communication. The back transformation of Q_1 to the resulting matrix from above, $Q_1 \times (Q_2 Z)$, involves efficient BLAS 3 kernels and it is done by using the GPU function magma_dormtr, which is the GPU implementation of the standard LAPACK function (dormtr).

1.6 Summary and Future Directions

In conclusion, GPUs can be used with astonishing success to accelerate fundamental linear algebra algorithms. We have demonstrated this on a range of algorithms, from the matrix–matrix multiplication kernel written in CUDA, to the higher level algorithms for solving linear systems, to eigenvalue and SVD problems. Further, despite the complexity of the hardware, acceleration was achieved at a surprisingly low software development effort using a high-level methodology of developing hybrid algorithms. The complete implementations and more are available through the MAGMA library. The promise shown so far motivates and opens opportunities for future research and extensions, e.g., tackling more complex algorithms and

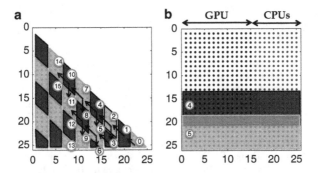

Fig. 1.16 Blocking technique to apply the Householder reflectors V_2 with a hybrid implementation on GPU and CPU. (**a**) Blocking for V_2; (**b**) eigenvectors matrix

hybrid hardware. Several major bottlenecks need to be alleviated to run at scale though, which is an intensive research topic. When a complex algorithm needs to be executed on a complex heterogeneous system, scheduling decisions have a dramatic impact on performance. Therefore, new scheduling strategies must be designed to fully benefit from the potential of future large-scale machines.

References

1. Anderson, E., Bai, Z., Bischof, C., Blackford, L.S., Demmel, J.W., Dongarra, J.J. Du Croz, J., Greenbaum, A., Hammarling, S., McKenney, A., Sorensen, D.: LAPACK Users' Guide. SIAM, Philadelphia (1992). http://www.netlib.org/lapack/lug/
2. Bientinesi, P., Igual, F.D., Kressner, D., Quintana-Ortí, E.S.: Reduction to condensed forms for symmetric eigenvalue problems on multi-core architectures. In: Proceedings of the 8th International Conference on Parallel Processing and Applied Mathematics: Part I, PPAM'09, pp. 387–395. Springer, Berlin/Heidelberg (2010)
3. Dongarra, J.J., Sorensen, D.C., Hammarling, S.J.: Block reduction of matrices to condensed forms for eigenvalue computations. J. Comput. Appl. Math. **27**(1–2), 215–227 (1989)
4. Gansterer, W., Kvasnicka, D., Ueberhuber, C.: Multi-sweep algorithms for the symmetric eigenproblem. In: Vector and Parallel Processing - VECPAR'98. Lecture Notes in Computer Science, vol. 1573, pp. 20–28. Springer, Berlin (1999)
5. Golub, G., Loan, C.V.: Matrix Computations, 3rd edn. Johns Hopkins, Baltimore (1996)
6. Haidar, A., Ltaief, H., Dongarra, J.: Parallel reduction to condensed forms for symmetric eigenvalue problems using aggregated fine-grained and memory-aware kernels. In: Proceedings of SC '11, pp. 8:1–8:11. ACM, New York (2011)
7. Haidar, A., Ltaief, H., Luszczek, P., Dongarra, J.: A comprehensive study of task coalescing for selecting parallelism granularity in a two-stage bidiagonal reduction. In: Proceedings of the IEEE International Parallel and Distributed Processing Symposium, Shanghai, 21–25 May 2012. ISBN 978-1-4673-0975-2
8. Haidar, A., Tomov, S., Dongarra, J., Solca, R., Schulthess, T.: A novel hybrid CPU-GPU generalized eigensolver for electronic structure calculations based on fine grained memory aware tasks. Int. J. High Perform. Comput. Appl. **28**(2), 196–209 (2014)

9. Haidar, A., Kurzak, J., Luszczek, P.: An improved parallel singular value algorithm and its implementation for multicore hardware. In: SC13, The International Conference for High Performance Computing, Networking, Storage and Analysis, Denver, CO, 17–22 November 2013

10. Lang, B.: Efficient eigenvalue and singular value computations on shared memory machines. Parallel Comput. **25**(7), 845–860 (1999)

11. Ltaief, H., Luszczek, P., Haidar, A., Dongarra, J.: Enhancing parallelism of tile bidiagonal transformation on multicore architectures using tree reduction. In: Wyrzykowski, R., Dongarra, J., Karczewski, K., Wasniewski, J. (eds.) Proceedings of 9th International Conference, PPAM 2011, Torun, vol. 7203, pp. 661–670 (2012)

12. MAGMA 1.4.1: http://icl.cs.utk.edu/magma/ (2013)

13. Nath, R., Tomov, S., Dong, T., Dongarra, J.: Optimizing symmetric dense matrix-vector multiplication on GPUs. In: 2011 International Conference for High Performance Computing, Networking, Storage and Analysis (SC), pp. 1–10. New York, NY, USAm 2011, ACM

14. Strazdins, P.E.: Lookahead and algorithmic blocking techniques compared for parallel matrix factorization. In: 10th International Conference on Parallel and Distributed Computing and Systems, IASTED, Las Vegas, 1998

15. Strazdins, P.E.: A comparison of lookahead and algorithmic blocking techniques for parallel matrix factorization. Int. J. Parallel Distrib. Syst. Netw. **4**(1), 26–35 (2001)

16. Tomov, S., Nath, R., Dongarra, J.: Accelerating the reduction to upper Hessenberg, tridiagonal, and bidiagonal forms through hybrid GPU-based computing. Parallel Comput. **36**(12), 645–654 (2010)

17. Yamazaki, I., Dong, T., Solcà, R., Tomov, S., Dongarra, J., Schulthess, T.: Tridiagonalization of a dense symmetric matrix on multiple GPUs and its application to symmetric eigenvalue problems. Concurr. Comput. Pract. Exp. (2013). doi:10.1002/cpe.3152

Chapter 2
A Guide for Implementing Tridiagonal Solvers on GPUs

Li-Wen Chang and Wen-mei W. Hwu

2.1 Introduction

The tridiagonal solver has been recognized as a critical building block for many engineering and scientific applications [3, 8, 9, 11, 17, 18] on GPUs. However, a general high-performance tridiagonal solver for GPU is challenging, not just because the number of independent, simultaneous matrices varies greatly among applications, but also because applications may require their tridiagonal solvers to have customized requirements, such as: data with different layouts, matrices with a certain structure, or execution on multi-GPUs. Therefore, although building a tridiagonal solver library is crucial, it is very difficult to meet all demands. In this chapter, guidelines are given for customizing a high-performance tridiagonal solver for GPUs.

A wide range of algorithms for implementing tridiagonal solvers on GPUs, including both sequential and parallel algorithms, was studied. The selected algorithms were chosen for the requirement of applications, and to take the advantage of massive data parallelism of GPU architecture. Meanwhile, corresponding optimizations were proposed to compensate for some inherent limitations of the selected algorithms. In order to achieve high performance on GPUs, workloads have to be partitioned and computed in parallel on stream processors. For the tridiagonal solver, the inherent data dependence found in sequential algorithms (e.g. the Thomas algorithm [5] and the diagonal pivoting method [10]), limits the opportunities for partitioning the workload. On the other hand, parallel algorithms (e.g. Cyclic Reduction (CR) [12], Parallel Cyclic Reduction (PCR) [12], or the SPIKE algorithm [16, 19]) allow the partitioning of workloads, but suffer from the required overheads of extra computation, barrier synchronization, or communication.

L.-W. Chang • W.-m.W. Hwu (✉)
University of Illinois, 1308 W Main St, Urbana, IL 61801, USA
e-mail: lchang20@illinois.edu; w-hwu@illinois.edu

V. Kindratenko (ed.), *Numerical Computations with GPUs*,
DOI 10.1007/978-3-319-06548-9__2, © Springer International Publishing Switzerland 2014

Two main kinds of components are recognized in most GPU tridiagonal solvers. (1) Partitioning methods are applied to divide workloads for parallel computing. Independent solvers compute massive independent workloads in parallel. In this chapter, we first review cutting-edge partitioning techniques for GPU tridiagonal solvers. Different partitioning techniques require different types of overheads, such as computation or memory overhead. (2) State-of-the-art optimization techniques for independent solvers are discussed. Different algorithms of independent solvers might require different optimizations. Optimization techniques might perform together for more robust independent solvers. Finally, a case study of a new algorithm, SPIKE-CR, which replaces part of the traditional SPIKE algorithm with Cyclic Reduction, is given to demonstrate how to systematically build a highly optimized tridiagonal solver by selecting the partitioning method, and by applying optimization techniques to the independent solver for each partition. *The main purpose of this chapter is to inspire readers building their own GPU tridiagonal solvers to meet their application requirement, instead of demonstrating high performance of SPIKE-CR.*

The rest of the sections in this chapter are organized as following. Section 2.2 briefly reviews the selected algorithms used by GPU tridiagonal solvers. Section 2.3 reviews and compares corresponding optimizations applied to the GPU tridiagonal solvers. Section 2.4 shows a case study of the new GPU tridiagonal solver, SPIKE-CR; discusses its partitioning and optimizations; and compares its performance to alternative methods. Section 2.5 concludes the chapter. In the following sections, we use NVIDIA CUDA [14] terminology.

2.2 Related Algorithms

In this section, we briefly cover the selected tridiagonal solver algorithms used for GPUs. Although, in general, most tridiagonal solvers may be used to solve multiple systems of equations each with its own tridiagonal matrix, for simpler explanation here, we only discuss the case of solving a single system with one tridiagonal matrix. The tridiagonal solver solves $Tx = d$, where T is a tridiagonal matrix with n rows and n columns, defined in Eq. (2.1), and x and d are both column vectors with n elements. Note that the first row of T is row 0, and the first element of x and d is element 0.

$$T = \begin{bmatrix} b_0 & c_0 & & & \\ a_1 & b_1 & c_1 & & \\ & a_2 & \ddots & \ddots & \\ & & \ddots & \ddots & c_{n-2} \\ & & & a_{n-1} & b_{n-1} \end{bmatrix} \tag{2.1}$$

2.2.1 Thomas Algorithm

The Thomas algorithm is a special case of Gaussian elimination without pivoting (or LU decomposition with LU solvers) for a tridiagonal matrix. It consists of two phases, a forward reduction and a backward substitution. The forward reduction sequentially eliminates the lower diagonal of the original matrix, while the backward substitution sequentially solves for unknown variables using known variables and the upper and main diagonals in the resultant matrix. For $Tx = d$, decompose $T = LU$ by LU decomposition, let $Ux = y$, solve $Ly = d$, and then solve $Ux = y$.

2.2.2 Diagonal Pivoting Algorithm

The diagonal pivoting algorithm for tridiagonal matrices was proposed by Erway et al. [10]. Although Gaussian elimination with partial pivoting is widely used for tridiagonal solvers on CPUs, it is not efficient on GPUs due to its inherent data dependence and expensive row interchange operations. Erway's diagonal pivoting method avoids row interchanges by dynamically selecting 1-by-1 or 2-by-2 pivots. The factorization is defined as follows:

$$T = \begin{bmatrix} P_h & B \\ C & T_r \end{bmatrix} = \begin{bmatrix} I_h & 0 \\ CP_h^{-1} & I_r \end{bmatrix} \begin{bmatrix} P_h & 0 \\ 0 & T_s \end{bmatrix} \begin{bmatrix} I_h & P_h^{-1}B \\ 0 & I_r \end{bmatrix} \tag{2.2}$$

where P_h is a 1-by-1 ($[b_0]$) or 2-by-2 pivoting block ($\begin{bmatrix} b_0 & c_0 \\ a_1 & b_1 \end{bmatrix}$), and

$$T_s = T_r - CP_h^{-1}B = \begin{cases} T_r - \frac{a_1 c_0}{b_0} e_1^{(n-1)} e_1^{(n-1)T}, & \text{for 1-by-1 pivoting} \\ T_r - \frac{a_2 b_0 c_1}{\Delta} e_1^{(n-2)} e_1^{(n-2)T}, & \text{for 2-by-2 pivoting} \end{cases} \tag{2.3}$$

where $\Delta = b_0 b_1 - a_1 c_0$ and $e_1^{(k)}$ is the first column vector of the k-by-k identity matrix. Since T_s is still tridiagonal (Eq. (2.3)), it can also be factorized by the same Eq. (2.2). Therefore, a tridiagonal matrix T can be recursively factorized in LBM^T, where B only contains either 1-by-1 or 2-by-2 blocks in its diagonal. After LBM^T factorization, the tridiagonal matrix T can be solved by solving L, B, and M^T sequentially.

$$\begin{bmatrix} b_0 & c_0 & & \\ a_1 & b_1 & c_1 & \\ & a_2 & b_2 & c_2 \\ & & a_3 & b_3 \end{bmatrix} \rightarrow \begin{bmatrix} b_0' & 0 & c_0' & \\ a_1 & b_1 & c_1 & \\ a_2' & 0 & b_2' & 0 \\ & & 0 & a_3 & b_3 \end{bmatrix} \Rightarrow \begin{bmatrix} b_0' & c_0' \\ a_2' & b_2' \end{bmatrix}$$

Fig. 2.1 One step CR forward reduction on a 4-by-4 matrix: a_2 and c_2 on row 2 are eliminated by row 1 and 3. Similarly, c_0 is eliminated by row 1. After that, row 0 and row 2 can form a smaller matrix

2.2.3 Cyclic Reduction

The Cyclic Reduction (CR) algorithm, also known as an odd-even reduction, contains two phases, forward reduction and backward substitution. In every step of the forward reduction, defined in Eq. (2.4), each odd (or even) equation is eliminated by using the adjacent two even (or odd) equations.

$$\alpha = a_i/b_{i-stride}, \quad \beta = c_i/b_{i+stride},$$
$$a_i' = -\alpha a_{i-stride}, \quad b_i' = b_i - \alpha c_{i-stride} - \beta a_{i+stride}, \quad (2.4)$$
$$c_i' = -\beta c_{i+stride}, \quad d_i' = d_i - \alpha d_{i-stride}\alpha - \beta d_{i+stride},$$

where the *stride* starts from 1 and increases exponentially step-by-step, and the domain of i starts from all odd and shrinks exponentially. The boundary condition can be simplified by using $a_i = c_i = 0$, and $b_i = 1$. Figure 2.1 shows a CR example for a 4-by-4 tridiagonal matrix. After a step of CR forward reduction, redundant unknown variables and zeros can be removed, and a half-size matrix is formed of the remaining unsolved equations. Each step of the backward substitution, defined in Eq. (2.5), solves for unknown variables by substituting solutions obtained from the smaller system.

$$x_i = \frac{d_i' - a_i' x_{i-stride} - c_i' x_{i+stride}}{b_i'} \quad (2.5)$$

where the stride decreases exponentially step-by-step, and the domain of i increases exponentially. The graph representation of CR for a 8-by-8 matrix is shown in Fig. 2.2, where each vertical line represents an equation, and each circle represents forward or backwards computation.

2.2.4 Parallel Cyclic Reduction

The PCR algorithm, different from CR, only performs the forward reduction, Eq. (2.4). Also, the PCR forward reduction is performed on all equations, instead of odd (or even). That means the domain of i does not decrease exponentially, but

Fig. 2.2 The CR access pattern on a 8-by-8 matrix: each *vertical line* represents an equation, each *circle* represents forward or backwards computation, and each *edge* represents communication between two equations

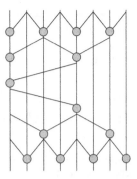

the *stride* still keeps increasing exponentially step-by-step. Figure 2.3 shows a PCR example for the same 4-by-4 tridiagonal matrix. After a step of PCR, two half-size matrices are formed of the resultant new equation by reorganizing unknown variables. It also illustrates how a matrix can be partitioned after each PCR (forward reduction) step.

$$
\begin{bmatrix}
b_0 & c_0 & & \\
a_1 & b_1 & c_1 & \\
& a_2 & b_2 & c_2 \\
& & a_3 & b_3
\end{bmatrix}
\rightarrow
\begin{bmatrix}
b_0' & 0 & c_0' & \\
0 & b_1' & 0 & c_1' \\
a_2' & 0 & b_2' & 0 \\
& a_3' & 0 & b_3'
\end{bmatrix}
\Leftrightarrow
\begin{bmatrix}
b_0' & c_0' \\
a_2' & b_2' \\
\hline
b_1' & c_1' \\
a_3' & b_3'
\end{bmatrix}
$$

Fig. 2.3 One step PCR forward reduction on a 4-by-4 matrix: a_i and c_i on each row i are eliminated by adjacent two rows. For example, a_2 and c_2 on row 2 are eliminated by row 1 and 3. After that, row 0 and 2 can form a smaller matrix, and row 1 and 3 can form another

2.2.5 Recursive Doubling

The Recursive Doubling (RD) algorithm [21] can be considered as a reformulation of a parallel tridiagonal solver into a second-order linear recurrence, Eq. (2.6). By solving the relationship between x_0 and d_{n-1}, all unknown variables, x_i's, can be solved.

$$
\begin{bmatrix}
1 & & & & & \\
b_0/c_0 & 1 & & & & \\
a_1/c_1 & b_1/c_1 & 1 & & & \\
& \ddots & \ddots & \ddots & & \\
& & a_{n-2}/c_{n-2} & b_{n-2}/c_{n-2} & 1 & \\
& & & a_{n-1} & b_{n-1} & 1
\end{bmatrix}
\begin{bmatrix}
x_0 \\
x_1 \\
x_2 \\
\vdots \\
x_{n-1} \\
0
\end{bmatrix}
=
\begin{bmatrix}
x_0 \\
d_0/c_0 \\
d_1/c_1 \\
\vdots \\
d_{n-2}/c_{n-2} \\
d_{n-1}
\end{bmatrix}
\tag{2.6}
$$

However, in the Recursive Doubling algorithm, huge numerical errors might be produced, even for a diagonally dominant matrix, since division operations are performed on upper diagonal elements (c_i's). Because of this shortcoming, we skip the discussion of RD in this chapter.

2.2.6   SPIKE Algorithm

The SPIKE algorithm was originally introduced by Sameh et al. [19] and the latest version described by Pollizi et al. [16]. It is a domain decomposition algorithm, that partitions a matrix into block rows containing diagonal sub-matrices, T_i, and off-diagonal elements, a_{hi} and c_{ti}. The original matrix, T, can be further defined as the product of two matrices, the block-diagonal matrix D and the spike matrix S, Fig. 2.4, where V_i and W_i of S can be solved by Eq. (2.7).

$$T_i V_i = \begin{bmatrix} 0 \\ \vdots \\ 0 \\ c_{ti} \end{bmatrix}, \quad T_i W_i = \begin{bmatrix} a_{hi} \\ 0 \\ \vdots \\ 0 \end{bmatrix}. \tag{2.7}$$

After the formation of the matrices D and S, the SPIKE algorithm solves $Dy = d$ for y, and then uses the special form of S to solve $Sx = y$ [16]. The spike matrix, S, can also be considered a specialized block tridiagonal matrix, and can be solved by a block tridiagonal solver algorithm, such as the block Cyclic Reduction algorithm [2].

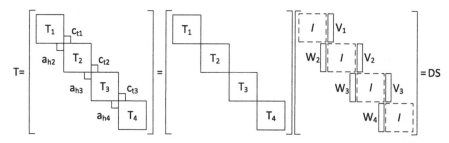

Fig. 2.4 A tridiagonal matrix T can be defined as $T = DS$, where D is a block diagonal matrix and S is a spike matrix (a specialized block tridiagonal matrix)

2.3   Optimization Techniques

As mentioned in Sect. 2.1, partitioning is necessary for high performance on GPUs. Although sequential algorithms inherently cannot be partitioned, they are widely applied to solving multiple independent systems in parallel. On the other hand, although parallel algorithms are capable of both partitioning individual systems and solving multiple independent systems, they might require high overheads. In this section, all existing optimization techniques for GPU tridiagonal solvers are examined. However, while not every optimization is discussed in detail, references are provided for each technique to satisfy readers who need more information.

2.3.1   Partitioning Method

Many of the early tridiagonal solvers on GPUs [6, 8, 11, 17, 20, 23] can only be applied to problems with multiple independent matrices. They simply assume no partitioning occurs, and exploit only the inherent parallelism from multiple independent matrices. This assumption works very efficiently, simply because parallelism is inherent and no partitioning overhead is required. However, when the number of independent matrices shrinks, the overall performance drops dramatically.

Partitioning is found in many studies of tridiagonal solvers for GPUs, and particularly, the PCR algorithm was widely applied to partitioning. Sakharnykh [18] first introduced PCR in his PCR-Thomas implementation to further extract more parallelism for a limited number of independent matrices. Kim et al. [13] and Davidson et al. [7] first recognized that partitioning is necessary for a tridiagonal solver to handle a single large matrix on GPUs, and they proposed PCR-Thomas tridiagonal solvers. In both papers, PCR was used to decompose one large matrix into many smaller independent matrices. The main limitation of PCR is its computation overhead. In order to minimize the computation overhead of PCR, only a few PCR steps are performed. Kim et al. further proposed the sliding window technique to reduce the requirement of scratchpad memory size for PCR, and to make PCR more efficient.

Compared to PCR, domain partitioning requires less computational overhead. The CR-PCR implementation for the non-pivoting tridiagonal solver in NVIDIA CUSPARSE [15] uses implicit domain partitioning by duplicating memory accesses between two adjacent partitions. By storing data back to global memory between two CR steps, the redundant equations of CR (see Fig. 2.1) can be removed to avoid unnecessary memory overhead. Although this naive partitioning method simplifies the source code, it may cost a large memory overhead, since each CR step requires reloading data from global memory.

Argüello et al. [1] proposed a split-and-merge method for CR by separating computation workloads into two sets, called split and merge sets. The split sets represent the independent workloads partitioned and are assigned to stream processors, while the merge sets represent computation workloads requiring data from two or more independent split sets. Figure 2.5a illustrates the graph representation for the split-and-merge method of CR forward reduction. The independent split sets can be simply computed in parallel, while the merge sets are postponed and computed in a separate kernel later. Compared to the NVIDIA CR-PCR implementation, Argüello's method dramatically reduces memory access overhead, since multiple steps of CR might be computed with shared data in a kernel. Chang et al. [2] further refined Argüello's split-and-merge CR to support the larger split sets. The corresponding illustration is shown in Fig. 2.5b.

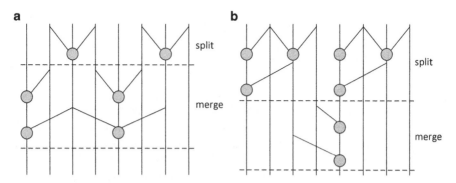

Fig. 2.5 The graph representation of a 8-by-8 matrix for CR using split-and-merge. (**a**) Argüello's split-and-merge method, which has smaller splits sets and larger merge sets; (**b**) Chang's split-and-merge method, which has larger splits sets and smaller merge sets

Chang et al. [4] and the pivoting tridiagonal solver in NVIDIA CUSPARSE applied the SPIKE algorithm to decompose a matrix into disjoint partitions. The SPIKE algorithm requires extra overhead for solving the spike matrix, $Sx = y$. The computation cost for solving the spike matrix is relatively small, compared to the cost for solving all of the independent partitions.

2.3.2 Algorithms and Optimizations for Independent Solver

After using a matrix partitioning method, or given multiple independent matrices, the multiple independent workloads can be computed in parallel. The Thomas algorithm was applied in [4, 7, 13, 17] simply for its low complexity and lack of warp divergence. Chang et al. [4] first introduced the diagonal pivoting method [10] for numerical stability, and the same method is also implemented in the CUSPARSE pivoting tridiagonal solver. With Chang's dynamic tiling technique, the overhead of warp divergence in the diagonal pivoting method is dramatically reduced.

Different from the sequential algorithms, the parallel algorithms, such as CR, require more optimization techniques to reduce possible overheads and to perform efficiently. Göddeke et al. [11] eliminated bank conflict caused by the strided access of CR, by marshaling data on scratchpad memory. Davidson et al. [6] proposed register packing for CR to hold more data in registers within a stream processor without increasing the size of scratchpad memory. Figure 2.6 illustrates an example of 4-equation register-packing CR forward reduction for an 8-by-8 matrix. A 4-equation CR forward reduction is computed locally in packed vector4 registers. The label S represents the data copied to scratchpad memory for communication among threads. Note that the needed scratchpad size is equal to the number of threads. Davidson's optimization can potentially increase the size of each partition, and further reduce the possible overhead of partitioning, though the benefits were not explicitly mentioned in Davidson's paper.

Fig. 2.6 The graph representation of a 8-by-8 matrix for Davidson's 4-equation register-packing CR forward reduction: the label S represents the data copied to scratchpad memory. Davidson's method can hold the number of thread times the number of register packing equations in a thread block. The needed scratchpad size is equal to the number of threads. In this illustration, the scratchpad size is only 2 equations

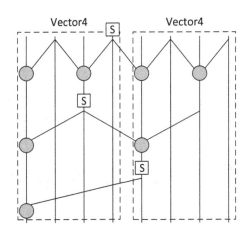

PCR can be used as an efficient independent solver for small-size matrices. Zhang et al. [23] first demonstrated it in their CR-PCR method, and CUSPARSE extended the CR-PCR method in a non-pivoting tridiagonal solver to support larger matrices. A high-performance warp-level PCR that has no barrier overhead is proposed in Sect. 2.4 and Listing 2.5.

Zhang et al. [23] first systematically introduced the hybrid methods for GPU tridiagonal solvers, by combining the Thomas, CR, PCR, and RD algorithms, to gain feasible complementary benefits. Although Zhang's idea only worked for small matrices, implementing an independent solver using his idea is extremely efficient when running on a stream processor. The reading of Zhang's paper is highly recommended.

2.3.3   Short Summary

Table 2.1 summarizes the above partitioning methods, and the corresponding limitation or overhead. Different applications may require different partitioning methods, and have different overheads. Another possible overhead for all methods is the data marshaling [22] overhead to glue two memory access patterns of the applied

Table 2.1 Summary of partitioning methods

Methods	Limitation or overhead
No partitioning	No overhead, but only for massive independent matrices
PCR	Heavy computation overhead
Naive domain partitioning	Heavy memory access overhead
SPIKE algorithm	Light computation/memory overhead
Split-and-merge	Light memory access overhead

Table 2.2 Optimization of independent solver

Optimization	Algorithms
Dynamic tiling	Diagonal pivoting method
Register packing	CR
Bank conflict elimination	CR
Warp-level computation	PCR
Hybrid method	All tridiagonal algorithms

partitioning method and independent solver. For example, in Chang's SPIKE-based tridiagonal solver [4], data marshaling is used to guarantee a coalesced memory access pattern in the independent solver. The data marshaling overhead is required only if the output pattern of the partitioning method is different from the input pattern of the independent solver.

Table 2.2 categorizes applicable optimizations for the algorithms used in independent solvers. Different optimization techniques might perform better together for more robust independent solvers. The concept of Zhang's hybrid method [23] can further enable more potential optimizations across different algorithms. For example, in the case study (Sect. 2.4), we use a hybrid of CR and PCR to enable optimizations in the both algorithms.

2.4 Case Study: SPIKE-CR

In this section, a new hybrid algorithm, SPIKE-CR, is used as a case study to demonstrate how to apply the optimization techniques that were summarized in Sect. 2.3. Using a systematic optimization analysis, the implementation of the SPIKE-CR tridiagonal solver conceptually interacts with the GPU architecture. Previous works did not discover the SPIKE-CR method. This is mainly because the previous works did not systematically analyze the partitioning methods.

In the SPIKE-CR, the SPIKE algorithm is applied to partitioning for its lower computation overhead than PCR and lower memory access overhead than the other domain partitioning methods. After the partitioning method is selected, CR is applied for the independent solver. Although the sequential algorithms are efficient with the SPIKE algorithm [4], the CR algorithm is chosen to avoid the potential data marshaling overhead from combining the SPIKE algorithm and the sequential algorithms. SPIKE-PCR is another potential direction for a GPU tridiagonal solver. However, the computation cost of PCR is much higher than CR.

In order to implement an efficient SPIKE-CR, the following optimization techniques are applied in this case study. First, Davidson's register packing [6] is applied to hold more equations in a partition. This optimization can potentially reduce partitioning overhead of the SPIKE algorithm by reducing the number of partitions. Second, Zhang's hybrid idea [23] of CR and PCR is used to avoid the potential low utilization of vector units in CR and to further enable more options of optimization in PCR. Third, a new warp-level PCR is proposed to remove barrier synchronization overheads in PCR. Last, another level partitioning using the SPIKE algorithm is applied to minimize communication between warps within a thread block. This strategy makes partitioning become hierarchical and further reduces communication overheads.

Listing 2.1 The baseline kernel of CR forward

```
1    ...
2    tx = threadIdx.x;
3    b_dim = blockDim.x;
4    ...
5    //CR iteration within a thread block
6    active_tx = b_dim;
7    for(int i=1;i<b_dim;i*=2)
8    {
9            active_tx/=2;
10           if(tx < active_tx)
11           {
12                   //CR forward computation using data in
13           scratchpad
14           }
15           __syncthreads();
16           if(tx < active_tx)
17           {
18                   //update scratchpad
19           }
20           __syncthreads();
21   }
22   ...
```

Listing 2.2 The optimized kernel of CR

```
1    ...
2    tx = threadIdx.x;
3    b_dim = blockDim.x;
4    lane_id = tx % warpSize;
5    warp_id = tx / warpSize;
6    ...
7    double2 a_reg,b_reg, c_reg, d_reg; //vectorize register
8    //load data into scratchpad using vector2
9    a_reg = a[id];
10   b_reg = b[id];
11   c_reg = c[id];
12   d_reg = d[id];
13   //code fragment 1, CR forward reduction
14   //code fragment 2, warp-level PCR
15   //code fragment 3, CR backward substitution
16   //store partial results and the spike matrix
17   ...
```

Since the source codes of the SPIKE algorithm have been provided by Chang et al. [4] at http://impact.crhc.illinois.edu, and the computation cost for solving the spike matrix is much smaller than the cost for solving all independent partitions, we only discuss the detailed source codes of the independent solver. The reading of Chang's paper and source codes is highly recommended. Listing 2.1 shows the simplified baseline of the CR forward reduction kernel, and Listing 2.2 shows the structure of our optimized CR. The code fragments are written for NVIDIA Fermi architecture, and possible changes for NVIDIA Kepler architecture are further discussed.

Listing 2.3 The code fragment 1: CR forward reduction

```
1    ...
2    //CR forward in register
3    sh_a[tx] = a_reg.y;
4    sh_b[tx] = b_reg.y;
5    sh_c[tx] = c_reg.y;
6    sh_d[tx] = d_reg.y;
7    //up side
8    {
9            k1=c_reg.x/b_reg.y;
10           b_reg.x -= a_reg.y*k1;
11           d_reg.x -= d_reg.y*k1;
12           c_reg.x = -c_reg.y*k1;
13   }
14   // down side
15   if(lane_id>=1)
16   {
17           k1=a_reg.x/sh_b[tx-1];
18           b_reg.x -= sh_c[tx-1]*k1;
19           d_reg.x -= sh_d[tx-1]*k1;
20           a_reg.x = -sh_a[tx-1]*k1;
21   }
22   sh_a[tx] = a_reg.x;
23   sh_b[tx] = b_reg.x;
24   sh_c[tx] = c_reg.x;
25   sh_d[tx] = d_reg.x;
26   ...
```

Listing 2.4 The code fragment 3: CR backward substitution

```
1    ...
2    //CR backward in register
3    k1 = a_reg.y/b_reg.x;
4    a_reg.y = 0.0;
5    if(lane_id<warpSize-1)
6    {
7            k2 = c_reg.y/sh_b[tx+1];
8            c_reg.y = -sh_c[tx+1]*k2;
9            a_reg.y = -sh_a[tx+1]*k2;
10           d_reg.y -= sh_d[tx+1]*k2;
11   }
12   c_reg.y -= c_reg.x*k1;
13   a_reg.y -= a_reg.x*k1;
14   d_reg.y -= d_reg.x*k1;
15   ...
```

Listing 2.3 shows the portion of 2-equation register-packing CR forward reduction using Davidson's technique [6], and Listing 2.4 shows the portion of corresponding CR backward substitution. Note that Listing 2.4 is the fragment 3, and is performed after the warp-level PCR. Here, we change the order of the listings for an easier discussion by putting the two fragments of CR together. The packed registers are defined in line 7 of Listing 2.2, and the computation of CR happens at line 3–14 of Listing 2.3 and all of Listing 2.4. Scratchpad memory, sh_a to sh_d, is used to communicate among threads only within a warp. Compared to the baseline of CR forward reduction, which contains at least two barrier synchronizations, line 14 and 19 of Listing 2.1, in a loop, the optimized CR requires no barrier synchronization, since communication only happens within a warp. Also, for NVIDIA Kepler architecture, shuffle instructions can replace those scratchpad memory accesses, since communication happens within a warp. Moreover, since Kepler provides larger register files, a larger size register packing can be applied to holding more data.

Listing 2.5 shows the warp-level PCR fragment of our CR-PCR hybrid. Similarly, since PCR only happens in a warp, no barrier synchronization is needed. Also, shuffle instructions can be used for Kepler by replacing scratchpad memory accesses. In these code fragments, our CR-PCR performs 1 CR forward reduction step, followed by 5 PCR steps in the warp-level PCR and 1 CR backward substitution step, without any barrier synchronization. After CR-PCR, the computed results are stored back to global memory, and also the formed spike matrix is explicitly stored in another space. Since each thread block is further partitioned into multiple warps, another level of domain partitioning using SPIKE algorithm is implicitly applied.

Listing 2.5 The code fragment 2: warp-level PCR

```
1     ...
2     //PCR for each warp, no barrier needed
3     for(int i=1;i<warpSize;i*=2)
4     {
5             // down side
6             if(lane_id>=i)
7             {
8                     k1=sh_a[tx]/sh_b[tx-i];
9                     b_reg.x -= sh_c[tx-i]*k1;
10                    d_reg.x -= sh_d[tx-i]*k1;
11                    a_reg.x = -sh_a[tx-i]*k1;
12            }
13            //up side
14            if(lane_id<warpSize-i)
15            {
16                    k1=sh_c[tx]/sh_b[tx+i];
17                    b_reg.x -= sh_a[tx+i]*k1;
18                    d_reg.x -= sh_d[tx+i]*k1;
19                    c_reg.x = -sh_c[tx+i]*k1;
20            }
21            sh_a[tx] = a_reg.x;
22            sh_b[tx] = b_reg.x;
23            sh_c[tx] = c_reg.x;
24            sh_d[tx] = d_reg.x;
25    }
26    ...
```

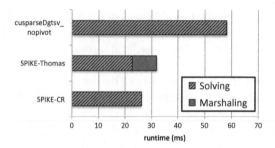

Fig. 2.7 Performance results for solving a 16M-equation double-precision matrix using CUS-PARSE non-pivoting tridiagonal solver(cusparseDgtsv_nopivot), Chang's SPIKE-Thomas, and SPIKE-CR on an NVIDIA Tesla C2050. The data marshaling overhead of Chang's SPIKE-Thomas implementation is shown in the *right portion* of the bar

2.4.1 Performance Comparison

Figure 2.7 shows the performance comparison for solving a 16M-equation (2^{24}) double-precision matrix using CUSPARSE non-pivoting tridiagonal solver (CR-PCR), Chang's SPIKE-Thomas [4], and SPIKE-CR on an NVIDIA Tesla C2050. Although the Thomas algorithm is extremely efficient as an independent solver, in Chang's SPIKE-Thomas, the overhead of data marshaling, required to maintain coalescing memory access for Thomas algorithm, causes Chang's SPIKE-Thomas performing slightly slower than SPIKE-CR. Compared to CUSPARSE CR-PCR, the domain partitioning using the SPIKE algorithm tends to have less memory access overhead than the naive domain partitioning used by CUSPARSE. The memory access overhead causes the main performance difference between the SPIKE-based methods and CUSPARSE CR-PCR. In the end, SPIKE-CR has 1.23× and 2.23× speedups over SPIKE-Thomas and CUSPARSE CR-PCR, respectively.

2.5   Conclusion

This chapter summarizes most cutting-edge optimization techniques, applied in both partitioning methods and independent solvers, for GPU tridiagonal solvers, and demonstrates how to apply optimization techniques for building a high-performance tridiagonal solver in our case study, SPIKE-CR. The case study, SPIKE-CR, shows 1.23× and 2.23× speedups, respectively, over Chang's SPIKE-Thomas [4] and CUSPARSE non-pivoting tridiagonal solver, since SPIKE-CR has no data marshaling overhead and less memory access overhead.

As mentioned in Sect. 2.1, the main purpose of this chapter is to give readers the current status of GPU tridiagonal solvers, and further to inspire readers to customize GPU tridiagonal solvers to meet their application requirements, instead of showing high performance of SPIKE-CR. Multiple partitioning methods, such as the split-and-merge method [1, 2] and the SPIKE algorithm, tend to have very low overheads for a limited number of large matrices, while no partitioning is required for a massive number of matrices. For independent solvers, the sequential methods usually perform very efficiently, while the parallel algorithms, such as CR, can also provide comparable performance after optimization. Therefore, the main concern of building a high-performance GPU tridiagonal solver is how the applied algorithm and its memory access pattern meet a given application.

Some unique properties, such as numerical stability, of a GPU tridiagonal solver for the application are also very critical. So far, only few previous works [4, 23] recognized the numerical stability issue of current GPU tridiagonal solvers, and even fewer ones [4] investigated it. Numerical stability becomes the most important future work for the research of GPU tridiagonal solvers.

Acknowledgements This project was partly supported by the STARnet Center for Future Architecture Research (C-FAR), the DoE Vancouver Project (DE-FC02-10ER26004/DE-SC0005515), and the UIUC CUDA Center of Excellence.

References

1. Argüello, F., Heras, D.B., Bóo, M., Lamas-Rodríguez, J.: The split-and-merge method in general purpose computation on GPUs. Parallel Comput. **38**(6–7), 277–288 (2012)
2. Chang, L.-W., Hwu, W.-m.W.: Mapping tridiagonal solvers to linear recurrences. Technical report, University of Illinois at Urbana-Champaign (2013)
3. Chang, L.-W., Lo, M.-T., Anssari, N., Hsu, K.-H., Huang, N.E., Hwu, W.-m.W.: Parallel implementation of multi-dimensional ensemble empirical mode decomposition. In: International Conference on Acoustics, Speech, and Signal Processing, pp. 1621–1624 (May 2011)
4. Chang, L.-W., Stratton, J.A., Kim, H.-S., Hwu, W.-m.W.: A scalable, numerically stable, high-performance tridiagonal solver using GPUs. In: Proceedings of the International Conference on High Performance Computing, Networking, Storage and Analysis, SC '12, pp. 27:1–27:11 (2012)
5. Conte, S.D., De Boor, C.W.: Elementary Numerical Analysis: An Algorithmic Approach, 3rd edn. McGraw-Hill Higher Education, New York (1980)
6. Davidson, A., Owens, J.D.: Register packing for cyclic reduction: a case study. In: Proceedings of the Fourth Workshop on General Purpose Processing on Graphics Processing Units (2011)
7. Davidson, A., Zhang, Y., Owens, J.D.: An auto-tuned method for solving large tridiagonal systems on the GPU. In: Proceedings of the 25th IEEE International Parallel and Distributed Processing Symposium (May 2011)
8. Egloff, D.: GPUs in financial computing part II: massively parallel solvers on GPUs. Wilmott, **50**, 50–53 (Nov 2010)
9. Egloff, D.: GPUs in financial computing part III: ADI solvers on GPUs with application to stochastic volatility. Wilmott, **52**, 51–53 (Mar 2011)
10. Erway, J.B., Marcia, R.F., Tyson, J.A.: Generalized diagonal pivoting methods for tridiagonal systems without interchanges. IAENG Int. J. Appl. Math. **40**(4), 269–275 (2010)
11. Göddeke, D., Strzodka, R.: Cyclic reduction tridiagonal solvers on GPUs applied to mixed-precision multigrid. IEEE Trans. Parallel Distrib. Syst. **22**, 22–32 (2011)
12. Hockney, R.W., Jesshope, C.R.: Parallel Computers: Architecture, Programming and Algorithms. Hilger, Bristol (1981)
13. Kim, H.-S., Wu, S., Chang, L.-W., Hwu, W.-m.W.: A scalable tridiagonal solver for GPUs. In: 2011 International Conference on Parallel Processing (ICPP), pp. 444–453 (2011)

14. NVIDIA Corporation: CUDA Programming Guide 5.5 (2013)
15. NVIDIA Corporation: CUSPARSE Library (2013)
16. Polizzi, E., Sameh, A.H.: A parallel hybrid banded system solver: the SPIKE algorithm. Parallel Comput. **32**(2), 177–194 (2006)
17. Sakharnykh, N.: Tridiagonal solvers on the GPU and applications to fluid simulation. In: NVIDIA GPU Technology Conference (September 2009)
18. Sakharnykh, N.: Efficient tridiagonal solvers for ADI methods and fluid simulation. In: NVIDIA GPU Technology Conference (September 2010)
19. Sameh, A.H., Kuck, D.J.: On stable parallel linear system solvers. J. ACM **25**(1), 81–91 (1978)
20. Sengupta, S., Harris, M., Zhang, Y., Owens, J.D.: Scan primitives for gpu computing. In: Graphics Hardware 2007, pp. 97–106 (2007)
21. Stone, H.S.: An efficient parallel algorithm for the solution of a tridiagonal linear system of equations. J. ACM **20**(1), 27–38 (1973)
22. Sung, I.-J., Stratton, J.A., Hwu, W.-M.W.: Data layout transformation exploiting memory-level parallelism in structured grid many-core applications. In: PACT '10: Proceedings of the 19th International Conference on Parallel Architectures and Compilation Techniques, pp. 513–522. ACM, New York (2010)
23. Zhang, Y., Cohen, J., Owens, J.D.: Fast tridiagonal solvers on the GPU. In: Proceedings of the 15th ACM SIGPLAN Symposium on Principles and Practice of Parallel Programming, PPoPP '10, pp. 127–136 (2010)

Chapter 3
Batch Matrix Exponentiation

M. Graham Lopez and Mitchel D. Horton

3.1 Introduction

Being the crucial component of numerical software packages such as LAPACK [3], ScaLAPACK [6], MUMPS [2], and SuperLU [13], the general dense matrix–matrix multiplication routine, GEMM,[1] is a common performance benchmark and a typical target of early optimization efforts for new computing architectures [33,34]. Major hardware vendors such as Intel, IBM, AMD, and NVIDIA maintain their own highly optimized GEMM implementations, which are included with their respective BLAS libraries: MKL [35], ESSL [28], ACML [1], and CUBLAS [8]. Non-vendor optimized implementations for various architectures are also available, examples being ATLAS [44] and GotoBLAS [24]. Autotuning efforts are now commonplace [33, 34], and GEMM is critical to the performance of the High Performance LINPACK Benchmark (HPL) [16], the official benchmark of the TOP500 list.

All of this importance attributed to GEMM is explained by the fact that many numerical algorithms, lower-upper (LU) factorization being one of several examples, can be expressed in terms of GEMM, or at least designed to partially use GEMM. This is achieved using delayed updates; the application of basic linear transformations expressed in terms of matrix–vector multiplications are delayed and accumulated, and then they are applied in aggregate as a GEMM [39]. LU is a canonical linear algebra procedure for solving linear systems of equations; improvements in the time to solution for LU has a direct impact on the execution time of applications in domains such as airplane wing design, radar cross-section

---

[1]In this work, we refer to general matrix–matrix multiplication as GEMM, in adherence with the Basic Linear Algebra Subroutines (BLAS) standard [5].

M.G. Lopez • M.D. Horton (✉)
Georgia Institute of Technology, Atlanta, GA 30332, USA
e-mail: graham.lopez@gatech.edu; mhorton9@mail.gatech.edu

V. Kindratenko (ed.), *Numerical Computations with GPUs*,
DOI 10.1007/978-3-319-06548-9__3, © Springer International Publishing Switzerland 2014

studies, flow around ships and other off-shore constructions, diffusion of solid bodies in a liquid, noise reduction, diffusion of light by small particles, etc. [14]

Besides its applicability to a wide variety of numerical algorithms and resulting application domains, GEMM has a high flops per memory access ratio and regular memory access pattern, which makes it well suited to a many-core architecture with a hierarchical memory such as the GPU [12]. Evaluation of GEMM is a well-studied problem, and it is the canonical GPU programming example [9]; double precision GEMM (DGEMM) can achieve 80 % of the peak theoretical performance on the Kepler architecture [32].

Consequently, batch GEMM, the matrix–matrix multiplication of a large number of relatively small matrices, is a growing area within dense linear algebra, and is relevant to various application areas such as phylogenetics [42], finite element modeling [29], image processing [11], fluid dynamics [11], and hydrodynamics [15]. NIVIDA began providing a batch GEMM routine with CUDA 4.1: `cublasXgemmBatched`, where X is one of S,D,C,Z [9]. With CUDA 5.5, NVIDIA provides batch LU, and batch matrix inversion [9].

3.2 Motivation

Our problem, matrix exponentiation based on batch GEMM, comes from the field of phylogenetics. Recent advances in sequencing technology (DNA sequencing, amino-acid and protein characterization, gene expression data, and whole-genome descriptions) are providing phylogenetics researchers with a plethora of biological sequence datasets [4, 18, 22, 23, 30, 36, 38, 40–42, 45]. Often, the goal is to infer a most probable phylogenetic history, which is represented as a tree. However, as the number of sequences increases, the number of trees that a brute force algorithm would evaluate to determine the most probable history quickly becomes prohibitive. The number of unrooted bifurcating trees, T, for n observed sequences is given by

$$T(n) = \prod_{i=3}^{n} (2i - 5), \tag{3.1}$$

and while the number of observed sequences can number in the hundreds or thousands, note that $T(50) \approx 2.84^{76}$ [19, 21, 23, 45].

As a result of this intractability, a number of techniques have emerged for reducing the number of trees that must be evaluated. Foremost among these is Markov chain Monte Carlo (MCMC), which has been enthusiastically embraced for phylogenetic inference [17, 26]. An MCMC based phylogenetics algorithm for inferring trees does its work in the following manner:

1. Randomly construct an initial tree. Call it the current tree.
2. Stochastically perturb the current tree (often this is simply a local regrouping and branch length modification).

3. Compute the acceptance ratio, R, of the probabilities of the modified tree and the current tree.
4. If $R \geq 1$, accept the new tree and make it the current tree. Otherwise, draw a uniform random number between 0 and 1. If it is less than R, accept the new tree and make it the current tree. Otherwise, reject the new tree.
5. Go to step 2.

It turns out that for a properly constructed and adequately run Markov chain, the proportion of the time that any tree is visited is a valid approximation of the probability of that tree [27, 43]. The tree that is visited the most would then also be the tree with the highest probability.

The probabilities given in step 3 above are computed using Felsenstein's algorithm for likelihood [19–21]. This evaluation is the most computationally intensive part of the algorithm, and is normally the prime candidate for GPU acceleration [42]. Briefly, Felsenstein's algorithm assumes independence of sites, independence of branches, and finite-time transition probabilities $P_{i,j}(t)$ that characterize how state i mutates to state j along a branch of length t; it then computes the probability of the given tree and set of branch lengths by summing across all possibilities for interior nodes and multiplying across all branches and sites.

For phylogenetic models, there are three common choices for the number of values a site can have: 4, 20, and 60 (nucleotide, amino acid, and codon model, respectively). For the nucleotide model, the finite-time transition probabilities are derived as follows [25]: a transition is a point mutation that changes a purine nucleotide base (A, G) to another purine, or a pyrimidine nucleotide base (C, T) to another pyrimidine. A transversion is a point mutation that changes a purine to a pyrimidine, or a pyrimidine to a purine. Each site evolves according to a Markov process in which a base $i \in \{T, C, A, G\}$ is replaced by another base j in an infinitesimally short interval of time, dt, with a probability $P_{ij}(dt)$ given by

$$
P_{ij}(dt) = \begin{cases} \alpha \pi_j dt & \text{(for transition)} \\ \beta \pi_j dt & \text{(for transversion)} \end{cases} \tag{3.2}
$$

where α is the proportion of mutations that are transitions, β is the proportion of mutations that are transversions, and π_j is the stationary composition of base j. The substitution probability matrix for an infinitesimally short interval of time can then be written as:

$$
P(dt) = \begin{array}{c} \\ T \\ C \\ A \\ G \end{array} \begin{array}{cccc} T & C & A & G \end{array} \left[\begin{array}{cccc} 1 - (\alpha \pi_C + \beta \pi_A + \beta \pi_G)dt & \alpha \pi_C dt & \beta \pi_A dt & \beta \pi_G dt \\ \alpha \pi_T dt & 1 - (\alpha \pi_T + \beta \pi_A + \beta \pi_G)dt & \beta \pi_A dt & \beta \pi_G dt \\ \beta \pi_T dt & \beta \pi_C dt & 1 - (\alpha \pi_G + \beta \pi_T + \beta \pi_C)dt & \alpha \pi_G dt \\ \beta \pi_T dt & \beta \pi_C dt & \alpha \pi_A dt & 1 - (\alpha \pi_A + \beta \pi_T + \beta \pi_C)dt \end{array} \right]
$$

$$
= I + A dt
$$

For an arbitrary time interval t, the function $P(t)$ satisfies the Chapman–Kolmogorov equation [25]

$$P(t + dt) = P(t)P(dt)$$
$$= P(t)(I + A dt) . \tag{3.3}$$

Therefore, we get

$$\frac{dP(t)}{dt} = P(t)A . \tag{3.4}$$

Since $P(0) = I$, we have

$$P(t) = e^{tA} . \tag{3.5}$$

The right hand side of Eq. (3.5) is matrix exponentiation. Matrix exponentiation is defined to be

$$e^X = \sum_{k=0}^{\infty} \frac{1}{k!} X^k, \tag{3.6}$$

where X is a matrix. For simple cases, matrix exponentiation can be computed explicitly. Otherwise, diagonalization is used. Given an eigendecomposition for X, the following holds:

$$X = EDE^{-1} \Rightarrow X^k = ED^k E^{-1} \Rightarrow e^X = E \left(\sum_{k=0}^{\infty} \frac{1}{k!} D^k \right) E^{-1} = E e^D E^{-1} . \tag{3.7}$$

Because raising a diagonal matrix to a power amounts to raising each diagonal entry to that power, Eq. (3.5) can be expressed as

$$P(t) = e^{tA} = E \times \mathrm{diag}(e^{t\lambda_1}, \ldots, e^{t\lambda_4}) \times E^{-1} = ED_t E^{-1}, \tag{3.8}$$

where $\lambda_1, \ldots, \lambda_4$ are the eigenvalues of A.

It is from Eq. (3.8) that our motivating batch GEMM arises. For each MCMC step, and for each tree branch, we must compute the finite time transition probability $P_{i,j}(t)$ that characterizes how state i mutates to state j along a branch of length t. Since an MCMC algorithm can run for hundreds of millions of steps, and tree branches can number in the tens of thousands, this computation is a good candidate for acceleration on the GPU. As part of the Keeneland project [31], optimizing the acceleration of this batched matrix exponentiation was undertaken as a contribution to the beast/beagle phylogenetics community code [4, 17].

3.3 Implementation

As can be seen from Eq. (3.8), the fundamental operations for calculating the matrix exponentiation involves two GEMMs, plus M floating point exponential operations[2] to construct the diagonal D_t matrix. Of course, the eigendecomposition of the transition matrix is needed as well. For our models, the transition matrix A does not change when a tree is modified. Only the branch lengths, t, of the trees in Eq. (3.5) change (across branches, and across MCMC steps), and so D_t in Eq. (3.8) must be recalculated at each MCMC step for each branch.

This implies that our two outer matrices E and E^{-1} referred to as A and B in the pseudocode examples also remain the same for every step in the algorithm. The pseudocode examples given throughout this section assume A and B are the same across steps, however, to generalize to unique matrices, the cublasSgemmBatched and cublasSgemm examples need no modification, and the handwritten CUDA needs only to be changed in how the input matrices are read from global device memory to shared memory. The size of memory transfers would also be different for the input A and B matrices being unique in the batched operation. However, all of the performance data shown here for comparison purposes excludes all memory transfers, since this cost is similar across implementation methods anyway.

3.3.1 NVIDIA Library Solutions

As pointed out before, each tree can have tens of thousands of internal branches. The exponentiation involves two GEMM operations per branch length evaluation, and the number of flops required by the GEMMs dominates that for the exponentiation of the diagonal matrix by $O(n^3)$ to $O(n)$. NVIDIA has provided a batched GEMM implementation, cublasXGemmBatched, since the release of CUDA 4.1, and so we examine how to use this implementation for batched matrix exponentiation and the resulting performance.

```
1  __global__ void kernelComputeD(float* D,
2                                  float* eigenvals,
3                                  float* lengths,
4                                  int n) {
5
6      float* position;
7      float length;
```

---

[2]Here, M is the dimension of the probability matrix and number of sites in the model. For example, $M = 4$ for the nucleotide model.

```
 8
 9    int bx=blockIdx.x;
10    int tx=threadIdx.x;
11
12    position=D+bx*n*n;
13    length=lengths[bx];
14
15    position[n*tx+tx]=__expf(eigenvls[tx]*length);
16
17  }
18
19  int main(int argc,char **argv) {
...
20    float **dAin=0;
21    float **dAin_d=NULL;
22    dAin=(float**)malloc(numLengths*sizeof(*dAin));
23    for (i=0;i<numLengths;i++) {
24     cudaMalloc((void**)&dAin[i],n*n*sizeof(float));
25    }
26    for (i=0;i<numLengths;i++) {
27      cudaMemcpy(dAin[i],hA,n*n*sizeof(float),
28                 cudaMemcpyHostToDevice);
29    }
30    cudaMalloc((void**)&dAin_d,
31               numLengths*sizeof(*dAin)));
32    cudaMemcpy(dAin_d,dAin,numLengths*sizeof(*dAin),
33               cudaMemcpyHostToDevice);
34
35    dim3 dimBlock(n,1,1);
36    dim3 dimGrid(numLengths,1,1);
37    kernelComputeD<<<dimGrid,dimBlock>>>(dDout,
38                                         dEigenvls,
39                                         dLengths,
40                                         n);
41    cudaMemcpy(hDout,
42               dDout,
43               numLengths*n*n*sizeof(float),
44               cudaMemcpyDeviceToHost);
45    for (i=0;i<numLengths;i++) {
46      cudaMemcpy(dinD[i],
47                 hDout+i*n*n,
48                 n*n*sizeof(float),
49                 cudaMemcpyHostToDevice);
50    }
51    cudaMemcpy(dinD_d,
```

```
52                dinD,
53                numLengths*sizeof(*dinD),
54                cudaMemcpyHostToDevice);
55    cublasSetStream(handle,streamArray[0]);
56    cublasSgemmBatched(handle,CUBLAS_OP_N,
57                CUBLAS_OP_N,n,n,n,
58                &alpha,
59                (const float**)dAin_d,n,
60                (const float**)dDin_d,n,
61                &beta, dOut1_d,n,numLengths);
...
```

Code 1: cublasSgemmBatched for Matrix Exponentiation.

Code 1 shows the main points of computation associated with using cublasSgemmBatched to evaluate the right hand side of Eq. (3.8) for a single tree, for each branch length of that tree. For the sake of clarity, no error checking is done. The variable holding the input matrix A in device memory, dA on Ln. 59, is declared, allocated, and initialized according to the pattern set out in batchCUBLAS.cpp from the NVIDIA SDK [10]. On Ln. 35, n is the dimension of D_t from Eq. (3.5), in this case, 4. On Ln. 36, numLengths is the number of branch lengths in a single tree and the number of matrices in the batch computation. On Ln. 38, dEigenvals is the matrix of four eigenvalues from Eq. (3.5). On Ln. 39, dLengths is an array of branch lengths. The second required batched GEMM, multiplying the result of the first one shown with the matrix B, is done in the same pattern as the one shown in Cd. 1, but is omitted here for brevity.

Another solution that is suggested by the CUBLAS documentation [8] for a batched situation involves making multiple calls to the normal cublasSgemm routine in separate streams. Accordingly, the solution for batched exponentiation is unchanged, except for replacing the cublasSgemmBatched routine with a loop which launches cublasSgemm in multiple streams, as shown in Cd. 2. In this example, the computation on Ln. 56 from Cd. 1 is replaced with Lns. 2–8 in Cd. 2.

```
1    float *dAin;
...
2    for (i=0;i<numLengths;i++) {
3      cublasSetStream(handle, streamArray[i]);
4      cublasSgemm (handle,CUBLAS_OP_N,
5                CUBLAS_OP_N,n,n,n,
6                &alpha,dAin,n,
7                dDin[i],n,&beta,dDout1,n);
8    }
```

Code 2: cublasSgemm with Streams.

The relative performance of these two solutions will be discussed in the following section.

3.3.2 Handwritten CUDA

As will be shown in Fig. 3.1, neither cublasSgemmBatched nor the streams solution achieves a very high percentage of the theoretical peak performance of the device. Hence, it makes sense to consider hand written CUDA for this algorithm. Code 3 shows a first attempt at a hand written CUDA kernel that implements all steps of the matrix exponentiation expressed on the right hand side of Eq. (3.8) into a single kernel, while keeping performance considerations in mind, for example by using shared memory where possible.

```
1  __global__ void exp4x4(float* output,
2                         float* A,
3                         float* D,
4                         float* B,
5                         float* lengths) {
6    __shared__ float* C;
7    __shared__ float length;
8    int bx = blockIdx.x;
9    int tx = threadIdx.x;
10   int ty = threadIdx.y;
11
12   if (tx == 0 && ty == 0) {
13     C = output + 4*4*bx;
14     length = lengths[bx];
15   }
16   __syncthreads();
17
18   float Csub = 0;
19   __shared__ float As[4][4];
20   __shared__ float Bs[4][4];
21   __shared__ float Ds[4];
22
23   if (ty == 0)
24     Ds[tx] = __expf(D[tx] * length);
25   __syncthreads();
26
27   As[ty][tx] = A[4 * ty + tx];
28   Bs[ty][tx] = B[4 * ty + tx];
29   __syncthreads();
30
31   for (int k = 0; k < 4; k++)
32     Csub += As[ty][k] * Ds[k] * Bs[k][tx];
33   __syncthreads();
34
35   C[ty*4 + tx] = Csub;
40 }
```

```
41
42 int main(int argc, char **argv) {
...
43    dim3 dimBlock(4,4,1);
44    dim3 dimGrid(numLengths,1,1);
45    exp4x4<<<dimGrid,dimBlock>>>(dOut,dA,dD,
46                                 dB,dLengths);
...
```

Code 3: Hand-written Kernel for Matrix Exponentiation.

The argument dA on Ln. 45 is E from Eq. (3.8), the argument dD on Ln. 45 is the matrix of eigenvalues referred to in Eq. (3.8), the argument dB on Ln. 45 is E^{-1} from Eq. (3.8), and the argument dLengths on Ln. 46 is an array of branch lengths, one to be used in each computation of the batched exponentiation. Note that on the Fermi architecture, there is a limit of $2^{16} - 1$ possible blocks along each grid dimension. Therefore, additional logic is required to launch a kernel with more than $2^{16} - 1$ matrices in the batch, using more than one dimension of the grid. On Kepler, this limit is larger at $2^{31} - 1$.

All of the input and output memory that holds the matrices is one-dimensional linear; Line 13 simply points the current block's threads to the right spot in memory to record their output, and similarly Ln. 27 and 28 have the threads divide up the work of reading input from global to shared memory. It is at Ln. 24 and 32 where the computation takes place. First, D_t from Eq. (3.8) is constructed. Next, rather than doing two full GEMMs, the fact that the inner matrix is diagonal is exploited to turn the computation into a matrix–vector combined with GEMM operation.

Figure 3.1 shows the performance[3] of the cublasSgemmBatched, streams, and handwritten CUDA solutions on Fermi and Kepler with threaded MKL as a baseline. The results for cublasSgemmBatched are consistent with those published by NVIDIA for CUBLAS batched GEMM [7].

There are a couple of points to note about the streams solution performance as shown in Fig. 3.1. First, the performance is not only less than that of cublasSgemm-Batched, but also that of using MKL on the CPU alone. This is consistent with NVIDIA's CUBLAS documentation[8] where it is pointed out that the performance of cublasSgemmBatched should be much greater than that of using multiple streams for matrices where $M < 100$. Secondly, it is perhaps at first unexpected that the Kepler architecture would underperform Fermi, especially in the case of using multiple streams, given that Kepler has been enhanced with 32 work queues

---

[3]We use the following flop count throughout this work, regardless of the algorithm, implementation, or architecture:

$$flops = n * (3m^3 + 2m) \tag{3.9}$$

where n is the number of branch lengths, and m is the dimension of the matrix E from Eq. (3.8). This count comes from Ln. 24 and 32 of Cd. 3.

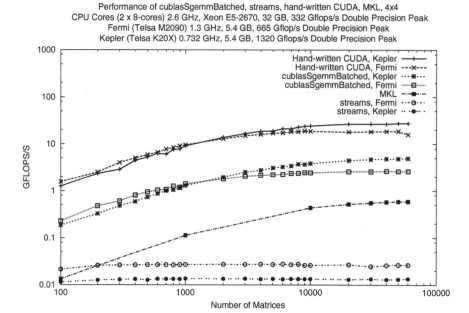

Fig. 3.1 Performance of NVIDIA library solutions vs. handwritten CUDA on Fermi and Kepler for $M = 4$

to Fermi's single work queue. A possible explanation is that due to the extra hardware involved in implementing the additional work queues for Kepler, there is a greater launch overhead cost. Consistent with the suggestion from the CUBLAS documentation, the matrices being used are too small to offset this extra overhead.

Figure 3.1 clearly indicates that the handwritten kernel from Cd. 3 outperforms `cublasSgemmBatched` and cublasSgemm when called from multiple streams. The simplification that turns two GEMMs into a GEMM plus a matrix–vector computation is one immediate advantage over the library-provided solutions. Additionally, because the GPU code is now accessible, there are further opportunities for optimization.

3.4 Tuning

Now that we have an algorithm in hand that performs reasonably well compared to the NVIDIA-provided solutions already described, we want to explore ways to further improve the performance of our hand-written CUDA kernel. Rather than exploring the performance effects of each possible optimization in isolation, the process will be presented incrementally, giving the results of each effort in the context of all previously successful changes to the algorithm. The examples in this section are for matrix sizes from our problem domain. However, the same principles that we show here for matrices of size $M = 4$, $M = 20$, and $M = 60$ can be applied to matrices of any size from $M \in \{2, \ldots, 8\}$, $M \in \{9, \ldots, 32\}$, and $M > 32$, respectively—with some caveats mentioned near the end of this section.

3.4.1 M = 4 Case

A first step is to adjust the kernel launch configuration. As seen in Cd. 3, each block of 16 threads is responsible for computing a single matrix in the batch. With only 16 threads in each block, only half of a single warp is being scheduled on each block. By having multiple matrices from the batch computed on a single block, more warps per block can make more efficient use of the hardware.

One way to achieve this configuration change is to introduce a third dimension of threads to address multiple matrices per block. We show how to do this using an adjustable parameter, `packsize`, because in our experience, the optimal number of warps per block can be hardware and problem-size dependent. For example, for the 4×4 matrix batch exponentiation, having a packsize of 8 which results in 128 threads or 4 warps per block gives the best performance. Code 4 shows the basic modifications needed to implement this change; a small amount of additional logic would be needed in the kernel if the total number of matrices in the batch is not evenly divisible by the packsize.

```
1    __global__ void exp4x4 (float* output,
2                            float* A,
3                            float* D,
4                            float* B,
5                            float* lengths) {
6
7      float* C;
8      float length;
9
10     int bx = blockIdx.x;
11     int tx = threadIdx.x;
12     int ty = threadIdx.y;
13
14     int tz = threadIdx.z;
15     int matrix_idx = bx*blockDim.z + tz;
16     int matrix_addr = 4*4*(bx*blockDim.z + tz);
17
18     C = output + matrix_addr;
19     length = lengths[matrix_idx];
20
21     float Csub = 0;
```

```
22    __shared__ float As[4][4];
23    __shared__ float Bs[4][4];
24    __shared__ float Ds[4][16];
25
26    if (ty == 0)
27      Ds[tx][tz] = __expf(D[tx] * length);
28    __syncthreads();
29
30    As[ty][tx] = A[4 * ty + tx];
31    Bs[ty][tx] = B[4 * ty + tx];
32    __syncthreads();
33
34    for (int k = 0; k < 4; k++)
35      Csub += As[ty][k] * Ds[k][tz] * Bs[k][tx];
40    __syncthreads();
41
42    C[ty*4 + tx] = Csub;
43 }
44
45 int main(int argc, char **argv) {
...
46    dim3 dimBlock(Nsize,Nsize,packsize);
47 dim3 dimGrid((numLengths+packsize-1)/packsize,1,1);
48    exp4x4<<<dimGrid,dimBlock>>>(dOut,dA,dD,
49                                 dB,dLengths);
...
```

Code 4: Adjusting Kernel Launch Configuration.

Two new variables are added to the kernel for convenience, matrix_idx and matrix_addr. In Lns. 18 and 19 of Cd. 4, these variables function in the same way as Lns. 13 and 14 of the original kernel in Cd. 3. However, now there are multiple matrices per block, so the z-dimension of the thread index indicates which matrix within the current block is being worked on. Since there are now multiple matrices being handled within a block, length and the address of the output matrix are no longer shared as they were in Cd. 3. Finally, the shared D matrix, Ds in Cd. 4, is now two-dimensional, with the second dimension being used to indicate which matrix within the block is being used by addressing it with the z-dimension of the thread index. Using these changes, we see a 4.5–5x speedup on both Fermi and Kepler for 50,000 matrices of size 4×4 when going from a single matrix per block to 8 matrices (packsize = 8) per block.

The next most obvious optimization that can be done is to remove all unnecessary branching and barriers from the kernel. In fact, the kernel from Cd. 4 is already quite compact, but a small amount of refactoring can have a nontrivial impact on performance. Code 5 shows an improved kernel.

```
1    __global__ void exp4x4(float* output,
2                           float* A,
3                           float* D,
4                           float* B,
5                           float* lenghts) {
6
7    float* C;
8    float lengths;
9
10   int bx = blockIdx.x;
11   int tx = threadIdx.x;
12   int ty = threadIdx.y;
13
14   int tz = threadIdx.z;
15   int matrix_idx = bx*blockDim.z + tz;
16   int matrix_addr = 4*4*(bx*blockDim.z + tz);
17
18   C = output + matrix_addr;
19   length = lenghts[matrix_idx];
20
21   float Csub = 0;
22   __shared__ float As[4][4];
23   __shared__ float Bs[4][4];
24   __shared__ float Ds[4][16];
25
26   As[ty][tx] = A[4 * ty + tx];
27   Bs[ty][tx] = B[4 * ty + tx];
28   Ds[tx][tz] = D[tx];
29   __syncthreads();
30
31   for (int k = 0; k < 4; k++)
32     Csub += As[ty][k] *
33               __expf(Ds[tx][tz]*length) *
34             Bs[k][tx];
35   __syncthreads();
36
37   C[ty*4 + tx] = Csub;
38
39 }
...
```

Code 5: Removing Barriers and Branches.

The main difference between Cd. 4 and 5 is the construction of the D matrix. Lns. 26–28 from Cd. 4 have been eliminated, and the entire computation is done at once on Lns. 32–34 of Cd. 5. Even though more threads will be doing redundant work to compute D in Cd. 5, these threads actually would be waiting due to the branch and barrier from Cd. 4, which has now been eliminated. Making this change in Cd. 5 gives a roughly 20 % performance increase on Fermi, and around an 8 % increase on Kepler over Cd. 4.

In certain situations, a performance improvement can result from reading the input data from texture memory instead of device global memory. In Cd. 6, we show how this is done for our problem.

```
 1 texture <float> textureA;
 2 texture <float> textureD;
 3 texture <float> textureB;
 4 texture <float> textureLengths;
 5
 6 __global__ void exp4x4(float* output,
 7                        float* A,
 8                        float* D,
 9                        float* B,
10                        float* lengths) {
11
12    float* C;
13    float length;
14
15    int bx = blockIdx.x;
16    int tx = threadIdx.x;
17    int ty = threadIdx.y;
18
19    int tz = threadIdx.z;
20    int matrix_idx = bx*blockDim.z + tz;
21    int matrix_addr = 4*4*(bx*blockDim.z + tz);
22
23    C = output + matrix_addr;
24    length = tex1Dfetch(textureLengths, matrix_idx);
25
26    float Csub = 0;
27    __shared__ float As[4][4];
28    __shared__ float Bs[4][4];
29    __shared__ float Ds[4][16];
30
31    As[ty][tx] = tex1Dfetch(textureA, 4 * ty + tx);
32    Bs[ty][tx] = tex1Dfetch(textureB, 4 * ty + tx);
33    Ds[tx][tz] = tex1Dfetch(textureD, tx);
34
```

```
35    __syncthreads();
40
41    for (int k = 0; k < 4; k++)
42    Csub += As[ty][k] *
43              __expf(Ds[tx][tz]*length) *
44            Bs[k][tx];
45    __syncthreads();
46
47    C[ty*4 + tx] = Csub;
48
49 }
...
50 int main(int argc, char **argv) {
...
51 cudaBindTexture(NULL,textureA,dA,16*sizeof(float));
52 cudaBindTexture(NULL,textureD,dD,4*sizeof(float));
53 cudaBindTexture(NULL,textureB,dB,16*sizeof(float));
54 cudaBindTexture(NULL,textureLengths,dLengths,
55                   numLengths*sizeof(float));
...
```

Code 6: Using Texture Memory.

Lines 1–4 of Cd. 6 declare the texture memory that will be used for the input data, while Lns. 51–55 bind the device memory to the texture memory. Note that dA in Ln. 51 is device memory that has the input already copied to it via cudaMemcpy(dA, ..., cudaMemcpyHostToDevice), and likewise for Lns. 52–55. While there is no significant overall performance improvement on Fermi or Kepler for small batches, there is a much larger improvement on Kepler with large batches, up to a 30 % increase over the kernel shown in Cd. 5.

Besides ensuring that each block has enough work allocated to it as was discussed with Cd. 4, a performance gain can be seen for some problems by adding to the amount of work that is assigned to each individual thread. Code 7 shows one way to go about giving each thread more work. The x-dimension of the grid is divided by some factor L, and each thread now steps through that grid dimension L times during each execution.

```
1 __global__ void exp4x4(float* output,
2                        float* A,
3                        float* D,
4                        float* B,
5                        float* lengths,
6                        int L) {
7
8    float* C;
9    float lengths;
```

```
10
11    int bx = blockIdx.x;
12    int tx = threadIdx.x;
13    int ty = threadIdx.y;
14
15    int tz = threadIdx.z;
16
17    float Csub;
18    __shared__ float As[4][4];
19    __shared__ float Bs[4][4];
20    __shared__ float Ds[4][16];
21
22    As[ty][tx] = tex1Dfetch(textureA, 4 * ty + tx);
23    Bs[ty][tx] = tex1Dfetch(textureB, 4 * ty + tx);
24    Ds[tx][tz] = tex1Dfetch(textureD, tx);
25
26    __syncthreads();
27
28    for (int l=0;l<L;l++) {
29
30      Csub=0.0;
31      int matrix_idx = l*gridDim.x +
32                       bx*blockDim.z + tz;
33      int matrix_addr = 4*4*(l*gridDim.x +
34                        bx*blockDim.z + tz);
35
36    length = tex1Dfetch(textureLengths, matrix_idx);
37
38      C = dMatrices + matrix_addr;
39
40      for (int k = 0; k < 4; k++)
41        Csub += As[ty][k] *
42                __expf(Ds[tx][tz]*length) *
43                Bs[k][tx];
44      __syncthreads();
45
46      C[ty*4 + tx] = Csub;
47
48    }
49
50 }
...
51 int main(int argc, char **argv) {
...
52 dim3 dimBlock(Nsize,Nsize,packsize);
```

```
53 dim3 dimGrid(((numLengths+packsize-1)/packsize/
   L,1,1);
54 exp4x4<<<dimGrid,dimBlock>>>(dOut,dA,dD,
55                               dB,dLengths,L);
```
. . .

Code 7: Adding Work for Each Thread.

On Ln. 53, the kernel launch parameters are changed so that the grid dimension as previously calculated is now divided by a factor L, and this parameter is passed into the kernel itself at Ln. 55. Then, in the kernel, a for loop is added at Ln. 28 which loops over chunks of the batch and calculates the stride matrix index and address at Lns. 31–34 using the width of the x-dimension of the grid as the stride length.

Similarly to the number of warps per block assignment, the optimal number of matrices to assign each thread execution can depend on the problem type and size. For our 4×4 batch matrix exponentiation, we found the greatest increase by setting L = 16, resulting in an increase of 32 and 48 % over the kernel shown in Cd. 6 for Fermi and Kepler, respectively.

Finally, there are a few things to point out regarding generalization of these kernels to handle problems with different parameters. First, in order to handle arbitrarily-sized batches of matrices, some hardware constraints must be kept in mind. Of course one must consider the total amount of device memory required to hold the input and output. However, for smaller matrices with constant A and B input matrices, as with the case of our motivating problem, the maximum number of blocks in a particular grid dimension must also be considered. This is especially true for Fermi, which has a limit of $2^{16} - 1$ for each dimension, whereas Kepler increases this to $2^{31} - 1$. To get around such a limitation, multiple grid dimensions can be used. However, caution and ideally prior knowledge about the problem specifics must be used, because the additional logic and kernel launch overhead required to accommodate the larger batch sizes can cause a performance decrease for smaller batch sizes versus using the simpler configuration.

Figure 3.2 shows the performance of the final kernel given by Cd. 7 versus the original kernel in Cd. 3 on both Fermi and Kepler for $M = 4$. An advantage for Cd. 7 can be seen at a batch size of 10,000 matrices, and the difference in performance grows significantly as the batch size increases from there. There is a dip in the Fermi curve at around 60,000 matrices for Cd. 7 due to the effect of the modified block allocation to allow for batch sizes larger than $2^{16} - 1$ matrices as discussed in the previous paragraph.

Another generalization that may be attempted is to allow for multiple matrix sizes with the same kernel code, replacing the hard-coded values of $M = 4$ at Lns. 18–20 and 42 of Cd. 7, for example. However, be aware that using a variable loop limit at Ln. 42 can impose a performance penalty because the compiler can no longer unroll this loop, which reduces performance by 35 and 48 % on Fermi and Kepler, respectively. This issue is more pronounced in later versions of CUDA, where the compiler has gotten better at increasing performance with automatic unrolling. Adding #pragma unroll to try to alleviate this problem only recovers a few percent of the performance that was lost.

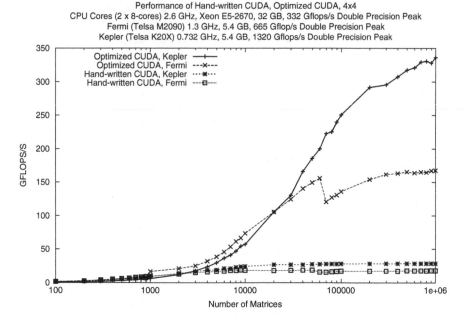

Fig. 3.2 Relative performance of handwritten kernels for $M = 4$ on Fermi and Kepler

3.4.2 $M = 20$ and $M = 60$ Cases

There are two other matrix sizes that are of interest to our motivating phylogenetics problem, $M = 20$ and $M = 60$, as described in Sect. 3.2. These additional sizes introduce a few considerations that can affect the performance of the kernels already shown.

The $M = 20$ case is similar to the $M = 4$ case already discussed. However, packing multiple matrices per block is not as important for the larger matrices. Assigning only a single matrix to each block, there are still 400 threads and 13 warps per block, which is enough to make efficient use of the hardware. Besides this difference, the kernel for the 20×20 case looks exactly like that in Cds. 3–7, except that the constants of '4' in the shared memory allocations and the loop limit for the computation should be changed to '20'. The enhancements discussed in Cds. 5 and 7 are also relevant to the 20×20 case and yield significant performance improvements.

The performance of the `cublasSgemmBatched` solution from Cd. 1, along with the original and optimized kernels presented in Cds. 3 and 7, as applied to the 20×20 case is shown in Fig. 3.3. Since the problem provides a denser computation, the performance of all methods increases relative to the $M = 4$ case. However, the ordering of relative performance of each method remains the same, with the optimized kernel of Cd. 7 yielding a nice performance advantage, even at smaller batch sizes.

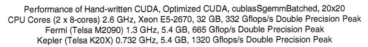

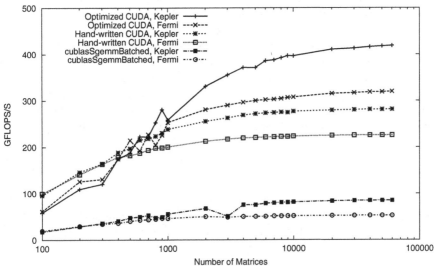

Fig. 3.3 Relative performance of handwritten kernels and cublasSgemmBatched for $M = 20$ on Fermi and Kepler

Since the beginning of this work, improvements have been made to NVIDIA's CUBLAS library and the `cublasSgemmBatched` implementation. In the most recent CUDA version 5.5 as of this writing, `cublasSgemmBatched` outperforms our kernel for the 60×60 case. However, since the eigendecomposition algorithm presented here outperforms `cublasSgemmBatched` for $M = 4$ and $M = 20$, and if a unified solution is desired, there are a couple of considerations to adapting the eigendecomposition method to the $M = 60$ case. On many devices the hardware limit on shared memory necessitates the use of the common tiling algorithm which brings the matrix into shared memory in smaller pieces to be worked on separately. This algorithm is taught in most introductory CUDA programming materials and is not repeated here. After tiling has been applied to the kernel as previously presented, then the remaining optimizations (excepting the multiple matrices per block, which has been replaced by tiling) can be applied, although the effect of increasing the computational work of each thread is expected to be less pronounced since each thread is already working on multiple tiles.

3.5   Alternative Methods for Matrix Exponentiation

Besides the eigendecomposition method presented here, there are many other ways
to numerically compute matrix exponentiation[37]. We chose to evaluate two of
these alternative methods which appeared to have a computational pattern that
would fare well for a batched application on the GPU.

The first method is Lagrange interpolation, which can be applied to matrix exponentiation as shown in Eq. (3.10). Note that for greatest efficiency, the eigenvalues
of the matrix A should be known and precomputed, as is the case of the matrix A
from the phylogenetics problem in Eq. (3.5) because it does not change throughout
the batch operation (only t changes from case to case within the batch).

$$e^{tA} = \sum_{j=1}^{n} e^{\lambda_j t} \prod_{k=1, k \neq j}^{n} \frac{(A - \lambda_k I)}{(\lambda_j - \lambda_k)} \tag{3.10}$$

Another viable option that was identified is Newton interpolation, which is shown
in Eqs. (3.11) and (3.12) for matrix exponentiation. This method also presumes
knowing the eigenvalues of the A matrix beforehand,

$$e^{tA} = e^{\lambda_1 t} I + \sum_{j=2}^{n} [\lambda_1, \cdots, \lambda_j] \prod_{k=1}^{j-1} (A - \lambda_k I), \tag{3.11}$$

where the $[\lambda_1, \cdots, \lambda_j]$ are functions of t and are recursively defined as [37],

$$[\lambda_1, \lambda_2] = (e^{\lambda_1 t} - e^{\lambda_2 t})/(\lambda_1 - \lambda_2),$$

$$[\lambda_1, \cdots, \lambda_{k+1}] = \frac{[\lambda_1, \cdots, \lambda_k] - [\lambda_2, \cdots, \lambda_{k+1}]}{\lambda_1 - \lambda_{k+1}} \quad (k \geq 2). \tag{3.12}$$

The performance of the Lagrange and Newton interpolation for the simplest
$M = 4$ case is shown in Fig. 3.4. While both the Lagrange and Newton
interpolation methods perform well compared to the `cublasSgemmBatched`
solution presented in Sect. 3.3.1, for much larger batch sizes, Newton interpolation
becomes preferable to Lagrange interpolation. However, neither of these methods
outperforms the eigendecomposition method as shown in Cd. 7.

3.6   Conclusions

We have focused here on batch matrix–matrix multiplication as applied to
matrix exponentiation, for the phylogenetics domain (fixed A and B, 4×4,
20×20, and 60×60), in single precision. The methods discussed also apply

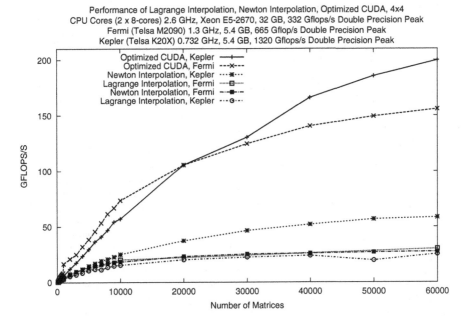

Fig. 3.4 Performance of Lagrange and Newton interpolation on Fermi and Kepler

to matrix exponentiation for variable A, B, and matrix size. The techniques can easily be applied to the general matrix–matrix multiplication problem: `C:=alpha*op(A)*op(B)+beta*C`, where `op(X)` is one of `op(X)=X` or `op(X)=X'`. Note that the matrices for the general problem can be rectangular. Our methods apply to any of the four precisions: single precision real, double precision real, single precision complex, and double precision complex. When all precisions are supported, it is common practice to maintain the code base in one precision only, say double precision complex, then generate the other three precisions automatically. Finally, our techniques clearly lend themselves to auto-tuning.

References

1. AMD Core Math Library (ACML): www.amd.com/acml. Cited 16 Dec 2013
2. Amestoy, P.R., Duff, I.S., L'Excellent, J.Y.: Multifrontal parallel distributed symmetric and unsymmetric solvers. Comput. Methods Appl. Mech. Eng. (2000). doi: 10.1016/S0045-7825(99)00242X
3. Anderson, E., Bai, Z., Bischof, C., Blackford, L.S., Demmel, J.W., Dongarra, J.J., Du Croz, J., Greenbaum, A., Hamarling, S., McKenney, A., Sorensen, D.: LAPACK Users' Guide. SIAM (1992). http://www.netlib.org/lapack/lug/. Cited 16 Dec 2013
4. Ayres, D.L., Darling, A., Zwickl, D.J., Beerli, P., Holder, M.T., Lewis, P.O., Huelsenbeck, J.P., Ronquist, F., Swofford, D.L., Cummings, M.P., Rambaut, A., Suchard, M.A.: BEAGLE: an application programming interface and high-performance computing library for statistical phylogenetics. Syst. Biol. **61**(1), 170–173 (2012)

5. Basic Linear Algebra Technical Forum: http://www.netlib.org/blas/blast-forum/blas-report. pdf. Cited 16 Dec 2013
6. Blackford, L.S., Choi, J., Cleary, A., D'Azevodo, E., Demmel, J., Dhillon, I., Dongarra, J.J., Hammarling, S., Henry, G., Petitet, A., Stanley, K., Walker, D., Whaley, R.C.: ScaLAPACK Users' Guide. SIAM (1997). http://www.netlib.org/scalapack/slug/. Cited 16 Dec 2013
7. CUBLAS: https://developer.nvidia.com/cuBLAS. Cited 16 Dec 2013
8. CUBLAS Documentation: http://docs.nvidia.com/cuda/cublas/. Cited 16 Dec 2013
9. CUDA C Programming Guide: http://docs.nvidia.com/cuda/cuda-c-programming-guide/ index.html. Cited 16 Dec 2013
10. CUDA Toolkit Documentation: http://docs.nvidia.com/cuda/cuda-samples/. Cited 16 Dec 2013
11. CULA Tools: http://www.culatools.com/blog/2011/12/09/batched-operations/. Cited 16 Dec 2013
12. Demmel, J., Volkov, V.: Benchmarking GPUs to tune dense linear algebra. In: Proceedings of the 2008 ACM/IEEE Conference on Supercomputing, vol. 31. IEEE Press, Piscataway (2008)
13. Demmel, J.W., Eisenstat, S.C., Gilbert, J.R., Li, X.S., Liu, J.W.H.: A supernodal approach to sparse partial pivoting. SIAM J. Matrix Anal. Appl. (1999). doi: 10.1137/S0895479895291765
14. Donfack, S., Dongarra, J., Faverge, M., Gates, M., Kurzak, J., Luszczek, P., Yamzaki, I.: LAPACK working note 280: On Algorithmic Variants of Parallel Gaussian Elimination: Comparison of Implementations in Terms of Performance and Numerical Properties. Innovative Computing Laboratory, University of Tennessee, Knoxville (2013)
15. Dong, T., Dovrev, V., Kolev, T., Rieben, R., Tomov, S., Dongarra, J.: Hydrodynamic Computation with Hybrid Programming on CPU-GPU Clusters. Innovative Computing Laboratory, University of Tennessee (2013)
16. Dongarra, J.J., Luszczek, P., Petitet, A.: The LINPACK benchmark: past, present and future. Concurr. Comput. Pract. Exp. (2003). doi: 10.1002/cpe.728
17. Drummond, A., Rambaut, A.: BEAST: Bayesian evolutionary analysis by sampling trees. BMC Evol. Biol. **7**, 214 (2007)
18. Drummond, A., Suchard, M., Xie, D., Rambaut, A.: Bayesian phylogenetics with BEAUti and the BEAST 1.7. Mol. Biol. Evol. **29**(8), 1969–1973 (2012)
19. Durbin, R., Eddy, S., Mitchison, G.: Biological Sequence Analysis: Probabilistic Models of Proteins and Nucleic Acids, 1st edn. Cambridge University Press, Cambridge (1997)
20. Felsenstein, J.: Evolutionary trees from DNA sequences: a maximum likelihood approach. J. Mol. Evol. **17**, 368–376 (1981)
21. Felsenstein, J.: Inferring Phylogenies. Sinauer Associates, Sunderland (2003)
22. Feng, X., Buell, D., Rose, J., Waddell, P.: Parallel algorithms for Bayesian phylogenetic inference. J. Parallel Distrib. Comput. **63**, 707–718 (2003)
23. Feng, X., Cameron, K., Sosa, C., Smith, B.: Building the tree of life on terascale systems. In: Parallel Distributed Processing Symposium (IPDPS 2007), Washington (2007)
24. GoToBLAS: Texas Advanced Computing Center. http://www.tacc.utexas.edu/. Cited 16 Dec 2013
25. Hasegawa, M., Kishino, H., Yano, T.: Dating of the human-ape splitting by a molecular clock of mitochondrial DNA. J. Mol. Evol. **22**(2), 160–174 (1985)
26. Huelsenbeck, J.P., Ronquist, F.: MrBayes: Bayesian inference of phylogenetic trees. Bioinformatics **17**, 754–755 (2001)
27. Huelsenbeck, J.P., Ronquist, F., Nielsen, R., Bollback, J.P.: Bayesian inference of phylogeny and its impact on evolutionary biology. Science **294**(5550), 2310–2314 (2001)
28. IBM: Engineering and Scientific Subroutine Library (ESSL) and parallel ESSL. http://www-03.ibm.com/systems/p/software/essl. Cited 16 Dec 2013

29. Jhurani, C., Mullowney, P.: A GEMM interface and implementation on NVIDIA GPUs for multiple small matrices. www.ices.utexas.edu/$\char126$chetan/preprints/2013-CJ-PM-GEMM.pdf. Cited 16 Dec 2013

30. Keane, T., Naughton, T., Travers, S., McInerney, J., McCormack, G.: DPRml: distributed phylogeny reconstruction by maximum likelihood. Bioinformatics 21, 969974 (2005)

31. Keeneland: http://keeneland.gatech.edu/. Cited 29 Jan 2014

32. Kepler Whitepaper: http://www.nvidia.com/content/PDF/kepler/NVIDIA-Kepler-GK110-Architecture-Whitepaper.pdf. Cited 16 Dec 2013

33. Kurzak, J., Tomov, S., Dongarra, J.: LAPACK Working Note 245: Autotuning GEMMs for Fermi. Innovative Computing Laboratory, University of Tennessee (2011)

34. Kurzak, J., Luszczek, P., Tomov, S., Dongarra, J.: LAPACK Working Note 267: Preliminary Results of Autotuning Gemm Kernels for the NVIDIA Kepler Architecture. Innovative Computing Laboratory, University of Tennessee (2012)

35. Math Kernel Library (MKL): Intel(R). http://www.intel.com/cd/software/products/asmo-na/eng.347757.htm. Cited 16 Dec 2013

36. Minh, B., Vinh, L., Haeseler, A., Schmidt, H.: pIQPNNI: parallel reconstruction of large maximum likelihood phylogenies. Bioinformatics 21, 3794–3796 (2005)

37. Moler, C., Van Loan, C.: Nineteen dubious ways to compute the exponential of a matrix, twenty-five years later. SIAM Rev. (2003). doi: 10.1137/S00361445024180

38. Moret, B., Badar, D., Warnow, T.: High-performance algorithm engineering for computational phylogenetics. J. Supercomput. 22, 99–11 (2002)

39. Nath, R., Tomov, S., Dongarra, J.: An improved MAGMA GEMM for Fermi GPUs. Int. J. High Perform. Comput. 24(4), 511–515 (2010)

40. Schmidt, H., Strimmer, K., Vingron, M., Haeseler, A.: TREE-PUZZLE: maximum likelihood phylogenetic analysis using quartets and parallel computing. Bioinformatics 18(2), 503–504 (2002)

41. Stamatakis, A., Meier, L.T.: RAxML-III: a fast program for maximum likelihood-based inference of large phylogenetic trees. Bioinformatics 21(4), 456–463 (2005)

42. Suchard, M., Rambaut, A.: Many-core algorithms for statistical phylogenetics. Bioinformatics 25, 1370–1376 (2009)

43. Tierney, L.: Markov chains for exploring posterior distributions. Ann. Stat. 22(4), 1701–1728 (1994)

44. Whaley, C.R., Petitet, A., Dongarra, J.: Automated empirical optimizations of software and the ATLAS project. Parallel Comput. 27(1–2), 3–35 (2001)

45. Zwickl, D.: Genetic algorithm approaches for the phylogenetic analysis of large biological sequence datasets under the maximum likelihood criterion. Ph.D. dissertation, University of Texas, Austin (2006)

Chapter 4
Efficient Batch LU and QR Decomposition on GPU

William J. Brouwer and Pierre-Yves Taunay

4.1 Batch LU Decomposition

While comparatively expensive, direct solvers based around matrix decomposition are used in various applications, for reasons of numerical stability, over iterative solvers. The implementation presented shortly was originally devised for the solution of many decoupled systems simultaneously [4], for what amounts to a domain decomposition approach [6]. The LU decomposition also provides a viable method for the calculation of the matrix determinant; after execution of an in-place implementation, the determinant is available from the product of the diagonal elements. This is particularly useful in condensed matter physics, specifically in studies of the fractional quantum Hall effect based on construction of the Pfaffian wave function, which requires $O(N!)$ determinant evaluations [9, 10].

4.1.1 Theory

The decomposition of matrix A into lower L (elements α_{ij}) and upper U (elements β_{ij}) matrix,

$$\mathbf{LU} = \mathbf{A}, \qquad (4.1)$$

has the advantage of permitting the solution of linear systems in two steps, comprised of forward and backward substitution procedures, for multiple right hand sides in $Ax = y$. Crout's approach to LU decomposition solves the set of equations implicit to Eq. (4.1); these are:

W.J. Brouwer (✉) • P.-Y. Taunay
The Pennsylvania State University, University Park, PA, USA
e-mail: wjb19@psu.edu; py.taunay@psu.edu

V. Kindratenko (ed.), *Numerical Computations with GPUs*,
DOI 10.1007/978-3-319-06548-9_4, © Springer International Publishing Switzerland 2014

$$\beta_{ij} = a_{ij} - \sum_{k=1}^{i-1} \alpha_{ik}\beta_{kj}, \tag{4.2}$$

and

$$\alpha_{ij} = \frac{1}{\beta_{jj}} \left(a_{ij} - \sum_{k=1}^{j-1} \alpha_{ik}\beta_{kj} \right). \tag{4.3}$$

Numerical stability relies on suitable choice of pivot, or dividing element in the solution for α_{ij}. Pivoting may be partial (a row interchange) or full (both row and column); the former is implemented in this chapter. Following the approach detailed in Numerical Recipes [5], the choice of the best pivot is made only after both Eqs. (4.2) and (4.3) are solved for a given column, and thereafter the row swap and a scaling performed. Recording the row permutations in a separate vector is required for use with the solution of linear equations, in order that the right hand side vector be subsequently rearranged to suit. Equations (4.2) and (4.3) give rise to $N^2 + N$ equations, whose overdetermined nature permits the setting of N elements arbitrarily. A popular choice is to set the diagonal elements of α to one, followed in this chapter. Crout's approach to LU decomposition is summarized in Algorithm 1.

4.1.2   GPU Implementation

With the foreknowledge that the decomposition will be applied in batch, the mapping of computational thread to matrix is a seemingly reasonable strategy for a GPU implementation. However, on the device this virtually eliminates the possibility of coalesced loads from global memory, and thread cooperation via shared memory, key requirements for good performance. At the other extreme, mapping thread to matrix element would introduce significant overhead in the form of synchronization, owing to dependencies between the loops described in Algorithm 1. In a compromise between the two extremes, $O(N)$ threads were assigned to the operations for each matrix, and individual CUDA thread blocks assigned one or more matrices to process. Referring to Algorithm 1, there are at least two key points at which threads must cooperate. The first is the determination of scaling information, lines 1–5, which may be considered a separate scope to lines 6 forward. This task is readily solved using parallel reduction, a well known primitive. Turning attention to the main steps of the algorithm, lines 7–13 perform updates to matrix elements above the diagonal, specifically column j. By assigning the index of the loop at line 7 to thread index, increasingly more threads in this scope work as the outer loop progresses; a brief summary of this scope as executed in CUDA is detailed in Table 4.1. Within a warp, one may rely on SIMD execution, and thus updated column elements are available to threads of higher indices when needed.

Algorithm 1 LU decomposition with partial pivoting

Input : **A**, Batch of $N \times N$ matrices
Output: **A**, In–place LU decomposed matrices

```
 1 for i ← 0 to N − 1 do
 2 │   for j ← 0 to N − 1 do
   │   │   // find largest element q
 3 │   end
 4 │   scale[i] = 1.0/q;
 5 end
 6 for j ← 0 to N − 1 do
 7 │   for i ← 0 to j − 1 do
 8 │   │   sum = A[i][j];
 9 │   │   for k ← 0 to i − 1 do
10 │   │   │   sum− = A[i][k] ∗ A[k][j];
11 │   │   end
12 │   │   A[i][j] = sum;
13 │   end
14 │   for i ← j to N − 1 do
15 │   │   sum = A[i][j];
16 │   │   for k ← 0 to j − 1 do
17 │   │   │   sum− = A[i][k] ∗ A[k][j];
18 │   │   end
19 │   │   A[i][j] = sum;
20 │   end
   │   // find index l of largest element q = scale[i]*fabs(sum)
21 │   if j != l then
   │   │   // swap rows j and l
   │   │   // update scale
   │   │   // save permutation details
22 │   end
23 │   if j! = N − 1 then
24 │   │   sum = A[j][j] ;
25 │   │   for k ← j to N − 1 do
26 │   │   │   A[i][j]/ = sum ;
27 │   │   end
28 │   end
29 end
```

As one might expect, matrices of side greater than a single warp require serialization of warp execution, due to the unpredictable way in which instructions are scheduled and dispatched within the Streaming Multiprocessor (SM), as illustrated in Fig. 4.1. Some parallelism is regained by mapping matrix to warp, for this scope alone.

No such limitations pervade lines 14–20, where loop index is also mapped to thread index, and column data is read from above the diagonal. Threads in this scope update from diagonal downwards; however, barrier synchronization is necessary before and after this scope. The particular column updated in a single iteration of

Table 4.1 Global memory read[], shared memory read(), write{ }, critical†
and arithmetic operations for several iterations and CUDA threads t_id of
algorithm lines 7–14

k	t_id	$j=2$	$j=3$	$j=4$	$j=5$
–	1	(1,2)	(1,3)	(1,4)	(1,5)
0	1	−[1,0]*(0,2)	−[1,0]*(0,3)	−[1,0]*(0,4)	−[1,0]*(0,5)
–	1	{1,2}	{1,3}†	{1,4}†	{1,5}†
–	2		(2,3)	(2,4)	(2,5)
0	2		−[2,0]*(0,3)	−[2,0]*(0,4)	−[2,0]*(0,5)
1	2		−[2,1]*(1,3)†	−[2,1]*(1,4)†	−[2,1]*(1,5)†
–	2		{2,3}	{2,4}†	{2,5}†
–	3			(3,4)	(3,5)
0	3			−[3,0]*(0,4)	−[3,0]*(0,5)
1	3			−[3,1]*(1,4)†	−[3,1]*(1,5)†
2	3			−[3,2]*(2,4)†	−[3,2]*(2,5)†
–	3			{3,4}	{3,5}†
–	4				(4,5)
0	4				−[4,0]*(0,5)
1	4				−[4,1]*(1,5)†
2	4				−[4,2]*(2,5)†
3	4				−[4,3]*(3,5)†
–	4				{4,5}

Fig. 4.1 An example of
instruction scheduling and
execution in a streaming
multiprocessor

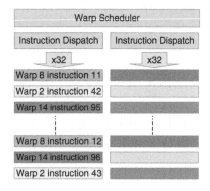

the outer loop is cached in shared memory before line 7, and written back to global
after line 20. Shared memory buffers used for communication are declared using the
volatile keyword, to ensure that write operations are not optimized out during
compilation. Once the column update is complete, and working threads have written
elements q before line 20 to another shared memory buffer, parallel reduction is
employed in order to find the index of the pivot. Should the condition at line 21 be
satisfied, then a row swap is completed by threads, storing temporary elements in
registers. Thereafter, row elements are scaled by diagonal elements; once again loop
index k is mapped to thread. Barrier synchronization is employed before the end of

Table 4.2 LU algorithm executed on K40c GPU device versus 16 Intel E5-2670 (Sandy Bridge) CPU threads

Batch size	Matrix size	K40c (s)	CPU(s)	mats./blk
800	256	1.5	1.5	1
1,600	128	0.33	0.45	1
8,000	64	0.20	0.30	2
16,000	32	0.05	0.11	4
64,000	16	0.03	0.15	8

the outer loop at line 29. An abbreviated listing of the main CUDA kernel is recorded in Appendix 1, based around the `float2` type, for processing complex data.

4.1.3 LU Results

An implementation of Algorithm 1 was written in C for execution on CPU, for use with row-major storage format matrices and complex (single precision) floating point data. This routine was compiled using a recent revision of the Intel compiler, with flags `-O3 -xHost` to ensure the highest degree of optimization, taking advantage of AVX hardware and instructions of the Sandy Bridge CPU. OpenMP was used to distribute matrices to separate threads for processing. The main GPU kernel as described and supporting routines including parallel reduction were compiled using `nvcc`, CUDA revision 5.5, for compute architecture 3.5 and with optimization flag `-O3`. Table 4.2 summarizes results, comparing execution times. Profiling using `nvvp` revealed a total global memory bandwidth of approximately 62 GB/s (54.5 GB/s read + 7.5 GB/s write). Both CPU and GPU routines were devoted to calculating the in-place LU decomposition alone. No permutations were stored; however, the sign of the permutation was recorded in memory, as is necessary for any subsequent calculation of matrix determinants. Crout's algorithm when executed on the K40c device experienced a 1.0–5.0x performance improvement over a single Sandy Bridge CPU socket, running 16 threads. The super-linear scaling of the CPU results was investigated further using tools from the Valgrind suite [8]. As expected, the effect had little correlation with cache performance; miss rates for both instructions and data were negligible for all matrix and batch sizes considered. However, profiling with `callgrind` did reveal that instructions devoted directly to the LU calculation itself steadily increased as a fraction of the total instructions, with matrix size. This fraction was as little as 60 % for a matrix of side 32, increasing to almost 100 % for matrices of side 256. Similarly, the percentage of instructions derived from other sources, particularly the Intel KMP interface for thread management and CPU affinity decreased to negligible contributions, for matrices of side 256.

4.2   QR Decomposition

While also a method that may be applied in the solution of systems of linear equations, the QR decomposition,

$$\mathbf{QR} = \mathbf{A},\tag{4.4}$$

generally takes preeminence in a popular approach to eigendecomposition, the QR algorithm. In numerical implementations of the QR decomposition algorithm, the upper diagonal matrix R is constructed by the action of operations on A. R can be produced by one of several means, the most popular being Householder reflections, or Givens rotations [3]. This chapter focuses on the latter, whereby successive rotations G_i are applied, selectively eliminating elements below the diagonal of A, and producing the upper diagonal matrix R. One such step for the first column of a 3×3 complex matrix is illustrated in Eq. (4.5), where * denotes the complex conjugate.

$$\begin{bmatrix} 1 & 0 & 0 \\ 0 & c & s \\ 0 & -s* & c \end{bmatrix} \begin{bmatrix} a_{11} & a_{12} & a_{13} \\ a_{21} & a_{22} & a_{23} \\ a_{31} & a_{32} & a_{33} \end{bmatrix} = \begin{bmatrix} a_{11} & a_{12} & a_{13} \\ a'_{21} & a'_{22} & a'_{23} \\ 0 & a'_{32} & a'_{33} \end{bmatrix}\tag{4.5}$$

4.2.1   Theory

4.2.1.1   Serial QR Decomposition

The kernel of rotation matrix G_i is a 2×2 matrix that operates on pairs of values $a = a_{i,j}$ and $b = a_{i+1,j}$ in A, where elements c and s are chosen to eliminate the lower element in the operation:

$$\begin{bmatrix} c & s \\ -s* & c \end{bmatrix} \begin{bmatrix} a \\ b \end{bmatrix} = \begin{bmatrix} r \\ 0 \end{bmatrix}.\tag{4.6}$$

Bindel et al. [1] give expressions for suitable c and s in a variety of contexts; the following are used in the remainder of this chapter for complex values, analogous to those for real values:

$$c = \pm \frac{|a|}{\sqrt{|a|^2 + |b|^2}},\tag{4.7}$$

$$s = \pm \mathrm{sgn}(a) \frac{b}{\sqrt{|a|^2 + |b|^2}},\tag{4.8}$$

where

$$\text{sgn}(a) = \begin{cases} a/|a| & \text{if } a \neq 0 \\ 1 & \text{if } a = 0 \end{cases}.$$
(4.9)

The concatenation of all orthogonal operations G_i comprises the transpose of the orthogonal matrix Q ie., using 0-based indexing,

$$Q^T A = \left[\prod_{j=0}^{j=N-2} \left\{ \prod_{i=j}^{i=N-2} G_i^j \right\} \right] A = R$$
(4.10)

where the superscript on G refers to the matrix column operated on during a particular iteration.

4.2.1.2 Parallel QR Decomposition

Sameh and Kuck [7] developed a parallel scheme dedicated to matrices of even side, in which the elimination process pictured in Eq. (4.5) can be carried in parallel across multiple rows and columns. Multiple independent Givens rotations $\tilde{Q}_{m,n}$ can be executed at the same time, where m and n refer to the row and column indices of the eliminated element. The product of these matrices constructs the matrix $\hat{Q}_i$, which is applied at the i-th step of the algorithm:

$$\hat{Q}_i = \prod \tilde{Q}_{m,n}.$$
(4.11)

For a given step i, the matrices $\tilde{Q}_{m,n}$ can be multiplied in any order to obtain $\hat{Q}_i$, as they are a direct sum of plane rotations [7]. As a result, $\hat{Q}_i$ is a block-diagonal matrix, with Givens rotations matrices G_i on the diagonal, as pictured in Eq. (4.12).

$$\hat{Q}_i = \begin{bmatrix} 1 \\ & \ddots \\ & & 1 \\ & & & c_{k,l} & s_{k,l} \\ & & & -s_{k,l}* & c_{k,l} \\ & & & & & \ddots \\ & & & & & & c_{m,n} & s_{m,n} \\ & & & & & & -s_{m,n}* & c_{m,n} \\ & & & & & & & & 1 \end{bmatrix}$$
(4.12)

The scheme from Sameh and Kuck is completed in $2N - 3$ steps, where N is the rank of the matrix. The i-th transform is obtained by eliminating an entry in A at the row m and column n, where m and n are given by

$$m = \begin{cases} \{N-i, N-i+1, \ldots, N-1-\delta(i)\} & 1 \le i \le N-1 \\ \{i-N+2, i-N+4, \ldots, N-1-\delta(i)\} & N \le i \le 2N-3 \end{cases}, \quad (4.13)$$

and

$$n = \begin{cases} \left\{1, 2, \ldots, \lceil \frac{i}{2} \rceil \right\} & 1 \le i \le N-1 \\ \left\{i-N+2, i-N+3, \ldots, \lceil \frac{i}{2} \rceil \right\} & N \le i \le 2N-3 \end{cases}, \quad (4.14)$$

with $\delta(i)$ defined as

$$\delta(i) = \begin{cases} 0 & i \text{ odd} \\ 1 & i \text{ even} \end{cases}. \quad (4.15)$$

Though other elimination patterns are possible, this approach has been proven to be one of the most efficient, both from a practical and mathematical point of view, as it is easy to implement and asymptotically optimal [2].

At each step of the process, the total number of rotations performed simultaneously, N_{rot}, is obtained by counting the total number of columns n and rows m affected:

$$N_{rot} = \begin{cases} \lceil i/2 \rceil & 1 \le i \le N-1 \\ \lceil i/2 \rceil - i + N - 1 & N \le i \le 2N-3 \end{cases}. \quad (4.16)$$

An example of the entries successively eliminated by this algorithm is shown in Fig. 4.2, for an 8×8 matrix. Numbers in the matrix correspond to the order in which the associated matrix element is eliminated in the algorithm.

Fig. 4.2 Illustration of the successive elimination scheme in the QR parallel decomposition algorithm, for an 8×8 matrix

$$\begin{bmatrix} * & * & * & * & * & * & * & * \\ 7 & * & * & * & * & * & * & * \\ 6 & 8 & * & * & * & * & * & * \\ 5 & 7 & 9 & * & * & * & * & * \\ 4 & 6 & 8 & 10 & * & * & * & * \\ 3 & 5 & 7 & 9 & 11 & * & * & * \\ 2 & 4 & 6 & 8 & 10 & 12 & * & * \\ 1 & 3 & 5 & 7 & 9 & 11 & 13 & * \end{bmatrix}$$

Algorithm 2 Outer loop of the parallel QR decomposition

Input: A, Batch of $N \times N$ matrices
Input: NMAT, Size of the batch
Input: NMPBL, Number of matrices to process per CUDA block
Input: NTH, Total number of CUDA threads per block
Input: blocks, CUDA Grid configuration
Output: A,Q, Upper diagonal batch of matrices R stored in place in A,
            batch of the transpose of the orthogonal matrices Q

1 blocks.x $\leftarrow \lceil \dfrac{\text{NMAT}}{\text{NMPBL}} \rceil$

2 $Q \leftarrow \mathcal{I}_N$

3 **for** $1 \leq i \leq 2*N\text{-}3$ **do**

4   **if** $i < N$ **then**

5   blocks.z $\leftarrow \lceil \dfrac{i}{2} \rceil$

6   **else**

7   blocks.z $\leftarrow \lceil \dfrac{i}{2} \rceil - i + N - 1$

8   **end**

9   QR_Kernel $<<<$ blocks,NTH $>>>$ (Q,A,i)

10 **end**

4.2.2 GPU Implementation

The previous observations made in Sect. 4.1.2 related to global and shared memory accesses are also valid for the QR decomposition; therefore, each CUDA thread block is assigned one or more matrices to process, and N threads operate on a single matrix. The parallel QR algorithm is driven by an outer loop executed on the CPU, as detailed in Algorithm 2. This routine calculates the number of CUDA blocks to run in the x-dimension of the CUDA grid, initializes the orthogonal matrix Q as the identity matrix, and calculates the total number of Givens rotations that can be executed in parallel, based on Eq. (4.16). This number sets the z-dimension of the CUDA grid, to ensure that a total of N_{rot} Givens rotations are applied in parallel to the same matrix, at each iteration of the outer loop. Finally, each iteration launches the CUDA kernel to be executed on the GPU, shown in Algorithm 3. Each CUDA block in the x-dimension performs operations on multiple matrices A, and accumulates the results in the corresponding matrix Q. All threads first calculate the indices m, n of the entry to eliminate in their corresponding matrix. Threads then load rows $m - 1$ and m, on lines 10 and 11, subsequently calculating their corresponding Givens rotation, on line 14. Algorithm 4 details this operation: multiple threads load the elements a and b defined in Eq. (4.6) through a shared memory broadcast on lines 1 and 2. The components of the Givens rotation kernel, c and s, are then evaluated on line 3 based on Eqs. (4.7) through (4.9). Turning attention back to Algorithm 3, the threads perform the Givens rotation on their

---

Algorithm 3 QR_KERNEL Core GPU kernel for the parallel QR decomposition

---

Input   : **A**, Batch of $N \times N$ matrices
Input   : **Q**, Batch of $N \times N$ matrices
Input   : **NMPBL**, Number of matrices to process per block
Input   : **i**, Kuck–Sameh iteration number
Input   : **upperRow, lowerRow**, Shared memory buffers

```
// Variables for convenience
```
1 tdx ← threadIdx.x
2 bdx ← blockIdx.x
3 bdz ← blockIdx.z

```
// Calculate the location of the matrix on which to act
```
4 myMatrix ← tdx / N
5 matrixLocation ← (bdx * NMPBL + myMatrix) * N * N
6 myIndex ← tdx % N

```
// Calculate the indices m and n on which to act
```
7 [m,n] ← CalcIndices (i,bdz)
```
// Calculate the memory location of rows m and n
```
8 memLocUp ← matrixLocation + (m-1)*N+myIndex
9 memLocLo ← matrixLocation + m*N+myIndex

```
// Load the rows m and m-1 of the matrix in the shared memory
```
10 upperRow[tdx] ← A[memLocUp]
11 lowerRow[tdx] ← A[memLocLo]
```
// Wait for all the data to be loaded
```
12 syncthreads ()

```
// Calculate the Givens rotation
```
13 smemIdx ← myMatrix*N + n
14 [c,s] ← CalcGivens(upperRow, lowerRow, smemIdx)

```
// Apply the Givens rotation to all A matrices
```
15 ApplyGivens(A,upperRow,lowerRow,c,s,memLocUp,memLocLo)

```
// Load the rows m and m-1 of Q in shared memory
```
16 upperRow[threadIdx.x] ← Q[memLocUp]
17 lowerRow[threadIdx.x] ← Q[memLocLo]

```
// Accumulate the Givens rotation for all Q matrices
```
18 ApplyGivens(Q,upperRow,lowerRow,c,s,memLocUp,memLocLo)

---

corresponding matrix with the APPLYGIVENS routine. The details of this function
are outlined in Algorithm 5. In the APPLYGIVENS routine, each thread within
a CUDA block operates on a single matrix element of the two rows loaded in
upperRow and lowerRow. The calculation presented in Eq. (4.6) is performed on
lines 5 and 6. The threads then store the data back in place, in global memory,

---

Algorithm 4 CALCGIVENS Calculate the [c,s] values of a Givens rotation

---

Input: upperRow, lowerRow, Shared memory buffers
Input: smemIdx, Location of the corresponding matrix in the shared
    memory
Output: c,s, Givens rotation kernel values
    // All threads calculate the Givens rotation for their
    corresponding matrix
1 a ← upperRow[smemIdx];
2 b ← lowerRow[smemIdx];
3 [c,s] ← Givens (a,b);

Algorithm 5 APPLYGIVENS Apply the [c,s] Givens rotation to an array of matrices

---

input   : **M,** Batch of $N \times N$ matrices to update
input   : **upperRow, lowerRow,** Shared memory buffers
input   : **c,s,** Givens rotation kernel values
input   : **memLocUp, memLocLo,** Per thread location to update in M
1 tdx ← threadIdx.x;
2 u ← upperRow[tdx];
3 l ← lowerRow[tdx];
    // Perform the rotation
4 syncthreads ();
5 upperRow[tdx] ← u*c+l*s;
6 lowerRow[tdx] ← u*(-s)+l*c;
    // Write the two modified rows back in global memory
7 M[memLocUp] ← upperRow[tdx];
8 M[memLocLo] ← (myIndex > n)*lowerRow[tdx];

---

on lines 7 and 8. Care is taken to introduce an exact zero for columns 1 through $n - 1$ with the boolean condition myIndex > n on line 8, in order to avoid floating point approximations. The remainder of the Algorithm 3—lines 16 through 18— accumulates the rotations in the matrix Q. Note that the boolean condition on line 8 of Algorithm 5 does not apply to matrix Q, as can be discerned from the last line of QR_Kernel in Appendix 2.

Memory optimizations are included in the QR kernel implementation. A few constants, for example the current iteration number and the total batch size are stored in constant memory to provide fast data access. The bandwidth-cost of copying the data from the CPU to the GPU through a call to cudaMemcpyToSymbol () does not impact the overall performance of the algorithm. Care is taken to avoid non-coalesced global memory accesses by providing contiguous indices for global memory loads and stores.

Table 4.3 QR parallel decomposition algorithm executed on K40c GPU device versus 16 Intel E5-2670 (Sandy Bridge) CPU threads, in ms

Matrix side	Batch size			Matrices per block
	1,000	10,000	100,000	
16	1.370 (14.3)	7.475 (6.03)	68.14 (3.26)	64
32	6.732 (4.82)	55.76 (2.86)	534.0 (1.83)	32
64	48.70 (2.5)	457.9 (1.73)	4,630 (0.87)	16
128	404.9 (1.69)	4,025 (0.81)	–	8
256	3,172 (0.76)	32,151 (0.57)	–	4

The number in parenthesis indicates the speedup over the QR serial decomposition executed on the CPU

4.2.3   QR Results

A serial implementation of the QR decomposition algorithm as described in the first paragraph of Sect. 4.2.1 was written in C for execution on the CPU. The source code was compiled with the latest AVX optimizations available for Intel processors, with flags -O3 -xHost. The core GPU kernel QR_Kernel was compiled with the CUDA 5.5 revision of nvcc for compute architecture 3.5, and with -O3 optimizations. The GPU method was tested on a Kepler K40c, while the CPU implementation was executed on a single Sandy Bridge CPU socket running 16 OpenMP threads. Benchmarking results are presented in Table 4.3. The GPU implementation of the QR algorithm as outlined here demonstrates a 0.6–14.3x performance improvement over a comparable CPU routine. The Nvidia profiler nvvp revealed a global memory bandwidth of 195 GB/s (97.5 GB/s read + 97.5 GB/s write).

Table 4.3 shows that the GPU results scale linearly at a constant matrix size. However, the scaling is not linear with the matrix size, at constant batch size; this effect can be attributed to a decreasing total number of matrices processed per block, as the size of the matrices increase. Therefore, more blocks are scheduled and executed on the GPU, resulting in a larger overhead. The QR GPU kernel as described was revealed to be *memory-bound* by the Nvidia profiler. Thus, additional optimizations to help the code scale with the matrix size may include increasing the total work performed by individual CUDA threads, in order to keep the total number of matrices processed per block constant. The super-linear behavior observed in the CPU scaling results was deduced to share similar origins as those of the CPU LU implementation.

4.3   Conclusion

This chapter has detailed new CUDA implementations of LU and QR decomposition, for large batches of matrices of side less than 1,024 elements. The kernels take advantage of several key GPU architectural features and display highly favorable performance and scaling as compared to comparable CPU implementations. However, QR decomposition was relatively more performant than LU decomposition, largely owing to the need for warp serialization and fairly excessive synchronization in the latter. Performance for initial kernels was improved significantly through introduction of several techniques guided by profiling. These techniques included configuring cache and shared memory in software, as well as optimizing thread blocksize and shared memory buffer size. Further optimizations and alternative kernels for these important methods are the subjects of ongoing work.

Acknowledgements The authors would like to thank Muhammed Kabiru Hassan and Sreejith Ganesh Jaya for bringing applications to their attention that benefit from the routines detailed here. The authors are also very grateful to reviewers from Nvidia for their comments and improvements to the manuscript.

Appendix 1

```
__global__ void luDecomposition ( float2 * inputMatrices, float * devSign ){

// NOTES:
// array indices have been simplified for readability
// eg.,
// #define buf_index        ( vectorIndex + myMatrix * MATRIX_SIDE )
// common tasks have been relegated to device functions
// temporary variables
float2 sum,dum,tmp,tmpr,tmpl;

// scratch space
__shared__ float          sign [ NUM_MATRICES ];
__shared__ volatile float  scale [ NUM_MATRICES * MATRIX_SIDE ];
__shared__ volatile float2 reduce [ NUM_MATRICES * MATRIX_SIDE ];
__shared__ volatile float2 vectors [ NUM_MATRICES * MATRIX_SIDE ];
__shared__ volatile int    indices [ NUM_MATRICES * MATRIX_SIDE ];
```

```
// index to matrix for processing
int myMatrix    = threadIdx.x / MATRIX_SIDE;

// index to vector for processing
int vectorIndex = threadIdx.x % MATRIX_SIDE;

// which warp
int myWarp      = vectorIndex / 32;


// initialize permutation signs
sign[threadIdx.x % NUM_MATRICES]=1.0f;

// initialize shared memory
initFloat2Buffer( vectors, FLOAT_MIN );


// determine scaling information
for (int i=0; i < MATRIX_SIDE; ++i){

        __syncthreads();

        // load shared memory

        vectors [ buf_index ].x = inputMatrices [ row_i_index ].x;
        vectors [ buf_index ].y = inputMatrices [ row_i_index ].y;

        __syncthreads();

        // find maxima by reduction
        findVectorMaxima ( vectors, vectorIndex, myMatrix );

        __syncthreads();
        // write scaling information
        if ( vectorIndex ==i ){

                // should test for singular
                scale [ scale_index ] = abs ( vectors [ buf_00 ].x );
        }
}

// initialize shared memory
initFloat2Buffer ( vectors, 0.0f );

for (int j=0; j<MATRIX_SIDE; j++){

        __syncthreads();

        // load the j column to shared
        vectors [ buf_index ].x        = inputMatrices [ col_j_index ].x;
        vectors [ buf_index ].y        = inputMatrices [ col_j_index ].y;

        __syncthreads();

        myMatrix=myWarp;
        if (( myMatrix < NUM_MATRICES ) && ( vectorIndex < j)){
         for (int i=0; i<WARPS_PER_MATRIX; i++){

                        sum.x = vectors [ buf_index ].x;
                        sum.y = vectors [ buf_index ].y;

                        for (int k=0; k< MATRIX_SIDE ; k++){

                                if (k>=vectorIndex) break;

                                tmpl = inputMatrices [ col_k_index ];

                                tmpr.x = vectors [ buf_k ].x;
                                tmpr.y = vectors [ buf_k ].y;

                                sum.x -= (tmpl.x * tmpr.x - tmpl.y * tmpr.y);
                                sum.y -= (tmpl.y * tmpr.x + tmpl.x * tmpr.y);

                                vectors [ buf_index ].x = sum.x;
                                vectors [ buf_index ].y = sum.y;

                        }

                }
        }
}
myMatrix = threadIdx.x / MATRIX_SIDE;

__syncthreads();
if ((vectorIndex >=j) && (vectorIndex < MATRIX_SIDE)){

        sum.x = vectors [ buf_index ].x;
        sum.y = vectors [ buf_index ].y;
```

```
                    for (int k=0; k< j; k++){

                            tmpl = inputMatrices [ col_k_index ];

                            tmpr.x = vectors [ buf_k ].x;
                            tmpr.y = vectors [ buf_k ].y;

                            sum.x -= (tmpl.x * tmpr.x - tmpl.y * tmpr.y);
                            sum.y -= (tmpl.y * tmpr.x + tmpl.x * tmpr.y);
                            vectors [ buf_index ].x= sum.x;
                            vectors [ buf_index ].y= sum.y;
                    }
            }
            __syncthreads();

            // write j column back to global
            inputMatrices [ col_j_index ].x = vectors [ buf_index ].x;
            inputMatrices [ col_j_index ].y = vectors [ buf_index ].y;

            // initialize shared memory
            initFloat2Buffer ( reduce, FLOAT_MIN );

            __syncthreads();

            if (vectorIndex >= j){

                    // init for pivot search by reduction
                    reduce [ buf_index - j ].x = abs ( vectors [ buf_index ].x ) \
                        / scale [ scale_index ];
                    indices [ buf_index - j ] = vectorIndex;
            }

            __syncthreads();
            findVectorMaximaKey ( reduce, indices, vectorIndex, myMatrix );
            __syncthreads();

            // possible row swap
            if (j != indices [ buf_00 ]){

                    int i = indices [ buf_00 ];

                    // each thread swaps one row element with another row element

                    sum                                 = inputMatrices [ row_i_index ];
                    inputMatrices [ row_i_index ]       = inputMatrices [ row_j_index ];
                    inputMatrices [ row_j_index ]       = sum;

                    if (vectorIndex==0){
                            scale [ buf_i ]             = scale [ buf_j ];
                            sign [ myMatrix ]           *= -1.0f;
                    }
            }

            __syncthreads();

            // final scaling
            if ( j != MATRIX_SIDE-1){

                    dum = inputMatrices [ diag_j_index ];

                    if (vectorIndex >= j+1){

                            tmp                                 = inputMatrices [ col_j_index ];
                            tmp                                 = divide ( tmp, dum );
                            inputMatrices [ col_j_index ]       = tmp;
                    }
            }

            __syncthreads();

    }// end j loops

    // write out sign
    if (vectorIndex == 0) devSign [ sign_ind ] = sign [ myMatrix ] ;


}
```

Appendix 2

```
// Iteration number and total batch size are stored in constant memory
__device__ __constant__ int cmem_k, cmem_size;

__global__ void QR_Kernel(float2 *matrices, float2 *q_complete) {

    //// Shared memory buffer rows
    // NOTE: This kernel as presented does not safeguard against buffer under/overflow
    __shared__ float2 upper_row[NMPBL*MATRIX_SIDE];
    __shared__ float2 lower_row[NMPBL*MATRIX_SIDE];

    //// Convenience indices
    // Index to matrix for processing
    int myMatrix   = threadIdx.x / MATRIX_SIDE;
    // Index to vector for processing
    int vectorIndex = threadIdx.x % MATRIX_SIDE;
    // Matrix offset for this block
    unsigned int memoryStride      = \
        ( blockIdx.x * NMPBL + myMatrix  ) * MATRIX_SIDE * MATRIX_SIDE ;

    //// Set column and line number we want to eliminate
    int my_i, my_j = 0; int iter = cmem_k;

    int dk = delta_k(iter);
    if( iter < MATRIX_SIDE ) {
        my_j = blockIdx.z;
        my_i = (MATRIX_SIDE-iter) + 2*my_j;
    } else {
        my_j = (iter-MATRIX_SIDE) + blockIdx.z + 1;
        my_i = (iter-MATRIX_SIDE) + 2*blockIdx.z + 2;
    }

    //// Load row data - if condition avoids out of bounds accesses
    if(memoryStride + my_i*MATRIX_SIDE + vectorIndex < cmem_size_kuck*MATRIX_SIDE*MATRIX_SIDE) {
        upper_row[threadIdx.x] = matrices[memoryStride + (my_i-1)*MATRIX_SIDE + vectorIndex];
        // Lower row w/ leading zero after rotation
        lower_row[threadIdx.x] = matrices[memoryStride + my_i*MATRIX_SIDE + vectorIndex];
    }

    //// Wait for all the data to be loaded first
    __syncthreads();

    float2 u,v,c,s;
    float f,g,den;

    u = upper_row[myMatrix*MATRIX_SIDE + my_j]; // broadcast operation from SMEM
    v = lower_row[myMatrix*MATRIX_SIDE + my_j]; // broadcast operation from SMEM

    //// Calculate c and s
    f = u.x*u.x + u.y*u.y;
    g = v.x*v.x + v.y*v.y;

    // Algorithm is provided in [BDK02]
    if( g < 2e-16 ) {

        c.x = 1.0f; c.y = 0.0f;
        s.x = 0.0f; s.y = 0.0f;

    } else if (f< 2e-16) {

        c.x = 0.0f; c.y = 0.0f;

        // s = conj(v)/g
        den = 1.0f/g;
        s.x = v.x*den; s.y = -v.y*den;

    } else {
        // r = sqrt(f + g)
        den = rsqrt(f + g);
        // c = f/r
        c.x = sqrt(f)*den; c.y = 0.0f;

        // s = x/f * conj(y) / r
        // den = -1/(f*r)
        den *= rsqrt(f);

        s.x = (u.x*v.x + u.y*v.y)*den;
        s.y = (u.y*v.x - u.x*v.y)*den;
    }

    //// Compute the two rows update
    // u*c + v*s
    // Load data
    u = upper_row[threadIdx.x];
    v = lower_row[threadIdx.x];
    // Perform product: real part
    f = (u.x*c.x - u.y*c.y) + (v.x*s.x - v.y*s.y);
```

```
        // Perform product: imaginary part
        g = (u.x*c.y + u.y*c.x) + (v.x*s.y + v.y*s.x);

        float2 tmp;
        tmp.x = f; tmp.y = g;

        // u*-conj(s) + v*c
        // Perform product: real part
        f = -(u.x*s.x + u.y*s.y) + (v.x*c.x - v.y*c.y);
        // Perform product: imaginary part
        g = (u.x*s.y - u.y*s.x) + (v.x*c.y + v.y*c.x);

        ///// Store: synchronization is necessary to avoid overwriting data ...
        ///// ...for warps that are still getting data from the broadcast operation
        __syncthreads();
        upper_row[threadIdx.x] = tmp;

        lower_row[threadIdx.x].x = f;
        lower_row[threadIdx.x].y = g;

        //// Write data in the original matrix
        if(memoryStride + my_i*MATRIX_SIDE + vectorIndex < cmem_size*MATRIX_SIDE*MATRIX_SIDE) {
                matrices[memoryStride + (my_i-1) * MATRIX_SIDE + vectorIndex ] = upper_row[threadIdx.x];

                matrices[memoryStride + my_i * MATRIX_SIDE + vectorIndex ].x = \
                        (vectorIndex > my_j) * lower_row[threadIdx.x].x;
                matrices[memoryStride + my_i * MATRIX_SIDE + vectorIndex ].y = \
                        (vectorIndex > my_j) * lower_row[threadIdx.x].y;
        }

        //// Load rows of Q to be updated
        if(memoryStride + my_i*MATRIX_SIDE + vectorIndex < cmem_size*MATRIX_SIDE*MATRIX_SIDE) {
                upper_row[threadIdx.x] = q_complete[memoryStride + (my_i-1) * MATRIX_SIDE + vectorIndex ];
                lower_row[threadIdx.x] = q_complete[memoryStride + my_i * MATRIX_SIDE + vectorIndex ];
        }

        //// Apply the Givens rotation
        u = upper_row[threadIdx.x];
        v = lower_row[threadIdx.x];

        // Q[i-1,k] = C*Q[i-1,k] + S*Q[i,k]
        // Perform product: real part
        f = (u.x*c.x - u.y*c.y) + (v.x*s.x - v.y*s.y);
        // Perform product: imaginary part
        g = (u.x*c.y + u.y*c.x) + (v.x*s.y + v.y*s.x);

        tmp.x = f; tmp.y = g;

        // Q[i,k] = -S'*Q[i-1,k] + C*Q[i,k]
        // Perform product: real part
        f = -(u.x*s.x + u.y*s.y) + (v.x*c.x - v.y*c.y);
        // Perform product: imaginary part
        g = (u.x*s.y - u.y*s.x) + (v.x*c.y + v.y*c.x);

        // No synchronization necessary here;
        // each thread operates independently on a single matrix element
        upper_row[threadIdx.x] = tmp;

        lower_row[threadIdx.x].x = f;
        lower_row[threadIdx.x].y = g;

        //// Write to global
        if(memoryStride + my_i*MATRIX_SIDE + vectorIndex < cmem_size*MATRIX_SIDE*MATRIX_SIDE) {
                q_complete[memoryStride + (my_i-1) * MATRIX_SIDE + vectorIndex ] = upper_row[threadIdx.x];
                q_complete[memoryStride + my_i * MATRIX_SIDE + vectorIndex ] = lower_row[threadIdx.x];
        }
}

// CPU driver loop
extern "C"{
        blocks.x = (int)ceil((float)size/(float)NMPBL);
        cudaMemcpyToSymbol(cmem_size,&size,sizeof(size));

        // Set the shared memory bank size to 8 bytes / 64 bits
        cudaDeviceSetSharedMemConfig(cudaSharedMemBankSizeEightByte);

        for(int k = 1;k<=2*MATRIX_SIDE-3;k++) {
                // Calculate the total number of rotations to apply at once
                // Launch blocks.z additional rows of CUDA blocks to compute the Nrot rotations
                if( k < MATRIX_SIDE ) {
                        blocks.z = (int)ceil((float)k/2.0f);
                } else {
                        blocks.z = (int)ceil((float)k/2.0f) - k + MATRIX_SIDE-1;
                }
                // Update constant memory
                cudaMemcpyToSymbol(cmem_k,&k,sizeof(k));
                // Launch the main kernel;
                // calculates the Givens rotations and places them in the temporary matrix Q_A
                QR_Kernel <<< blocks,NTH >>> (matrices,q_complete);
        }//end main loops
}
```

References

1. Bindel, D., Demmel, J., Kahan, W., Marques, O.: On computing givens rotations reliably and efficiently. ACM Trans. Math. Softw. **28**(2), 206–238 (2002)
2. Cosnard, M., Robert, Y.: Complexity of parallel QR factorization. J. Assoc. Comput. Machinery **33**, 712–723 (1986)
3. Golub, G.H.: Matrix Computations, 3rd edn. Johns Hopkins University Press, Baltimore (1996)
4. Lucente, E., Monorchio, A., Mittra, R.: An iteration-free MoM approach based on excitation independent characteristic basis functions for solving large multiscale electromagnetic scattering problems. IEEE Trans. Antennas Propag. **56**(4), 999–1007 (2008)
5. Press, W.H., Flannery, B.P., Teukolsky, S.A., Vetterling, W.T.: Numerical Recipes in C: The Art of Scientific Computing, 2nd edn. Cambridge University Press, Cambridge (1993)
6. Saad, Y.: Iterative Methods for Sparse Linear Systems, 2nd edn. SIAM, Philadelphia (2003)
7. Sameh, A.H., Kuck, D.J.: On stable parallel linear system solvers. J. Assoc. Comput. Machinery **25**, 81–91 (1978)
8. Seward, J., Nethercote, N., Weidendorfer, J.: Valgrind 3.3: Advanced Debugging and Profiling for GNU/Linux Applications. Network Theory Ltd., Bristol (2008)
9. Sreejith, G.J., Jolad, S., Sen, D., Jain, J.K.: Microscopic study of the $\frac{2}{5}$ fractional quantum Hall edge. Phys. Rev. B **84**, 245104 (2011)
10. Sreejith, G.J., Toke, C., Wójs, A., Jain, J.K.: Bipartite composite fermion states. Phys. Rev. Lett. **107**, 086806 (2011)

Chapter 5
A Flexible CUDA LU-Based Solver for Small, Batched Linear Systems

Antonino Tumeo, Nitin Gawande, and Oreste Villa

5.1 Introduction and Motivations

Many simulation models for hydrology, combustion and atmospheric modeling require solvers that operate on a large amount of small, independent systems of equations. These models typically operate by computing, at each time step of the simulation, the flow and then the chemical reactions of fluids and solids in elements over a large number of locations (a.k.a. physical grid nodes). The chemical reactions are usually described through a set of non-linear equations. The profiling of typical codes shows that these models can spend up to 95 % of the overall computation time to solve the chemical reactions [1]. Typical simulations involve computing tens to few hundreds chemical reactions, in tens of thousands up to millions of uniform or non-uniform grid nodes, depending on the geometry and the resolution of the problem to solve. The Newton–Raphson method is one of the most used approaches for obtaining a solution for such systems of non-linear equations. The technique involves the linearization of the systems by computing the Jacobian matrix and a residual vector for each set of equations that represent the reactions for a grid node. The method solves the linearized systems iteratively, performing Gaussian elimination with LU factorization until achieving the desired convergence. The method allows computing the LU factorization either with partial or full pivoting, depending on the numerical characteristics of the problem, time-step of the simulation and, ultimately, accuracy of the result. The Jacobian matrix is generated from the chemical reactions, and its size is typically a square function of the number of equations involved in the process. For example, the simulation of kinetic chemical

A. Tumeo (✉) • N. Gawande
Pacific Northwest National Laboratory, Richland, WA, USA
e-mail: antonino.tumeo@pnnl.gov; nitin.gawande@pnnl.gov

O. Villa
NVIDIA, Santa Clara, CA, USA
e-mail: ovilla@nvidia.com

V. Kindratenko (ed.), *Numerical Computations with GPUs*,
DOI 10.1007/978-3-319-06548-9_5, © Springer International Publishing Switzerland 2014

reactions in combustion modeling [2] typically involves matrices up to $\approx 40 \times 40$ in sizes, and is usually numerically stable by just using partial pivoting for the LU decomposition. Reactive transport models for fluids through the Earth's crust over multiple phases, instead, require matrices with sizes up to $\approx 100 \times 100$ and traditionally use LU decomposition with full pivoting to increase numerical stability. STOMP [3], HydroGeoChem [4], PRFLOTRAN [5], and TOUGH [6] use some of these models. All of these applications require solving the chemical reactions for at least thousands of grid nodes for the smallest problems they tackle.

There are many effective implementations of linear solvers for Graphic Processing Units (GPUs) [7]. However, GPUs are more efficient when they perform a large number of operations with respect to the amount of data involved in the operations (flop/byte ratio). In fact, although new GPUs keep providing higher and higher memory bandwidths, computational power is much higher and there still are strict requirements to reach the peak memory transfer rates. For these reasons, many of the available libraries focus on linear solvers for single, very large, matrices. Conventional solvers, such as MAGMA [8] or those provided by the CUDA library [9], target large matrices with several thousand of elements per dimension, achieving speedups of one order of magnitude when compared to CPUs. They exploit parallelism at the level of a single matrix solver and, in some cases, they also make use of heterogeneity, by assigning the diagonal blocks and interchange of row and columns to CPU cores and the reduction and scaling of large sub-matrices to GPUs [10]. The combination of increased parallelism and of solutions to increase bandwidth (e.g., through more effective and larger caches, bigger register files, larger on-chip memories), recently made GPUs much more interesting for operating in parallel on a large number of small matrices. Indeed, the latest versions of the CUBLAS library [11] include support for batched LU factorization. A software developer can use it to construct a solver operating on a set of small, independent matrices.

In this book chapter we present the CUDA implementation of a batched linear solver that operates on large numbers of small matrices, ranging from size 2×2 to 128×128. The presented implementation exploits, somehow counterintuitively, thread level parallelism, exploiting a employs GPU thread for each matrix. With respect to other existing implementations, the benefit of our approach resides in the support of matrices with sizes over the 100×100 elements, and the support of both partial and complete pivoting for the LU factorization. These are mandatory requirements for reactive transport simulators, which historically use complete pivoting, trading off some of the performance for higher accuracy. We discuss our implementation in comparison to other currently available solutions, which instead only integrate partial pivoting and support sizes up to 76×76 elements. Beside presenting the code of our implementation, we also discuss the performance tradeoffs, enabling a developer to choose the best implementation for his target applications.

The remainder of this chapter is organized as follows. Section 5.2 provides some preliminaries on linear solvers. Section 5.3 presents the commented code of

our implementation. Section 5.4 briefly introduces other existing approaches and
Sect. 5.5 discusses the performance and flexibility tradeoffs with respect to our
solution. Finally, Sect. 5.6 concludes the chapter.

5.2 Preliminaries on Solvers and LU Decomposition

The Newton–Raphson method is a technique for solving nonlinear equations
numerically. It is an iterative technique, which works by linearly approximating the
equations until convergence. A typical problem gives N nonlinear equations to be
zeroed, involving variables $x_i, i = 1, 2, \ldots, N$:

$$F_i(x_1, x_2, \ldots, x_N) = 0 \quad i = 1, 2, \ldots, N$$

Denoting with $\mathbf{x}$ the vector of values x_i and with $\mathbf{F}$ the vector of functions F_i, we
can expand each of the functions F_i in the neighborhood of X in Taylor series:

$$F_i(\mathbf{x} + \delta\mathbf{x}) = F_i(\mathbf{x}) + \sum_{j=1}^{N} \frac{\partial F_i}{\partial x_j} \delta x_j + O(\delta\mathbf{x}^2)$$

where:

$$J_{ij} \equiv \frac{\partial F_i}{\partial x_j}$$

is the Jacobian matrix $\mathbf{J}$.

In matrix notation:

$$\mathbf{F}(\mathbf{x} + \delta\mathbf{x}) = \mathbf{F}(\mathbf{x} + \mathbf{J}\delta\mathbf{x} + O(\delta\mathbf{x}^2)$$

By neglecting terms of order $\delta\mathbf{x}^2$ and higher and by setting $\mathbf{F}(\mathbf{x} + \delta\mathbf{x}) = 0$, we
obtain a set of equations for the corrections $\delta\mathbf{x}$ that move each function closer to
zero simultaneously:

$$\mathbf{J}\delta\mathbf{x} = -\mathbf{F}$$

This matrix equation can be solved by LU decomposition. The corrections are
then added to the solution vector:

$$\mathbf{x}_{new} = \mathbf{x}_{old} + \delta\mathbf{x}$$

And the process is iterated to convergence.

5.2.1 LU-Based Linear Solvers

A linear solver is a procedure that, given a system of linear equation described in matricial form as $\mathbf{Ax} = \mathbf{b}$, finds the solution vector $\mathbf{x}$. One of the most efficient method for dense and semi-dense matrices is finding a decomposition of the matrix $\mathbf{A}$ such that the solution is then obtained by back substitution. LU decomposition (also called LU factorization) factorizes a matrix $\mathbf{A}$ as the product of a lower triangular matrix $\mathbf{L}$ and an upper triangular matrix $\mathbf{U}$ such that $\mathbf{LU} = \mathbf{A}$. There are two basic approaches for arriving at an LU decomposition:

- simulate Gaussian elimination by using row operations to zero elements in A until an upper triangular matrix exists. Save the multipliers produced at each stage of the elimination procedure as L;
- use the definition of matrix multiplication to solve directly $LU = A$ for the elements of L and U.

Approaches that exploit Gaussian elimination mainly differs in the order in which $\mathbf{A}$ is forced into upper triangular form. The most common alternatives are to eliminate subdiagonal parts of $\mathbf{A}$ either one row at a time or one column at a time. The calculations required to zero a complete row or a complete column are referred as one stage of the elimination process.

At the kth stage of Gaussian elimination:

$$a_{ij}^{(k+1)} = a_{ij}^{(k)} - \left(\frac{a_{ik}^{(k)}}{a_{kk}^{(k)}}\right) a_{ij}^{(k)}, \text{ where } i, j > k$$

The term $\frac{a_{ik}^{(k)}}{a_{kk}^{(k)}}$ (referred as multiplier) describes the effect of eliminating element a_{ik} on the other entries in row i during the kth stage of the elimination. These multipliers are the elements of the lower triangular matrix $\mathbf{L}$, i.e.:

$$l_{ik} = \frac{a_{ik}^{(k)}}{a_{kk}^{(k)}}$$

Considering the Gaussian elimination procedure, LU decomposition fails when the value $a_{kk}^{(k)}$ (called the pivot element) is zero. Furthermore, Gaussian elimination is numerically unstable even if there are no zero pivot elements, because of the errors in approximation in finite precision representation of real numbers. A solution to numerical instability is to interchange the rows and columns of A to avoid zero and unstable pivot elements. These interchanges do not affect the solution of the approximated linear equations of the system as long as the permutations are logged and taken into account during the substitution process. The choice of pivot elements is referred as pivot strategy. There is not an optimal pivot strategy, but two common heuristics are:

- *Partial pivoting*: at the kth stage of the computation, select the largest element in column k as the pivot. When using partial pivoting, the factorization produces matrices **L** and **U** that satisfy the equation:

$$\mathbf{LU} = \mathbf{PA}$$

where **P** is a permutation matrix. Initially, **P** is initialized to **I**, then each row interchange that occurs during the decomposition of **A** causes a corresponding row swap in **P**. Starting from the linear system of equations $\mathbf{Ax} = \mathbf{b}$ and premultiplying both sides by **P**, we obtain $\mathbf{PAx} = \mathbf{Pb}$. Substituting **PA** with **LU**, we obtain $\mathbf{LUx} = \mathbf{Pb}$. Thus, we can achieve a solution for **A** by the sequential solution of two triangular systems: $\mathbf{y} = \mathbf{Pb}, \mathbf{Lc} = \mathbf{y}, \mathbf{Ux} = \mathbf{c}$.

- *Complete pivoting*: at the kth stage of the computation, choose the largest remaining element in **A** as the pivot. If pivoting has proceeded along the diagonal in stages 1 through $k - 1$, this implies that the next pivot should be the largest element $a_i^{(k-1)}j$ where $k \leq i \leq n$ and $k \leq j \leq n$. When using complete pivoting, factorization produces matrices **L** and **U** that satisfy the equation:

$$\mathbf{LU} = \mathbf{PAQ}$$

where **P** is a row permutation matrix and **Q** is a column permutation matrix. **Q** is derived from column interchanges in the same way **P** is derived from row interchanges. The linear system of equations $\mathbf{Ax} = \mathbf{b}$ can be solved by the sequential solution of two triangular systems: $\mathbf{y} = \mathbf{Pb}, \mathbf{Lc} = \mathbf{y}, \mathbf{Uz} = \mathbf{c}, \mathbf{x} = \mathbf{Qz}$.

The computational complexity of the LU factorization is $\mathbf{O(2/3 * n^3)}$. Partial pivoting contributes for a further $\mathbf{O((n^2 + n)/2)}$, while full pivoting adds $\mathbf{O(2/3 * n^3 + 1/2 * n^2 + 1/6 * n)}$. Once the matrix is decomposed, each triangular solver has computational complexity $\mathbf{O(n^2)}$. Asymptotically, a solver with partial pivoting has computational complexity of $\mathbf{O(2/3 * n^3)}$, while with full pivoting complexity is $\mathbf{O(4/3 * n^3)}$.

5.3   Proposed Implementation: CUDA Code and Comments

This section presents the implementation of our LU-based solver with complete pivoting. Listing 5.1 shows both the kernel invocation and the code of the kernel. The Jacobian matrix A has $n \times n$ elements, and the residual vector b has n elements. n is the size of the system. Our approach employs a single CUDA thread to find the solution for a single system (matrix). Because complete pivoting involves both a row and a column permutation, the procedure is difficult to parallelize effectively for a single matrix. Parallelization is achieved by batching multiple systems together. By using a single thread per matrix, and by directly accessing and storing the matrices and the residual vectors in memory, this implementation can potentially manage matrices of arbitrary sizes. In our experiments with Tesla M2090 boards based on the Fermi architecture and integrating 6 GB of memory, we easily reached sizes up to 128×128 elements, which allow solving a typical simulation integrated in a reactive transport simulator such as STOMP. At first glance, this implementation violates basic GPU programming principles, because it assigns a different "task" to different threads inside a warp. Generally applying this technique can lead to very poor performance due to thread divergence within the warp. However, in our code, each thread is exactly performing the same operations on all the independent matrices, except when it discovers pivot elements and swaps rows. The key observation is that, when the matrix is larger than 32×32 elements, the cost of these operations is much smaller than the cost of updating the lower matrix and back substituting in the triangular systems. Vice-versa, when the matrix is smaller than 32×32 elements, the cost of pivoting and row interchange can be compared to the cost of updating the matrices and performing back-substitution, resulting in possible thread divergence. However this is true with any other implementation, because with matrices smaller than 32×32 elements warps are not fully utilized. Another important issue of this approach is that the input matrices A and vectors b and x are stored as arrays of structures, meaning that big arrays contain all the elements of the different matrices and vectors. If each thread is accessing its own matrix, the threads in a warp are accessing elements that are strides of the number of elements in the matrix: i.e., they access elements at a distance of $n \times n$ memory locations. When accessing arrays b and x, data are instead at distance n. This results in un-coalesced accesses to memory, which are a main cause of performance degradation. To alleviate this problem, our code performs a transformation of the matrix A and of the arrays b and x before and after the solver phase, such that the resulting data structure is a structure of arrays, meaning that a given element (i,j) of a matrix is rearranged together in memory with those of the other matrices. This operation may appear quite expensive, in particular for the matrices, because the transformation must access their elements at least once in un-coalesced manner. However, the transformation has cost $O(N^2)$, while the entire computational complexity of the algorithm is $O(N^3)$.

Given the iterative nature of the solver, we want our code to preserve the original matrices A, without transforming them back after the solver completes. Thus, we need to store the transformed matrices in a temporary space. Unfortunately, the on-chip shared memory is not big enough. In fact, we need at least space for a number of matrices equal to the thread block size. To minimize divergence, the minimum effective thread block size obviously is 32 (warp size). Consequently, for a thread block of 32 threads, simultaneously operating on 32 matrices of size 100×100 with double precision elements (64 bits), we would need at least 2.5 MB. For this reason, although this may again seem counterintuitive for usual GPGPU programming, we utilize another portion of GPU memory that is allocated and deallocated on a thread block basis by a single thread in the block. These allocations are performed on a heap space that is set during initialization of the device by using the CUDA library call *cudaThreadSetLimit(cudaLimitMallocHeapSize, bytes)*. Allocations and de-allocations inside the heap space are performed with *_malloc/_free* primitives, which wrap the standard malloc/free that align to 128 bytes inside in the heap. As the heap space is reused across thread blocks that are executed on the same Streaming Multiprocessor, we do not need to have a heap space as large as the total dataset.

Finally, since the proposed implementation does not use the shared memory, we set the architecture to employ as much as possible the 64 kB of on-chip memory as L1 cache (i.e., 48 kB on both Fermi and Kepler architectures), exploiting the *cudaFuncSetCacheConfig* primitive.

Listing 5.1 GPU implementation of LU solver with complete pivoting

```
#include <stdio.h>
#include <assert.h>                                                              2
#include <cuda.h>
#include <cuda_runtime.h>                                                        4

#define BLOCKSIZE  96                                                            6

#define T(id) (threadIdx.x + blockDim.x * (id))                                  8

int axb_solve_d_gpu_batch(double * d_A, double * d_B,                            10
  double * d_X, int n, int batch) {
  cudaFuncSetCacheConfig(_axb_solve_d_gpu_batch,cudaFuncCachePreferL1);          12
  int gridDim = batch / BLOCKSIZE + 1;
  _axb_solve_d_gpu_batch<<<gridDim, BLOCKSIZE>>>(d_A, d_B, d_X, n, batch);        14
  cudaError_t err = cudaGetLastError();
  if (cudaSuccess != err) {                                                      16
    printf("ERROR__%d\n", err);
    return -1;                                                                   18
  }
  return 0;                                                                      20
}
                                                                                 22
__global__  void _axb_solve_d_gpu_batch(double * d_A,
    double * d_B, double * d_X, int n, int batch)                               24
{
  int matrixId = blockIdx.x * blockDim.x + threadIdx.x;                          26
  if ( matrixId >= batch) return;
  int i, j, k;                                                                   28

  __shared__   double * A;                                                       30
  __shared__   double * B;
  __shared__   double * X;                                                       32
  __shared__   int     * pivot;
                                                                                 34
  if(threadIdx.x == 0)
  {                                                                              36
    A = (double*)   malloc(blockDim.x * n * n * sizeof(double));
    B = (double*)   malloc(blockDim.x * n *     sizeof(double));                 38
    X = (double*)   malloc(blockDim.x * n *     sizeof(double));
    pivot = (int * ) malloc(blockDim.x * n *     sizeof(int));                   40
  }
  __syncthreads();                                                              42

  // Check for failure                                                          44
  if ( A == NULL || B == NULL || X == NULL || pivot == NULL ) {
    printf("Error_allocating_inside_kernel\n");                                 46
    return;
  }                                                                             48

  /* coalescing A and B */                                                      50
  for (j = 0; j < n; j++)
  {                                                                             52
    B[T(j)] = d_B[ matrixId * n + j];
    for (i = 0; i < n; i++)                                                     54
      A[T(j * n + i)] = d_A[ matrixId * n * n + j * n + i ];
  }                                                                             56
```

```
// For each row and column, k = 0, ..., n-1,                    58
for (k = 0; k < n; k++) {
                                                               60
  // find the pivot row
  int col = k;                                                 62
  double max = fabs( A[T(k * n + k)] );
  for (j = k + 1; j < n; j++) {                                64
    if ( max < fabs(A[T(j * n + k)]) ) {
      max = fabs(A[T(j * n + k)]);                             66
      col = j;
    }                                                          68
  }
                                                               70
  // and if the pivot row differs from the current row, then
  // interchange the two rows.                                 72
  if (col != k) {
    for (j = 0; j < n; j++) {                                  74
      double max = A[T(k * n + j)];
      A[T(k * n + j)] = A[T(col * n + j)];                     76
            A[T(col * n + j)] = max;
    }                                                          78
  }
                                                               80
  // and if the matrix is singular, return error
  if ( A[T(k * n + k)] == 0.0 ) {                              82
    printf("Inside_Kernel:_Matrix_singular!!\n");
    return;                                                    84
  }
                                                               86
  // otherwise find the lower triangular matrix elements for column k.
                                                               88
  for (i = k+1; i < n; i++)
    A[T(i * n + k)]  /= A[T(k * n + k)];                       90

  // update remaining matrix                                   92
  for (i = k+1; i < n; i++)
    for (j = k+1; j < n; j++)                                  94
      A[T(i * n + j)] -= A[T(i * n + k)] * A[T(k * n + j)];
                                                               96
  pivot[T(k)] = col;
}                                                              98

// Solve the linear equation Lx = B for x, where L is a lower  100
// triangular matrix with an implied 1 along the diagonal.
                                                               102
for (k = 0; k < n; k++) {
  if (pivot[T(k)] != k) {                                      104
    double dum = B[T(k)];
    B[T(k)] = B[T(pivot[T(k)])];                               106
    B[T(pivot[T(k)])] = dum;
  }                                                            108
  X[T(k)] = B[T(k)];
  for (i = 0; i < k; i++)                                      110
    X[T(k)] -= X[T(i)] * A[T(n * k + i)];
}                                                              112

// Solve the linear equation Ux = y, where y is the solution   114
// obtained above of Lx = B and U is an upper triangular matrix.
for (k = n-1; k >= 0; k--) {                                   116
  if (pivot[T(k)] != k) {
    double dum = B[T(k)];                                      118
    B[T(k)] = B[T(pivot[T(k)])];
    B[T(pivot[T(k)])] = dum;                                   120
  }
  for (i = k + 1; i < n; i++)                                  122
    X[T(k)] -= X[T(i)] * A[T(k * n + i)];
```

```
      if (A[T(k * n + k)] == 0.0) {                                         124
        printf("Inside_Kernel:_Matrix_singular!!\n");                       126
        return;
      }                                                                     128
      X[T(k)] /= A[T(k * n + k)];
    }                                                                       130

    /* un-coalescing X */                                                   132
    for (j = 0; j < n; j++)
      d_X[ matrixId * n + j] = X[T(j)];                                     134

    __syncthreads();                                                        136
    if(threadIdx.x == 0)
    {                                                                       138
      free(A);
      free(B);                                                              140
      free(X);
      free(pivot);                                                          142
    }
    return;                                                                 144
}
```

5.4   Other Implementations

There are two other implementations we are aware of that try to address the problem
of solving a set of small systems in a batch. The first one requires the batched
interfaces provided in the CUBLAS library [11] starting from CUDA 5.0. It exploits
parallelism at the warp-level. The second one [12], provided by NVIDIA on its
developer site, exploits parallelism at the thread block level. We briefly discuss
the features of these solutions, and present a tradeoff analysis with respect to our
proposed thread parallel implementation.

5.4.1   Warp Parallel Implementation

A software developer can implement a batched LU-based solver by exploiting the
batched interfaces of the CUBLAS library [11], provided in CUDA 5.0. Such
a solver performs a sequence of four GPU kernel calls for all the matrices, as
follows:

1. LU decomposition of A ($PA = LU$);
2. permutation of the array b with the array of pivots P ($y = Pb$);
3. solution of the triangular lower system ($Lc = y$);
4. solution of the upper system to obtain the final solution ($Ux = c$).

The library directly provides batched functions for three kernel calls: *cublas-
DgetrfBatched* for step 1 (batched LU decomposition), and *cublasDtrsmBatched*
for steps 3 and 4. A developer can implement a simple kernel that performs

step 2 to complete the solver. Compared to the other steps, step 2 has negligible execution time. With respect to our proposed implementation, this implementation has several limitations in terms of flexibility. First of all, it can perform the batched LU decomposition only with partial pivoting, because it is the only method provided in CUBLAS. Our approach supports complete pivoting, and can easily use partial pivoting by just substituting the related code. Second, the batched functions assign a warp (32 threads) per matrix, and they are limited to matrices at most of 32 × 32 elements. Thus, this solver can deal with a subset of the problems that our implementation can support solve. Because matrices are small, the implementation exploits shared memory (a matrix with 64 bit values occupies 8 kB). However, CUDA does not preserve the content of shared memory across subsequent kernel calls. Thus, every kernel has to reload the data in shared memory, with a performance penalty for the operation.

5.4.2 Thread Block Parallel Implementation

This implementation [12] is available on the NVIDIA developer site. It selects among three mutually exclusive kernels, depending on the size of the input matrices. The implementation exposes a *dsolve_batch()* function that, in turn, calls a single templatized function. This templatized function is parametrized by data type and architecture. The implementation loads the entire system to solve in shared memory, thus the size of the matrices it can handle are limited. For Fermi architectures, the maximum size is 76 × 76 double precision elements. When the solver loads the matrix into the shared memory, it augments the matrix on the right with the right hand side vector, allowing parallel manipulation. The two-dimensional shared memory layout of the matrix uses padding to minimize bank conflicts. The configuration class allows optimizing the padding for each matrix size. Each thread block solves a single system, so the number thread blocks in the launch configuration is identical to the batch size. The implementation exploits two-dimensional thread blocks: the x dimension is configured by the template class, the y dimension corresponds to the number of columns of the augmented matrix. In this way, each thread "row" handles one row of the augmented matrix in parallel. The three kernels used are gauss_jordan1 (for dimensions 2 through 9), gauss_jordan2 (used for dimension 10), and gauss2 (for dimensions 10 through 76), with switch-over points empirically determined. The first two kernels implement the Gauss–Jordan algorithm with partial pivoting, while the third implements Gaussian elimination with partial pivoting. In the first Gauss–Jordan kernel the number of thread rows is identical to the number of rows in the matrix (i.e., each thread handles one element of the augmented matrix), while in the second there are fewer thread rows than the number of matrix rows (i.e., each thread handles more than one element). The implementation performs the maximum search for partial pivoting as a two-stage process. In the first stage, a small number of threads search a maximum for a subset of column elements. In the second stage, a single thread finds the overall maximum. Row swapping is implemented by physical exchange.

5.5   Trade-Offs Evaluation

In the following, we refer to the CUBLAS-based solution as the *Warp* parallel one and to the custom NVIDIA solution as the *Thread block* parallel one. We refer to our custom implementation as the *Thread* parallel one. To perform a fair comparison, we execute our implementation with both the partial pivoting and the full pivoting implementation.

We present a brief performance evaluation of the three different implementations on two different GPUs. The objective of this analysis is to provide to developers an informed view of which implementation to prefer, depending on the requirements of their target applications. For certain applications, it may also be useful to integrate a switching logic able to select the best implementation depending on the size of the systems to solve, on the number of systems and on the required numerical accuracy (choosing between complete or partial pivoting). For this analysis, we selected a Fermi-based and a Kepler-based Tesla board. The Fermi-based solution is a Tesla M2090 board, which includes the Fermi T20a GPU, with 16 Streaming Multiprocessors (SMs), providing a total of 512 Streaming Processors (SPs) at 1.3 GHz, and 6 GB of GDDR5 at 1.85 GHz connected through a 384-bit interface. The peak memory bandwidth is 177 GB/s. The Kepler-based solution is a Tesla K20 board with a GPU that implements 13 SMXes (2,496 SPs). The GPU works at 706 MHz, and the board includes 5 GB of GDDR5 at 2.6 GHz, connected through a 320-bit bus with 5 memory controllers. The peak memory bandwidth is 208 GB/s.

We underline that the architectures of the two GPUs are radically different. The Fermi architecture exploits a set of Streaming Multiprocessors (SMs) that include 32 Streaming Processors (SPs), 4 Super Function Units (SFUs), 16 Load/Store Units and 64 kB of on-chip memory configurable either as 48 kB of L1 cache and 16 kB of shared memory or as 16 kB of L1 cache and 48 kB of shared memory. A Fermi's SM can simultaneously execute two single precision Warps (group of 32 threads) or one double precision Warp in a minimum of 2 clock cycles. Thus, peak double precision is half of the single precision. Each SM includes a total of 32,768 registers and can maintain up to 1,536 threads in-flight. All the SMs in a chip interface to a L2 cache of 768 kB. The SMs access the global memory through a crossbar connected to several 64 bits memory controllers. In Fermi, the SMs run at higher clocks (double) than the rest of the chip. In Kepler, instead, a SM, now called SMX, includes 192 single precision streaming processors, 64 double precision streaming processors, 32 SFUs, 32 Load/Store Units. The number of threads and of registers per SMX is, respectively, 2,048 and 65,536. Kepler can dispatch 8 instructions (2 independent instructions from 4 Warps) simultaneously and can pair double precision instructions with other instructions. Each SMX still has 64 kB of configurable shared memory, but now there is also a 32/32 kB split.

This results in a higher number of warps competing for the same shared memory. An SMX also includes a new 48 kB cache for read-only data. Kepler doubles the L2 cache both in terms of size (1,536 kB) and bandwidth with respect to Fermi.

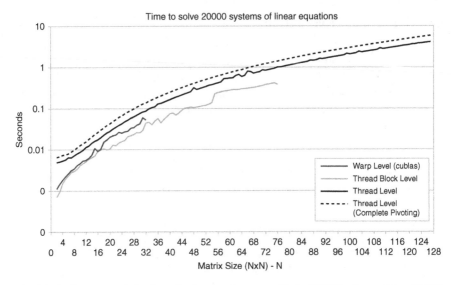

Fig. 5.1 Performance of the three implementations on a Tesla M2090 when solving 20,000 systems of linear equations while increasing the number of double precision elements in each matrix

Figure 5.1 shows the performance of the three implementations on a Tesla M2090 board when simultaneously solving 20,000 systems of variable size (the matrices have from 2 × 2 to 128 × 128 elements). The Thread block parallel implementation results the fastest. However, the performance varies a lot by changing the matrix size. Performance significantly degrades with matrices larger than 56 × 56 elements. This happens because increasing the size of the matrices increases shared memory occupation. Because shared memory is a limited resource allocated per thread block, over a certain threshold there is a reduction in the number of thread blocks that are simultaneously active, determining resulting in under-utilization. The Warp parallel implementation is the second fastest. Up to 16 × 16 elements, it provides performance very near to the Thread block parallel implementation, but over 16 × 16 elements its performance significantly reduces. The reason is the use of multiple kernels, which does not allow to fully exploit the increased bandwidth provided by the shared memory. The only implementation that manages matrices bigger than 76 × 76 elements is our proposed Thread parallel solution. It presents the lowest performance of the three implementations, but it is also the most stable: the execution time almost linearly increases with the size the systems. Our implementation is mainly limited by the number of registers used by each thread, which limits utilization of the SMs. Our proposed implementation is also the only one that supports full pivoting. Full pivoting is slower, for the higher computational complexity, than single pivoting. However, it follows the same behavior of linearly increasing its execution time with the size of the systems.

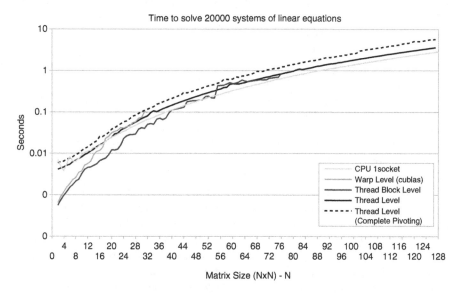

Fig. 5.2 Performance of the three implementations on a Tesla K20 when solving 20,000 systems of linear equations while increasing the number of double precision elements in each matrix

Figure 5.2 proposes the evaluation on the Tesla K20 (Kepler). The Thread parallel implementation still results the slowest in average. However, with Kepler, it is faster than both the Warp level and the Thread block parallel implementations for certain dimensions. The switch points are dimensions of 16×16 for the Warp parallel implementation and 56×56 for the Thread block parallel implementations. The Thread parallel implementation utilizes almost all the available global memory bandwidth, thus it benefits of its increase in Kepler. At the opposite, Kepler provides less bandwidth to the on-chip shared memory for each active warp. This limits the effectiveness of the Warp parallel and Thread block parallel implementations. The performance spread between the Warp parallel and the Thread block parallel implementation with matrices over sizes of 16×16 is more significant. The Thread parallel implementation with full pivoting is the slowest, but it still follows the same general behavior of when partial pivoting is used. The figure also shows that, for small matrices, the performance of Kepler with the Thread parallel implementation is comparable to a reference x86 implementation (Xeon X5650 at 2.67 GHz, 6 Nehalem cores with 12 threads at 12 MB of L3 cache), while for larger matrices it becomes slower. Therefore, it may be useful to provide heterogeneous implementations able to distribute the workload across GPUs and CPUs, depending on the characteristics of the applications and the systems to solve.

5.6 Conclusions/Summary

In this chapter we presented and discussed the CUDA implementation of a batched linear solver based on LU factorization for small matrices. These matrices are usually generated from small sets (up to 100) of non linear equations, typical in reactive transport simulators, that are then solved through the Newton–Raphson iterative technique. The code presented in this chapter exploits a thread parallel implementation (a matrix is assigned to a CUDA thread), does not exploit the on-chip shared memory and employs dynamic allocation inside the kernel. Although these approaches may appear counterintuitive, our code can manage bigger matrices (well over 100×100 elements) than other currently available solutions, which can only reach 76×76 elements. Furthermore, our approach supports both partial and complete pivoting for the LU decomposition. The support of larger matrices and full pivoting are strict requirements for certain reactive flow transport simulators for fluids through the Earth's crust over multiple phases, such as STOMP from Pacific Northwest National Laboratory (PNNL). We also presented an evaluation of our implementation against the other solutions, discussing tradeoffs in performance and flexibility. This may allow a developer to select and then integrate in its target application the best approach, depending on the requirements, or even implementing dynamic switching solutions among the different methods to maximize performance, depending on the characteristics and sizes of the problems to solve.

References

1. Tang, G., D'Azevedo, E.F., Zhang, F., Parker, J.C., Watson, D.B., Jardine, P.M.: Application of a hybrid MPI/OPENMP approach for parallel groundwater model calibration using multi-core computers. Comput. Geosci. **36**, 1451–1460 (2010)
2. Higham, N.J.: Gaussian elimination. Comput. Stat. **3**, 230–238 (2011)
3. White, M.D., Oostrom, M.: STOMP Subsurface Transport Over Multiple Phase: User's Guide. Technical report, Pacific Northwest National Laboratory, Richland (2006). PNNL-15782
4. Yeh, G.T., Tripathi, V.S., Gwo, J.P., Cheng, H.P., Chend, J.-R.C., Salvage, K.M., Li, M.H., Fang, Y., Li, Y., Sun, J.T., Zhang, F., Siegel, M.D.: HYDROGEOCHEM: a coupled model of variably saturated flow, thermal transport, and reactive biogeochemical transport, on laptops to leadership-class supercomputers. In: Groundwater Reactive Transport Models. Bentham Science Publishers, Sharjah (2012)
5. Hammond, G.E., Lichtner, P.C., Lu, C., Mills, R.T.: Pflotran: reactive flow and transport code for use on laptops to leadership-class supercomputers. In: Groundwater Reactive Transport Models. Bentham Science Publishers, Sharjah (2012)
6. Zhang, K., Wu, Y., Pruess, K.: User's Guide for TOUGH2-MP - A Massively Parallel Version of the TOUGH2 Code. Technical report, Lawrence Berkeley National Laboratory, Berkeley (2008). LBNL-315E
7. Tomov, S., Nath, R., Ltaief, H., Dongarra, J.: Dense linear algebra solvers for multicore with gpu accelerators. In: IPDPSW'10: IEEE International Symposium on Parallel Distributed Processing, Workshops and Phd Forum, pp. 1–8 (2010)
8. Agullo, E., Augonnet, C., Dongarra, J., Faverge, M., Langou, J., Ltaief, H., Tomov, S.: Lu factorization for accelerator-based systems. In: AICCSA: 9th IEEE/ACS International Conference on Computer Systems and Applications, pp. 217–224 (2011)

9. NVIDIA Corporation. Nvidia CUDA C Programming Guide, Version 5.0 (2012)
10. Song, F., Tomov, S., Dongarra, J.: Enabling and scaling matrix computations on heterogeneous multi-core and multi-GPU systems. In: ICS '12: The 26th ACM International Conference on Supercomputing, pp. 365–376 (2012)
11. NVIDIA Corporation. Nidia CUBLAS Library, Version 5.0 (2012)
12. NVIDIA custom batched LU Decomposition. NVIDIA. Available at http://developer.nvidia.com (2013)

Chapter 6
Sparse Matrix-Vector Product

Zbigniew Koza, Maciej Matyka, Łukasz Mirosław, and Jakub Poła

6.1 Introduction

The sparse matrix-vector (SpMV) multiplication is one of the key kernels in scientific computing. Efficient SpMV is crucial for the performance of several popular algorithms of computational linear algebra, especially sparse linear solvers and sparse eigenvalue solvers. The former are common, for example, in codes that solve partial differential equations like those governing the air flow round an airplane or propagation of seismic waves through the Earth. The latter are essential, for example, in quantum physics, but also in the PageRank algorithm used by Google in its web search engine to rank websites.

From a mathematical point of view, the aim of the SpMV kernel is to calculate the product $\mathbf{y} = \hat{A}\mathbf{x}$, where $\hat{A}$ is a large sparse matrix and $\mathbf{x}$, $\mathbf{y}$ are dense vectors. While SpMV operation belongs to the most simple operations of linear algebra, it is rather surprising—and instructive—to realize to what extent its efficient implementation on GPUs requires a deep understanding of the hardware. The SpMV kernel can thus serve as a good illustration of the GPU programming principles. Moreover, the ideas behind this kernel turn out to be helpful in other GPU kernels that deal with sparse data structures.

Z. Koza (✉) • M. Matyka
Faculty of Physics & Astronomy, University of Wrocław, Wrocław, Poland
e-mail: zkoza@ift.uni.wroc.pl; maq@ift.uni.wroc.pl

Ł. Mirosław
Institute of Informatics, Wrocław University of Technology, Wrocław, Poland
e-mail: lukasz.miroslaw@vratis.com

J. Poła
Institute of Physics, University of Silesia, Katowice, Poland
e-mail: jakub.pola@gmail.com

V. Kindratenko (ed.), *Numerical Computations with GPUs*,
DOI 10.1007/978-3-319-06548-9__6, © Springer International Publishing Switzerland 2014

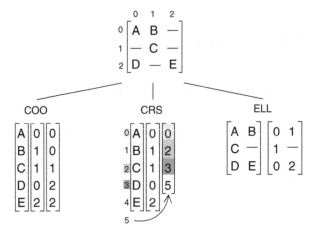

Fig. 6.1 A simple matrix encoded in various sparse formats

We start the chapter with a short introduction into two main issues each SpMV designer must cope with: sparse matrix storage and architecture-specific aspects of the problem.

6.1.1 Sparse Matrix Formats

To perform numerical algebra on matrices, we have to keep them in computer memory. With sparse structures this is a non-trivial task. The straightforward usage of a two-dimensional array is impractical as sparse data written this way could easily exceed the computer memory. Moreover, reading all these zeroes would increase the execution time by several orders of magnitude. Therefore, compact formats for sparse matrices that omit an unnecessary storage of zeroes are required. Below we focus on three most representative formats used in the context of the GPU (cf. Fig. 6.1):

(a) COO (coordinate format), in which non-zero matrix elements together with their row and column indices are stored in separate arrays;
(b) CRS (compressed row storage), in which references to the first nonzero element within each matrix row are stored instead of the row indices;
(c) ELL, in which relatively small two-dimensional dense arrays are used to store the nonzero matrix elements and their column indices.

The simplest sparse matrix format is the coordinate (COO) format, in which the information about the row index, column index, and the value of each non-zero matrix element is stored in three 1D arrays, RowInd, ColInd, and Val, respectively. As an example, consider a matrix:

$$\hat{M} = \begin{bmatrix} 1 & 0 & 0 & 2 & 0 \\ 0 & 3 & 0 & 0 & 4 \\ 0 & 0 & 5 & 0 & 6 \\ 0 & 0 & 7 & 8 & 9 \\ 0 & 0 & 0 & 0 & 10 \end{bmatrix}. \tag{6.1}$$

Its COO representation (with zero-based indexing) reads

$$\text{Val} = [\, 1 \ 2 \ 3 \ 4 \ 5 \ 6 \ 7 \ 8 \ 9 \ 10 \,],$$

$$\text{ColInd} = [\, 0 \ 3 \ 1 \ 4 \ 2 \ 4 \ 2 \ 3 \ 4 \ 4 \,],$$

$$\text{RowInd} = [\, 0 \ 0 \ 1 \ 1 \ 2 \ 2 \ 3 \ 3 \ 3 \ 4 \,].$$

To complete the definition of a sparse matrix representation, one also needs to supply three integers: the number of matrix rows (rows), columns (cols) and non-zero elements (nnz).

In the above example the row-major ordering was used, i.e., the arrays were first sorted by row indices and then by column indices. In such a case array RowInd will typically contain sequences of many identical entries. This property is utilized in the compressed row storage (CRS, also known as compressed sparse row, CSR) format to reduce the memory footprint by replacing array RowInd with a shorter array RowPtr. This array has exactly rows+1 elements and is defined by the requirement that RowPtr[j] be equal to the number of non-zero elements in all the rows preceding the j-th row ($j = 0, \ldots,$ rows $- 1$) and RowPtr[rows] = nnz. If the matrix contains no empty rows, RowPtr[j] gives the index into Val corresponding to the first non-zero element in the j-th matrix row. The CRS representation of $\hat{M}$ reads

$$\text{Val} = [\, 1 \ 2 \ 3 \ 4 \ 5 \ 6 \ 7 \ 8 \ 9 \ 10 \,],$$

$$\text{ColInd} = [\, 0 \ 3 \ 1 \ 4 \ 2 \ 4 \ 2 \ 3 \ 4 \ 4 \,],$$

$$\text{RowPtr} = [\, 0 \ 2 \ 4 \ 6 \ 9 \ 10 \,].$$

Note that arrays Val and ColInd are the same as in COO format.

In the ELLPACK/ITPACK (ELL) format an $n \times m$ sparse matrix is represented by two $n \times k$ dense arrays, Val and ColInd, where k is the maximum number of non-zero elements per row. Array Val is constructed from the original matrix by removing all zeros, while ColInd holds column indices into Val. The rows with less than k non-zero elements are padded in Val and ColInd arrays with 0 and -1, respectively. The ELL representation of $\hat{M}$ is thus:

$$
\mathrm{Val} = \begin{bmatrix} 1 & 2 & 0 \\ 3 & 4 & 0 \\ 5 & 6 & 0 \\ 7 & 8 & 9 \\ 10 & 0 & 0 \end{bmatrix}, \quad \mathrm{ColInd} = \begin{bmatrix} 0 & 3 & -1 \\ 1 & 4 & -1 \\ 2 & 4 & -1 \\ 2 & 3 & 4 \\ 4 & -1 & -1 \end{bmatrix}. \tag{6.2}
$$

6.1.2 Architecture-Specific Issues

The ideal sparse matrix representation should

- store only the non-zero matrix elements,
- require no extra storage space,
- require no additional computation (e.g. sorting),
- allow for efficient utilization of the hardware,
- allow for a single-kernel implementation of the SpMV product.

Unfortunately, this ideal representation does not exist and one has to strike a balance between conflicting requirements. Here we briefly present the hardware-related issues that must be taken into account in designing an efficient SpMV kernel running on the GPU.

Matrix size. Modern GPUs are massively parallel, throughput-oriented architectures that need to process at least tens of thousands of threads to hide a high latency of the off-chip memory. Moreover, one should take into account a relatively high kernel launch time, $\approx 4\,\mu s$ for Tesla K20M. Note that during $4\,\mu s$ a GPU with the memory bandwidth of 200 GB/s can read $\approx 8 \times 10^5$ bytes, or 10^5 numbers in double precision. This means that for today's hardware the minimal number of nonzero matrix elements that could saturate the GPU and amortize the kernel launch time is of order of 10^5. Thus, the sparse matrices that can be processed efficiently on GPUs need to be large.

Rows are preferred to columns. If different threads were allowed to write to the same entry in the output vector, they would have to be synchronized, e.g., by atomic operations, which would diminish the performance. Therefore GPU implementations of the SpMV kernel are based on matrix formats that facilitate accessing the matrix elements by rows (CRS and ELL) or assume the row-major ordering of the data (COO). This, in turn, hinders the development of implementations of algorithms where a sparse matrix should be traversed along its columns, e.g., multiplication by a matrix transpose.

Memory boundedness. Calculation of a sparse matrix-vector product essentially reduces to many "multiply and add" operations, which in modern GPUs are implemented as a single fused multiple-add (FMA) instruction. Since SpMV multiplication involves several memory accesses per arithmetic instruction, the SpMV kernel is inherently memory-bound. For example, a server-class Tesla K20X GPU can perform $\approx 6.5 \times 10^{11}$ FMA operations per second and can access its main memory at $\approx 2.5 \times 10^{11}$ B/s, which yields approximately 2.5 operations per byte. For the SpMV kernel this sets the upper bound for the processor computational efficiency to $\approx 2\,\%$ of its peak theoretical value. It is therefore of utmost importance to focus on the memory utilization as well as on reducing the memory footprint of the SpMV kernel.

Storage overhead. CRS generally needs less memory than COO. As for ELL, the situation is far more complex. If each matrix row contains exactly the same number of nonzero elements (we shall call this parameter "a row length"), then ELL is the most storage-efficient format of the three. However, if row lengths vary, shorter rows must be padded with explicit zeroes. For example, if the length of the first matrix row is 10 and all other rows contain only 1 nonzero element, the ELL format imposes a huge, ten-fold memory overhead. On the one hand, this makes ELL impractical as a general sparse matrix format. On the other hand, for regular matrices ELL is really fast on GPUs, approximately three times faster than COO, so many attempts have been made to reduce its potentially unacceptable storage overhead. Two main ideas have been used to achieve this goal. The first one consists in partitioning the matrix into a regular part, stored in ELL, and an irregular part, stored in a storage-efficient format, e.g., COO. This is exactly the thought behind the HYB format from NVIDIA:

$$HYB = ELL + COO.$$

The second idea is to use some kind of matrix transformation, e.g., permutation of rows according to their size and then to divide the matrix into several slices, each represented separately in ELL, to reduce padding. This approach has led to the development of several ELL-based formats, e.g., sliced-ELL and sliced ELLR-T.

Coalescing memory transfers. Once the storage has been optimized, we still have to make sure that the off-chip memory can be read from or written to efficiently. In the case of the GPU, this is equivalent to requiring that the accesses to the data stored in the computer representations of $\hat{A}$, $\mathbf{x}$ and $\mathbf{y}$ can be coalesced. This is a key condition for the SpMV performance: failure to coalesce global data transfers can decrease the kernel performance by an order of magnitude.

The output vector, $\mathbf{y}$, can be coalesced quite easily; besides, it often contributes only a small fraction of all data transfers involved in SpMV. The input vector, $\mathbf{x}$, has a much serious impact on the kernel performance, as its elements are requested as often as the matrix values. As this is the only data array whose elements can be used many times during an SpMV kernel invocation, it would be very advantageous to have it buffered on-chip. However, the access pattern for the elements of $\mathbf{x}$ is completely unpredictable and so an SpMV kernel designer has a very limited control over the way $\mathbf{x}$ can be cached. The most common strategy is to bind it to a texture cache (on devices supporting OpenCL or CUDA with compute capability cc < 3.5) or to the 48 kB read-only cache (only for CUDA cc ≥ 3.5). Perhaps the best thing that can be done regarding $\mathbf{x}$ is to reduce the so called matrix bandwidth, i.e., permute matrix rows and columns so that to move the nonzero elements towards its main diagonal. We have seen physics simulations in which renumbering the mesh cells with the Cuthill–McKee algorithm accelerated the SpMV by a factor of six.

Let's now consider the matrix data. In ELL all data accesses for both the value and column index arrays can be fully coalesced by assigning consecutive threads to consecutive matrix rows. One point worth noticing is that the data in these arrays are stored in column-major order, so in order to ensure the same coalescing for each column, the number of rows must be a multiply of a warp size (currently: 32). This can be achieved by padding the matrix with up to 31 empty rows.

The data transfer coalescence is much harder to achieve in COO and CRS because in these formats the row lengths can vary and, consequently, there is no simple *functional* mapping between the position in the internal array holding the nonzero matrix values, val, and the matrix row number. Hence, the data in COO and CRS must be stored in a row-major order. In CRS, the simplest choice is to assign consecutive threads to consecutive matrix rows. This is the essence of the so called scalar CRS kernel. In this approach threads process matrix elements in essentially the same order as in ELL, however, in contrast to ELL, the data are now arranged in a row-major order. This precludes any data coalescing except when the row lengths are extremely short. Thus, for most sparse matrices the scalar kernel is easy to write but very slow to run. Another option is to process matrix rows using whole warps, which leads to the so called vector CRS kernel. This mapping allows for good global memory coalescing and results in a kernel that is very efficient for matrices in which the mean row length is quite high, 100 or more. To understand this phenomenon, consider a matrix with row lengths equal to 5. In this case a warp of 32 threads would read only 5 data items per clock tick. Moreover, these 5 data items are quite likely to be located in different 128-byte-long global memory segments and hence two data transfers may be necessary to complete the read request. In this particular case the CRS vector loses to the ELL kernel at least 5:32 and perhaps even 5:64. This problem can be mitigated by making a warp process several consecutive rows, an idea that has led to the development of several CRS-based, GPU-oriented formats, e.g., CRS SIC and CMRS.

As for COO, this format allows for an elegant implementation based on a segmented reduction, an algorithm which is, however, beyond the scope of the present study. While the data transfers turn out to be well coalesced, several kernels must be launched sequentially to complete the job, each transferring data from or to the global memory. The COO kernel is not very fast for regular matrices, but since its computational performance is largely independent of the matrix structure, it can be found useful for matrices with a very irregular pattern of nonzero elements.

Work imbalance and thread divergence. Some matrices, e.g., those describing the WWW connections, exhibit a high variability of row lengths. If a block of threads is assigned one long and many short rows in the "vector" CRS kernel, then the warps processing short rows will quickly finish their job and stay idle waiting for the warp processing the long row to finish (work imbalance). If a warp is assigned a short row, than only a few of its threads will be active (thread divergence). Both of these problems can be mitigated by reordering the rows and processing more than one row per warp.

In summary, ELL excels in data coalescing, CRS in reduction of the storage overhead, and COO is a good alternative for the most irregular sparsity patterns. The main drawback of ELL is the memory overhead related to zero padding, CRS is inefficient for matrices with short rows, and COO is too slow for regular matrices. These problems can be mitigated in ELL either by combining it with a storage-efficient format or by some kind of matrix preprocessing involving row permutation, whereas disadvantages of CRS can be counteracted by processing several matrix rows per warp.

6.2 SpMV for Everyday Usage

Whenever we have to use a nontrivial piece of code, our first thought is to use a ready-made library. Many implementations of the SpMV kernel are already available to download from the Internet, both for the CUDA and OpenCL platforms. Among them, the cuSPARSE, CUSP and Paralution libraries are certainly worth recommendation.

6.2.1 CuSPARSE

The NVIDIA CUDA Sparse Matrix library (cuSPARSE) is a highly-optimized C/C++ library of basic linear algebra subroutines used for handling sparse matrices on the NVIDIA GPUs. It is freely available as part of the CUDA toolkit and contains implementations for several sparse matrix formats, including CRS and HYB. All functions are thread-safe and can be called from many host threads. Moreover, they are executed asynchronously with respect to the CPU and may return control to the application on the host before they complete their job.

The SpMV product for matrices in CRS format is handled by a family of functions cusparse[S,D,C,Z]csrmv, where exactly one of the upper-case letters in the square brackets must be selected to indicate whether the function accepts real data in single (S) or double (D) precision or perhaps complex data in single (C) or double (Z) precision. Each of these functions performs a general matrix-vector operation

$$y = \alpha * \operatorname{op}(A) * x + \beta * y,$$

where x, y are vectors, A is a sparse matrix stored in CRS, α, β are some constants, and op is one of three operators that can modify A: either the identity operator $(\operatorname{op}(A) = A)$ or the matrix transpose operator $(\operatorname{op}(A) = A^T)$, or the conjugate transpose operator $(\operatorname{op}(A) = A^H)$. This operation reduces to the SpMV product for $\alpha = 1, \beta = 0$, and $\operatorname{op}(A) = A$.

Performing the SpMV operation in HYB is a bit more complicated. CuSPARSE implements HYB in a opaque data type that can only be manipulated by calling appropriate subroutines. The first step is to create and initialize an internal data structure by calling cusparseCreateHybMat. Then one has to fill it with data by converting a matrix from CRS format using an appropriate subroutine from a cusparse[S,D,C,Z]csr2hyb family. Now it is possible to perform the SpMV operation by calling a cusparse[S,D,C,Z]hybmv function.

Tests show that NVIDIA HYB often yields better performance than NVIDIA CRS. However, HYB requires more storage, especially during conversion from CRS format, as at this stage a matrix is stored in two disjoint representations.

6.2.2 CUSP

CUSP is a C++ template library for sparse linear algebra operations on the CUDA platform. Its distinguishing feature is a flexible, high-level interface for manipulating sparse matrices and solving sparse linear systems. CUSP provides various linear solvers, preconditioners, sparse linear algebra and graph computation subroutines and can handle matrices in various sparse formats, including COO, CRS, ELL and HYB. While its SpMV routines are not as efficient as those available in NVIDIA cuSPARSE, CUSP is an open-source project based on a liberal Apache 2.0 licence, which makes it an excellent starting point for any CUDA-based software project that exploits sparse linear algebra. The library is available from https://github.com/cusplibrary.

Listing 6.1 shows an example of how simple and elegant can programming with CUSP be. This complete program declares a matrix in HYB format, loads its elements from a file stored in the MatrixMarket file format (*.mtx), allocates and initializes storage for the input and output vectors, performs the SpMV operation (cusp::multiply), and finally prints the result out.

Listing 6.1 An SpMV example in CUSP

```
#include <cusp/hyb_matrix.h>
#include <cusp/multiply.h>
#include <cusp/io/matrix_market.h>
#include <cusp/print.h>

int main()
{
    cusp::hyb_matrix<int, float, cusp::device_memory> A;
    cusp::io::read_matrix_market_file(A, "1.mtx");
    cusp::array1d<float, cusp::device_memory> x(A.num_rows, 1);
    cusp::array1d<float, cusp::device_memory> y(A.num_rows, 0);
    cusp::multiply(A, x, y);
    cusp::print(y);
}
```

6.2.3 Paralution

Paralution is another open-source C++ library for sparse linear algebra. Its unique feature is a high-level hardware and software abstraction, which enables its users to develop a portable software that can be compiled for various hardware accelerator and software backend configurations, including NVIDIA GPUs (CUDA, OpenCL), AMD GPUs (OpenCL), Intel Xeon Phi (OpenCL, OpenMP) and multicore CPUs (OpenMP). The target backend can be set at compile time by defining an appropriate preprocessor macro: SUPPORT_CUDA, SUPPORT_OCL or SUPPORT_MIC for the CUDA, OpenCL or Intel Xeon Phi, respectively.

An exemplary Paralution code, a direct counterpart of the program from Listing 6.1, is presented in Listing 6.2.

Listing 6.2 An SpMV example in Paralution

```
#include <paralution.hpp>

int main(int argc, char* argv[])
{
    paralution::init_paralution();

    paralution::LocalMatrix<float> mat;
    paralution::LocalVector<float> x, y;

    mat.ReadFileMTX("1.mtx");
    mat.ConvertToHYB();

    x.Allocate("x", mat.get_nrow());
    y.Allocate("y", mat.get_nrow());
    x.Ones();
    y.Zeros();

    mat.MoveToAccelerator();
    x.MoveToAccelerator();
    y.MoveToAccelerator();

    mat.Apply(x, &y); // y = A*x

    paralution::stop_paralution();
}
```

As can be seen, the matrix and the vectors are first allocated on the host. The matrix is read from a file (default format: CRS). Next, its format is converted to HYB. All the data are then moved to the accelerator, if the library can detect one; otherwise member functions MoveToAccelerator return immediately. If no accelerator is attached to the host, the data will remain on the CPU and Paralution will attempt to use the OpenMP backend (it is also possible to use the Intel MKL library instead). The SpMV operation is executed with the Apply member function. The library is available from http://www.paralution.comontheGPL-3licence.

6.3   Custom SpMV Kernels

Sometimes the structure of our sparse matrices exhibit some characteristic patterns or the problems we are solving require that some other operations, besides SpMV, should be implemented efficiently. In both cases a solution may consist in designing a special matrix format and writing an appropriate SpMV kernel. Such "custom" sparse formats are usually derived from simpler ones, especially ELL and CRS. Below we present CUDA and OpenCL implementations for these two basic formats and discuss the way these formats (and corresponding SpMV implementations) could be extended to improve the SpMV eperformance.

6.3.1   SpMV for ELL and ELL-Based Sparse Matrix Formats

As it was already stated, ELL belongs to the most efficient formats for sparse matrices in which all rows have the same lengths. Moreover, as we shall see, writing an efficient SpMV implementation for ELL is relatively simple, which makes this format a good starting point for our further discussion.

6.3.1.1   ELL

Complete SpMV kernels for matrices stored in ELL, written in OpenCL and CUDA, are shown in Listings 6.3 and 6.4, respectively.

As might be expected, the two implementations share a lot of features. They both take eight identical arguments: four integers that define the size of the original matrix (`rows`, `cols`), and the size of its ELL representations (`ell_rows`, `ell_cols`) followed by four pointers to 1D arrays that hold the column indices (`col_ind`), values (`val`), input (`x`) and output (`y`) vectors. Note that while in theory ELL holds the values and column indices in dense 2D arrays, in practice they are implemented as 1D arrays, which leads to instructions like

```
const int index = i * ell_rows + row;
```

that transform indices from 2D to 1D representation. Since rows are processed by individual threads, this formula ensures that *all* accesses to the output vector (`y`), matrix values (`val`) and column indices (`col_ind`) are *fully coalesced* provided that `ell_rows` is a multiply of the warp size. This, in turn, explains why the number of rows in the internal ELL representation need not be equal to the number of rows in the matrix and must be passed as a separate argument.

The bodies of the OpenCL and CUDA implementations are almost identical, the main difference being the mapping of a current thread id into a matrix row. In OpenCL this is achieved by calling `get_global_id`, whereas CUDA utilizes a more cumbersome method

```
const int row = blockDim.x * (gridDim.x * blockIdx.y + blockIdx.x) + threadIdx.x;
```

Listing 6.3 An OpenCL kernel for ELL

```
1  #pragma OPENCL EXTENSION \
2        cl_khr_fp64 : enable
3  #define T double
4
5  __kernel void ell_spmv_d(
6        const int rows,
7        const int cols,
8        const int ell_rows,
9        const int ell_cols,
10       __global const int *col_ind,
11       __global const T *val,
12       __global const T *x,
13       __global T *y)
14 {
15    const int row = get_global_id(0);
16
17
18    if (row >= rows)
19        return;
20
21    T sum = (T)0;
22    for (int i=0; i<ell_cols; ++i)
23    {
24        const int index = i * ell_rows + row;
25        const int column = col_ind[index];
26        if (column >= 0)
27            sum += val[index] * x[column];
28    }
29
30    y[row] = sum;
31 }
```

Listing 6.4 A CUDA kernel for ELL

```
template <typename T, size_t BLOCK_SIZE>
__launch_bounds__(BLOCK_SIZE,1)

__global__ void ell_spmv(
      const int rows,
      const int cols,
      const int ell_rows,
      const int ell_cols,
      const int * __restrict__ col_ind,
      const T * __restrict__ val,
      const T * __restrict__ x,
      T * __restrict__ y)
{
   const int row = blockDim.x * (gridDim.x *
             blockIdx.y + blockIdx.x) + threadIdx.x;

   if (row >= rows)
       return;

   T sum = (T)0;
   for (int i = 0; i < ell_cols; ++i)
   {
       const int index = i * ell_rows + row;
       const int column = col_ind[index];
       if (column >= 0)
           sum += val[index] * LOAD(x[column]);
   }

   y[row] = sum;
}
```

which contains a typical expression for a thread id in CUDA kernels invoked on 2D grids of thread blocks. Some implementations, e.g., CUSP, use 1D grids with a simplified expression for the thread id,

```
const int thread_id = blockDim.x * blockIdx.x + threadIdx.x;
```

This, however, brings about a problem on pre-Kepler architectures, where the maximum number of threads that can be launched in a 1D grid configuration is limited to $\approx 2^{26} \approx 6.7 \times 10^7$. If matrices with a larger number of rows are to be processed by the SpMV kernel, the implementation must be modified to allow a thread to process several matrix rows (see the CUSP source code for details).

The differences between OpenCL and CUDA versions are mostly technical. An important advantage of CUDA is that it fully supports C++ templates. This facilitates writing a generic code that can be used for single or double precision kernels. In contrast to this, OpenCL requires that a separate function be written for each data type. Listing 6.4 shows also how template arguments can be used in CUDA to pass to the compiler some additional bits of information to help it optimize the code. Function qualifier __launch_bounds__(BLOCK_SIZE,1) asserts that the kernel will never be launched with more than BLOCK_SIZE threads per block, which the compiler can use to optimize the register usage. Another interesting feature of the CUDA code is that all the pointers are marked with the __restrict__ qualifier to assert to the compiler that the pointers are not aliased and writing through y will never overwrite elements of other arrays. This helps the compiler to cache the read-only data (pointed to by const pointers) in the read-only data cache introduced in devices of CUDA compute capability $\geq$ 3.5. We may also explicitly demand that some data be fetched via this cache using the __ldg function. In Listing 6.4 we do it through a macro LOAD, which is defined in Listing 6.5.

Listing 6.5 A macro to speed up loading of read-only data on the newest CUDA-capable hardware

```
#if __CUDA_ARCH__ < 350
# define LOAD(x) x
#else
# define LOAD(x) __ldg(&x)
#endif
```

6.3.1.2 ELL-Based Formats

ELL is a great starting point for devising new sparse matrix formats tailored to our needs. Here we only list several extensions of ELL that were recently examined in the context of GPUs. The details can be found in the original research papers.

- ELL-R: This is ELL with an additional 1D array that stores the actual matrix row lengths.
- ELLR-T: This is ELL-R in which a warp processes w/T rows, where w is a warp size and $T = 1, 2, 4, \ldots, w$.
- Sliced ELL: The matrix is partitioned into strips of S adjacent rows, and each strip is stored in ELL. Further performance improvement can be achieved by reordering matrix rows according to their length. For $S = 1$ this ELL-based format reduces to CRS.
- Sliced ELLR-T: The matrix is partitioned into slices and each slice is stored in ELLR-T.

6.3.2 SpMV for CRS and CRS-Based Sparse Matrix Formats

While GPU-oriented extensions of ELL focus on reducing its storage overhead, CRS-based formats concentrate on mitigating problems with thread divergence and memory access coalescence. Writing an efficient SpMV kernel for vector architectures, like GPUs, is more challenging if the matrix is stored in CRS. However, as we shall see, CRS-based kernels can compete with or even surpass ELL-based kernels. To further improve CRS kernels, some extensions were suggested, i.e.:

- CRS SIC (CRS with segmented interleave combination) format: The matrix is partitioned into many strips of a constant height $h \geq 2$ and the matrix values are interleaved within each strip, with zero-padding of shorter rows. For example, if $h = 2$ and the two rows in a strip have the nonzero values $[6, 4, 2]$ and $[1, 5, 3, 7]$, they are interleaved to form a "longer row" $[6, 1, 4, 5, 2, 3, 0, 7]$. These "longer rows" are then stored in CRS. To reduce zero padding and work imbalance, the matrix is reordered according to row lengths and may be further partitioned into several segments containing rows with approximately equal lengths. A separate CRS SIC kernel is then launched for each segment. Implementation of the SpMV product for CSR SIC is thus quite complex.

- CMRS (compressed multi-row storage) format: The matrix is partitioned into strips of hight h. Strips are stored in CRS. An additional array is then used to identify the actual row index within a strip.

Below we present in a greater detail CUDA implementations of the SpMV kernel for the CRS and CMRS formats.

6.3.2.1   CRS

One of the problems in designing SpMV kernels for CRS format is how to map GPU threads into matrix rows. Listing 6.6 shows a "vector" implementation in which each matrix row is processed by all threads in a corresponding warp.

The kernel begins with a definition of a shared memory array `shared`. This array is marked with the `volatile` keyword to inform the compiler that the array will be used as a communication vehicle between threads of a block so that the compiler should never buffer its elements in registers. The size of the array is equal to the number of threads in a block of threads plus `WARP_SIZE`/2 ($=16$) additional elements to avoid buffer overrun. Here `WARP_SIZE` has of course the same value as that in CUDA's `warpSize` register, but the latter cannot be used as a compile-time constant.

The expression for the thread id,

```
const int thread_id = blockDim.x · blockIdx.x + threadIdx.x;
```

is written with the assumption that the kernel will be launched in a 1D grid configuration. This implies that a warp may be forced to process more than one matrix row on devices of cc < 3.5, which explains the outer `for` loop that runs over all rows assigned to the current warp. However, there is a deeper thought behind the outer loop: an attempt to balance warp load. The implementation assumes that a fixed number of warps has been launched that persist over the duration of the computation. This approach tends to even out moderate imbalances in per-row workload related to the variability of the matrix row lengths. Another strategy to reduce work imbalance is to use small, but not too small blocks. Tests show that `BLOCKSIZE = 128` is a good choice for modern NVIDIA devices.

Each thread accumulates the partial sum it has been assigned to compute and stores it in register `sum`, which is private to a thread. Once the whole row has been processed, all threads in the warp use the shared memory buffer `shared` to reduce these values to the actual sum, which is then written to the output vector `y`. Since they work in parallel, it suffices to perform only 5 instruction to work out the sum of

Listing 6.6 A CUDA SpMV kernel for CRS

```
template<typename T, int BLOCK_SIZE = 128>
__launch_bounds__(BLOCK_SIZE,1)
__global__ void spmv_crs(const int * __restrict__ row_ptr,
                         const int * __restrict__ col_ind,
5                        const T * __restrict__ val,
                         const T * __restrict__ x,
                               T * __restrict__ y,
                         const int rows)
{
10  const int WARP_SIZE = 32;
    __shared__ volatile T shared[BLOCK_SIZE + WARP_SIZE/2];

    const int thread_id = blockDim.x * blockIdx.x + threadIdx.x;
    const int warp_id = thread_id / WARP_SIZE;
15  const int thread_lane = threadIdx.x % WARP_SIZE;
    const int num_warps = ( (blockDim.x + WARP_SIZE − 1) / WARP_SIZE ) * gridDim.x;

    for(int row = warp_id; row < rows; row += num_warps)
    {
20    const int row_start = row_ptr[row];
      const int row_end = row_ptr[row+1];
      T sum = T(0);
      for(int j = row_start + thread_lane; j < row_end; j += warpSize)
        sum += val[j] * LOAD(x[col_ind[j]]);

25
      shared[threadIdx.x] = sum;
      shared[threadIdx.x] = sum += shared[threadIdx.x + 16];
      shared[threadIdx.x] = sum += shared[threadIdx.x + 8];
      shared[threadIdx.x] = sum += shared[threadIdx.x + 4];
30    shared[threadIdx.x] = sum += shared[threadIdx.x + 2];
      sum += shared[threadIdx.x + 1];

      if (thread_lane == 0)
        y[row] = sum;
35  }
}
```

32 numbers. Note that the parallel reduction code in Listing 6.6 explicitly assumes that the warp size is 32, which may change in future GPU architectures.

Listing 6.6 shows only a basic implementation of the "vector" CRS kernel. It can be still improved by several techniques, at the cost of increased complexity. For example, the CUSP library can virtually divide each warp into 2, 4, 8 or 16 smaller parts and assign them to different rows. This can improve the performance for matrices with short rows. Another technique, applicable for matrices with long rows, is to first process the unaligned part of each row to ensure fully coalesced accesses for the remaining part. It is also possible to speed up the parallel reduction by using shuffle instructions introduced in CUDA-capable devices of compute capability ≥ 3.5. These instructions allow to exchange data between the threads of a warp directly, bypassing the shared memory.

6.3.2.2   CMRS: A CRS-Based Format for Multi-Row Matrix Processing

The main drawback of the CRS SpMV kernel discussed in Sect. 6.3.2.1 is its low performance for matrices with relatively short rows. A natural solution to this problem is to group some rows into strips and process them in parallel, which is the key idea behind compressed multi-row storage (CMRS). In Fig. 6.2 an exemplary 3×3 matrix is encoded both in CRS and CMRS.

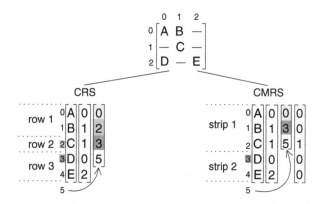

Fig. 6.2 A simple matrix encoded in CRS and CMRS formats, the latter with strip height $= 2$

CMRS uses four arrays to encode the sparse matrix, the first two of them being exactly the same as in CRS, while two arrays are specific to CMRS:

Val—a list of non-zero elements;
ColInd—column indices of all entries in Val;
StripPtr—locates the first elements of each strip (indices into Val);
RowInStrip—locates rows within a strip (for each element in Val).

If we assume the constant height of all strips (denoted by height or HEIGHT in the following text) then the number of strips is equal to $\lceil \text{rows/height} \rceil$ (the smallest integer greater than or equal to the ratio rows/height). As an example, let us consider matrix $\hat{M}$ introduced in Sect. 6.1.1. Assuming height $= 2$, the CMRS representation of $\hat{M}$ reads:

$$\text{Val} = [\, 1 \ 2 \ 3 \ 4 \ 5 \ 6 \ 7 \ 8 \ 9 \ 10 \,],$$

$$\text{ColInd} = [\, 0 \ 3 \ 1 \ 4 \ 2 \ 4 \ 2 \ 3 \ 4 \ 4 \,],$$

$$\text{StripPtr} = [\, 0 \ 4 \ 9 \ 10 \,],$$

$$\text{RowInStrip} = [\, 0 \ 0 \ 1 \ 1 \ 0 \ 0 \ 1 \ 1 \ 1 \ 0 \,].$$

Conversion between the CRS and CMRS formats is trivial and easy to parallelize. In particular, StripPtr[j] = RowPtr[j $*$ height] for $j <$ strips and

StripPtr[strips] = nnz, whereas RowInStrip[k] is the remainder of the row number divided by height. It is also clear that both formats are equivalent if height = 1, hence CMRS can be regarded as a generalization of the CRS format.

By introducing strips, we enlarged the number of contiguous data items processed by a warp at the cost of an extra array, without any need for zero-padding or row permutation. This extra array turns out to be a minor problem: it contains small integers, as height is assumed to be ≤16. Consequently, they can be stored on 4 bits of array ColInd—the remaining 28 bits are enough to identify column indices of the matrices that fit into 12 GB of modern GPUs. The main problem with CMRS is that while in the CRS kernel shown in Listing 6.6 we reserved in the shared memory only one word (float or double) per thread, with CMRS one should reserve height such words per thread. As the size of the shared memory is limited, for large height this will reduce the occupancy (i.e., the ratio of the number of resident threads to the maximum number of resident threads) and, consequently, kernel efficiency. For the Kepler-class architecture this sets the upper bound for height to 4. This problem can be mitigated by the fact that array RowInStrip allows to dynamically assign threads to rows. In other words, the order of the items within a strip is essentially arbitrary. In particular, we can try to order them in such a way that each warp will be assigned to process, at a given time, at most M values from the same matrix row, where $1 \leq M \leq$ WARPSIZE, all arranged in a contiguous manner. It turns out that for such arrangement of data items within a strip, the shared memory per thread is proportional to M. Fortunately, for most sparse matrices from real applications, one can assume M = 8, a number far smaller than the warp size. This allows to increase HEIGHT to 16. For most other matrices for which M = 8 cannot be achieved, the solution is. . . zero padding. In a vast majority of cases the resulting storage overhead is negligible. With height = 16 the mean number of nonzeroes per strip is 16 times larger than for CRS, usually ≳100, and hence we can apply one more optimization: pad each strip with zeroes to make sure its length is a multiply of the warp size. This final optimization ensures the *full coalescence* of memory accesses to the matrix data at a price of an acceptable additional storage overhead (usually below 10 %).

The SpMV kernel for the CMRS format is shown in Listing 6.7. Its general structure resembles that of the CRS "vector" kernel (Listing 6.6): after the current thread identifies itself, a big for loop is executed that runs over several different strips and consists of three main parts: the inner for loop running over all elements of a given strip, the parallel reduction and, finally, the storage of the results. For convenience, the shared memory buffer is allocated dynamically at run time, as its size depends on various parameters. A warp has an access to part of the shared memory buffer via ptr pointer. As this buffer is local to a warp, no explicit synchronization of different warps is necessary, which allows for massively parallel processing of strips.

The inner loop is a bit more complicated, as the row and column indices have to be decoded from a single value stored in col_ind. At this point the row index (r) is local to the strip and its value is between 0 and HEIGHT − 1. The parallel reduction is modified to account for the fact that now M × HEIGHT values must be reduced to HEIGHT values. Finally, the results are written out in a coalesced way by HEIGHT contiguous threads.

Tests performed on NVIDIA K20M GPU in double precision show that the speed-up of the CMRS kernel over both NVIDIA HYB and NVIDIA CRS can be as high as three-fold, although for some matrices the former implementation yields the shortest SpMV times—there is no such a thing as a single, universal SpMV kernel for all sparse matrices. The speedup over the "vector" CRS kernel shown in Listing 6.6 turns out to be the largest for matrices with short rows and can be as high as ten-fold (!).

6.4 Further Reading

Perhaps the best way to improve one's skills in designing SpMV kernels is to consult the source codes of high quality open-source GPU libraries for sparse linear algebra. These include:

– CUSP, http://cusplibrary.github.io (CUDA)
– Paralution, http://www.paralution.com (CUDA, OpenCL)

A thorough presentation of many sparse matrix formats and SpMV optimization techniques on traditional, cache-based processor designs can be found in Vuduc's thesis [11]. As for SpMV on GPUs, the primary source of information is a paper by Bell and Garland [1], which discusses the implementations of several SpMV kernels that can be found in the CUSP library. The ELL-R format in the context of SpMV on GPUs was discussed by Vázquez et al. [9]. For Sliced-ELL see Monakov et al. [6], for ELLR-T see Vázquez et al. [10], for Sliced ELLR-T see Dziekonski et al. [3], for CRS-T see Yoshizawa and Takahashi [13], and for CSR SIC see Feng et al. [4]. Blocked sparse formats are a separate class of matrix formats not covered here, see Choi et al. [2] for an example of such a format and the corresponding SpMV kernel.

Listing 6.7 A CUDA SpMV kernel for matrices stored in CMRS. For conciseness HEIGHT = 16 and $2 \le M \le 8$ is assumed

```
template<typename T, int M>
__global__ void
cmrs_multiply( const int * __restrict__ strip_ptr,
                const int * __restrict__ col_ind,
                const T * __restrict__ val,
                const T * __restrict__ x,
                        T * __restrict__ y,
                            int rows )
{
    assert (M >=2 && M <=8);
    const int CMRS_BITS = 4;
    const int CMRS_MASK = (1 << CMRS_BITS) − 1; // 15
    const int WARP_SIZE = 32;
    const int HEIGHT = 16; // parts of the code below rely implicitly on this particular value
    const int asize = HEIGHT*M; // size of warp−owned array in shared memory

    // shared memory is assigned dynamically at kernel invocation
    extern __shared__ char cdata[];
```

```
     // the buffer in shared memory actually contains T's
20   T volatile * sdata = reinterpret_cast<T volatile *>(cdata);
     // ptr points to the warp−owned buffer in shared memory
     T volatile * ptr = &sdata[(threadIdx.x / WARP_SIZE)*asize];

     const int thread_id = blockDim.x * blockIdx.x + threadIdx.x;
25   const int warp_id = thread_id / WARP_SIZE;
     const int thread_lane = threadIdx.x % WARP_SIZE;
     const int num_warps = ( (blockDim.x + WARP_SIZE − 1) / WARP_SIZE) * gridDim.x;

     for(int strip = warp_id; strip*HEIGHT < rows; strip += num_warps)
30   {
         // let's zero the buffer local to the current warp
     #pragma unroll
         for(int k = 0; k < M/2; k++)
             ptr[thread_lane + WARP_SIZE*k] = 0;
35
         const int strip_start = strip_ptr[strip];
         const int strip_end = strip_ptr[strip + 1];
         for(int j = strip_start + thread_lane; j < strip_end; j += WARP_SIZE)
         {
40           int c = col_ind[j];
             int r = c & CMRS_MASK;
             c >>= CMRS_BITS;
             r += HEIGHT*(thread_lane % M);
             ptr[r] += LOAD(x[c]) * val[j];
45       }

         // Now the parallel reduction of the data pointed by ptr.
         T z = 0;
         if (M == 2)
50           z = ptr[thread_lane];
         if (M > 4)
         {
             ptr[thread_lane] += ptr[thread_lane + HEIGHT*4];
             ptr[thread_lane + 32] += ptr[thread_lane + HEIGHT*4 + 32];
55       }
         if (M > 2)
             z = ptr[thread_lane] += ptr[thread_lane + HEIGHT*2];
         if (thread_lane < HEIGHT)
             z += ptr[thread_lane + HEIGHT];
60
         // write the results to y
         int row = strip*HEIGHT + thread_lane;
         if (thread_lane < HEIGHT && row < rows)
             y[row] = z;
65   }
     }
```

Special optimization techniques necessary for sparse matrices with a power-law distribution of row lengths were studied by Yang et al. [12]. Optimization techniques for Kepler-class GPUs were discussed by Mukunoki and Takahashi [7]. Autotuning of the parameters for the SpMV kernels on GPUs was discussed, for example, by Choi et al. [2] and Su and Keutzer [8]. Finally, the CMRS format with two SpMV implementations was described by Koza et al. [5] and examples of its application are available at http://speedit.vratis.com.

References

1. Bell, N., Garland, M.: Implementing sparse matrix-vector multiplication on throughput-oriented processors. In: SC'09: Proceedings of the Conference on High Performance Computing Networking, Storage and Analysis, pp. 1–11. ACM, New York (2009). doi:http://doi.acm.org/10.1145/1654059.1654078
2. Choi, J.W., Singh, A., Vuduc, R.W.: Model-driven autotuning of sparse matrix-vector multiply on GPUs. SIGPLAN Not. **45**(5), 115–126 (2010). http://doi.acm.org/10.1145/1837853.1693471
3. Dziekonski, A., Lamecki, A., Mrozowski, M.: A memory efficient and fast sparse matrix vector product on a GPU. Prog. Electromagn. Res. **116**, 49–63 (2011). http://www.jpier.org/PIER/pier116/03.11031607.pdf
4. Feng, X., Jin, H., Zheng, R., Hu, K., Zeng, J., Shao, Z.: Optimization of sparse matrix-vector multiplication with variant CSR on GPUs. In: 2011 IEEE 17th International Conference on Parallel and Distributed Systems (ICPADS), pp. 165–172. IEEE, Tainan (2011)
5. Koza, Z., Matyka, M., Szkoda, S., Mirosław, Ł.: Compressed multirow storage format for sparse matrices on graphics processing units. SIAM J. Sci. Comput. 36(2), 219–239 (2014). http://dx.doi.org/10.1137/120900216
6. Monakov, A., Lokhmotov, A., Avetisyan, A.: Automatically tuning sparse matrix-vector multiplication for GPU architectures. In: Patt, Y., Foglia, P., Duesterwald, E., Faraboschi, P., Martorell, X. (eds.) High Performance Embedded Architectures and Compilers. Lecture Notes in Computer Science, vol. 5952, pp. 111–125. Springer, Heidelberg (2010). http://dx.doi.org/10.1007/978-3-642-11515-8_10.
7. Mukunoki, D., Takahashi, D.: Optimization of sparse matrix-vector multiplication for CRS format on NVIDIA Kepler architecture GPUs. In: Murgante, B., Misra, S., Carlini, M., Torre, C.M., Nguyen, H.Q., Taniar, D., Apduhan, B.O., Gervasi, O. (eds.) Computational Science and Its Applications – ICCSA 2013. Lecture Notes in Computer Science, vol. 7975, pp. 211–223. Springer, Heidelberg (2013). http://dx.doi.org/10.1007/978-3-642-39640-3_15
8. Su, B.Y., Keutzer, K.: clSpMV: a cross-platform opencl spmv framework on GPUs. In: Proceedings of the International Conference on Supercomputing, ICS '12 (2012)
9. Vázquez, F., Garzón, E.M., Martınez, J.A., Fernández, J.: Accelerating sparse matrix vector product with GPUs. In: Proceedings of the International Conference on Computational and Mathematical Methods in Science and Engineering (CMMSE 2009), pp. 1081–1092. CMMSE, Gijón (2009)
10. Vázquez, F., Fernández, J.J., Garzón, E.M.: Automatic tuning of the sparse matrix vector product on GPUs based on the ELLR-T approach. Parallel Comput. **38**(8), 408–420 (2012). doi:10.1016/j.parco.2011.08.003. http://www.sciencedirect.com/science/article/pii/S0167819111001050
11. Vuduc, R.W.: Automatic performance tuning of sparse matrix kernels. Ph.D. thesis, University of California (2003)
12. Yang, X., Parthasarathy, S., Sadayappan, P.: Fast sparse matrix-vector multiplication on GPUs: implications for graph mining. Proc. VLDB Endowment **4**(4), 231–242 (2011)
13. Yoshizawa, H., Takahashi, D.: Automatic tuning of sparse matrix-vector multiplication for CRS format on GPUs. In: 2012 IEEE 15th International Conference on Computational Science and Engineering, pp. 130–136 (2012). doi:http://doi.ieeecomputersociety.org/10.1109/ICCSE.2012.28

Part II
Differential Equations

Chapter 7
Solving Ordinary Differential Equations on GPUs

Karsten Ahnert, Denis Demidov, and Mario Mulansky

7.1 Introduction

One of the most common problems encountered in Physics, Chemistry, Biology, but also Engineering or Social Sciences, is to find the solution of an initial value problem (IVP) of an ordinary differential equation (ODE). In fact, many physical laws are written in terms of ODEs, for example the whole classical mechanics, but ODEs also emerge from discretization of partial differential equations (PDEs) or in models of granular systems or when studying networks of interacting neurons. In the most cases one faces ODEs that are too complicated to be solved with analytic methods and one has to rely on numerical techniques to find at least an approximate solution. Of course, there exists a wide range of numerical algorithms to find such solutions of IVPs of ODEs. An introduction to both the mathematical background and the numerical implementation can be found in the textbooks from Hairer, Nørsett and Wanner [14,15]. The standard work for numerical programming, the "Numerical Recipes" [29] also contains detailed sections on solving ODEs.

K. Ahnert (✉)
Ambrosys GmbH, Albert-Einstein-Str. 1-5, 14469 Potsdam, Germany
e-mail: karsten.ahnert@gmx.de

D. Demidov
Kazan Branch of Joint Supercomputer Center, Russian Academy of Sciences,
Lobachevsky st. 2/31, 420011 Kazan, Russia
e-mail: dennis.demidov@gmail.com

M. Mulansky
Max-Planck Institute for the Physics of Complex Systems, Nöthnitzer Str. 38,
01187 Dresden, Germany

TU-Dresden, Institute for Theoretical Physics, Zellescher Weg 17, 01069 Dresden, Germany
e-mail: mulansky@pks.mpg.de

V. Kindratenko (ed.), *Numerical Computations with GPUs*,
DOI 10.1007/978-3-319-06548-9_7, © Springer International Publishing Switzerland 2014

There are also several special classes of ODEs that require specific numerical methods, e.g. the Hamiltonian systems in physics which are typically solved using symplectic routines [21].

Obviously, there is a variety of numerical tools and libraries dedicated to solving ODEs. All mathematical software packages, like Matlab, Maple, Mathematica, or even R [30, 34] contain routines for integrating ODEs. However, the focus here lies on the direct implementation of ODE simulations. For this task, one also finds a vast selection of numerical libraries, typically with Fortran or C/C++ bindings. Most prominent are probably the codes shipped with the "Numerical Recipes" book [29] containing several sophisticated explicit and implicit routines. The GNU scientific library (GSL) also provides ODE functionality [10], and finally the SUNDIALS suite [16] offers a modern implementation of all important algorithms. Unfortunately, none of those libraries supports GPU devices. However, there exists a highly flexible C++ library dedicated to ODEs: Boost.odeint,[1] which is designed in such a generic way that the algorithms are implemented completely independent from the computational backend. Thus, by providing a computational backend that employs GPUs one immediately gets a GPU implementation of the ODE solver. Boost.odeint already includes several backends for GPU computations: for the NVIDIA CUDA-framework based on the Thrust[2] library or the CUDA MTL4[3] [8] and for the OpenCL-framework based on VexCL,[4] ViennaCL,[5] or Boost.Compute.[6] In this text we will show how to implement ODE algorithms in such a generic way that separates the computational backend and thus greatly simplifies the portability to GPUs. Furthermore, we present two such backends, based on CUDA and OpenCL and develop several example simulations using these ODE codes. However, the most difficult part when writing an ODE simulation is the implementation of the right-hand-side (RHS) of the ODE, as it will be explained later. Hence, although Boost.odeint provides all the functionality to find a numerical solution of a given ODE, implementing the RHS of the ODE remains a non-trivial task.

The examples presented later will use modern C++ techniques and thus require the reader to be familiar with several advanced C++ concepts, e.g. we will make heavy use of templates to write generic code. Moreover, knowledge of the C++ Standard Library is also useful, specifically containers, iterators and algorithms. For the ODE algorithms implementation we make use of the C++03 standard only, but in some of the examples we employ the new C++11 and even C++14 abilities.

In the following sections we will give a short introduction to ODEs and the basic numerical schemes for finding approximate solutions (Sect. 7.2), followed by

---

[1]http://www.odeint.com.

[2]http://thrust.github.com.

[3]http://www.simunova.com/gpu_mtl4.

[4]https://github.com/ddemidov/vexcl.

[5]http://viennacl.sourceforge.net/.

[6]https://github.com/kylelutz/compute.

a description of the generic implementation of those algorithms in Sect. 7.3. Then in Sect. 7.4 we will specifically describe how to use the various GPU backends and how they are implemented. The Boost.odeint library is introduced in Sect. 7.5 and Sect. 7.6 contains several examples on how to efficiently implement the RHS of different ODE problems together with a discussion of the performance implications of possible implementations. Finally, Sect. 7.7 contains a short summary and conclusions.

7.2 Numerical Schemes

Before describing the generic implementation of ODE solvers and how to adapt them for GPU usage we will give a short introduction to ODEs and some mathematical background about the numerical schemes. This is mainly to familiarize the reader with our notation; for a more detailed description of the mathematics behind ODE integration we refer to standard textbooks, e.g. [14, 15].

7.2.1 Ordinary Differential Equations

Generally, an ODE is an equation containing a function $x(t)$ of an independent variable t and its derivatives $x', x'', \ldots$:

$$F(x, x', x'', \ldots, x^{(n)}, t) = 0. \tag{7.1}$$

This is the most general form, including implicit ODEs. However, we will here only consider *explicit* ODEs, which are of the form $x^{(n)} = f(x, x', x'', \ldots, x^{(n-1)})$ and are much simpler to be addressed numerically. The highest derivative n that appears in the ODE is called the *order* of the ODE. But any ODE of order n can be easily transformed into an n-dimensional ODE of first order. Therefore, it is sufficient to consider only first order differential equations where $n = 1$. The numerical routines presented later will all deal with initial value problems (IVP) where additionally to the ODE one has also given the value for x at a starting point $x(t = t_0) = x_0$. Thus, the mathematical formulation of the problem that will be numerically addressed throughout the following pages is:

$$\frac{d}{dt}\mathbf{x}(t) = \mathbf{f}(\mathbf{x}(t), t), \qquad \mathbf{x}(t = t_0) = \mathbf{x}_0. \tag{7.2}$$

Here, we use bold face $\mathbf{x}$ to indicate a possible vector character. Typically, the ODE is defined for real-valued variables, i.e. $\mathbf{x} \in \mathbb{R}^N$, but it is also possible to consider complex valued ODEs where $\mathbf{x} \in \mathbb{C}^N$. The function $\mathbf{f}(\mathbf{x}, t)$ is called the right-hand-side (RHS) of the ODE. The most simple physical example for an ODE is probably the *harmonic oscillator*, e.g. a point mass connected to a spring. Newton's equation of motion for such a system is:

$$\frac{d^2}{dt^2} q(t) = -\omega_0^2 q(t),$$ (7.3)

where $q(t)$ denotes the position of the mass and ω_0 is the oscillation frequency, a function of the mass m and the stiffness of the spring k: $\omega_0 = \sqrt{k/m}$. This can be brought into form (7.2) by introducing $p = dq/dt$, using $\mathbf{x} = (q, p)^T$ and defining some initial conditions, e.g. $q(0) = q_0$, $p(0) = 0$. Using the short-hand $\dot{\mathbf{x}} = d\mathbf{x}/dt$ and omitting explicit time dependencies we get:

$$\dot{\mathbf{x}} = \mathbf{f}(\mathbf{x}) = \begin{pmatrix} p \\ -\omega_0^2 q \end{pmatrix}, \qquad \mathbf{x}(0) = \begin{pmatrix} q_0 \\ 0 \end{pmatrix}.$$ (7.4)

Note, that $\mathbf{f}$ in Eq. (7.4) does not depend on the variable t, which makes Eq. (7.4) an *autonomous* ODE. Also note that in this example the independent variable t denotes the time and $\mathbf{x}$ a point in phase spaces, hence the solution $\mathbf{x}(t)$ is the *trajectory* of the harmonic oscillator. This is a typical situation in physical ODEs and the reason behind our choice of variables t and $\mathbf{x}$.[7]

For the harmonic oscillator in Eq. (7.4), one can easily find an analytic solution of the IVP: $q(t) = q_0 \cos \omega_0 t$ and $p(t) = -q_0 \omega_0 \sin(\omega_0 t)$. For more complicated, non-linear ODEs it is often impossible to find an analytic solution and one has to employ numerical methods to at least find an approximate solution. One specific example are systems exhibiting *chaotic dynamics* [26], where the trajectories can not be described in terms of analytic functions. One of the first models where this has been explored is the so-called Lorenz-system [35], a three-dimensional ODE given by the following equations for $\mathbf{x} = (x_1, x_2, x_3)^T \in \mathbb{R}^3$:

$$\dot{x}_1 = \sigma(x_2 - x_1)$$
$$\dot{x}_2 = Rx_1 - x_2 - x_1 x_3$$ (7.5)
$$\dot{x}_3 = x_1 x_2 - bx_3,$$

where σ, R, $b \in \mathbb{R}$ are parameters of the system. Although the solution might be impossible to find analytically, there are mathematical proofs about its *existence* and *uniqueness* under some conditions on the RHS $\mathbf{f}$, e.g. the Picard-Lindelöf theorem which requires $\mathbf{f}$ to be Lipschitz continuous [37]. Provided that this condition is fulfilled and a unique solution does exist, as it is the case for almost all practical problems, one can apply a numerical algorithm to find an approximate solution.

---

[7]In Mathematics, the independent variable is often called x and the function is $y(x)$.

7.2.2 Runge-Kutta Schemes

The most common general-purpose schemes for solving initial value problems of ordinary differential equations are the so-called *Runge-Kutta* (RK) methods [14]. We will focus on the *explicit* RK-schemes as those are easier to implement and well-suited for GPUs. They are a family of iterative one-step methods that rely on a temporal discretization to compute an approximate solution of the IVP. Temporal discretization means that the approximate solution is computed at time points t_n. So we use $\mathbf{x}_n$ for the numerical approximation of the solution $x(t_n)$ at time t_n. In the simplest, but most frequently used case of an equidistant discretization with a constant step size Δt, one writes for the numerical solution:

$$\mathbf{x}_n \approx \mathbf{x}(t_n), \quad \text{with} \quad t_n = t_0 + n \cdot \Delta t. \tag{7.6}$$

The approximate points $\mathbf{x}_n$ are obtained sequentially using a numerical algorithm that can in the most general form be written as:

$$\mathbf{x}_{n+1} = \mathbf{F}_{\Delta t}(\mathbf{x}_n). \tag{7.7}$$

The mapping $\mathbf{F}_{\Delta t}$ here represents the numerical algorithm, i.e. the Runge-Kutta scheme, that performs one iteration from $\mathbf{x}_n$ to $\mathbf{x}_{n+1}$ with the time step Δt. The numerical scheme is said to have the order m if the solution it generates is exact up to some error of order $m + 1$:

$$\mathbf{x}_1 = \mathbf{x}(t_1) + O(\Delta t^{m+1}), \tag{7.8}$$

where $\mathbf{x}(t_1)$ here is the exact solution of the ODE at t_1 starting from the initial condition $\mathbf{x}(t_0) = \mathbf{x}_0$. Hence, m denotes the order of accuracy of a *single step* of the scheme.

The most basic numerical algorithm to compute such a discrete trajectory $x_1, x_2, \dots$ is the *Euler scheme*, where $F_{\Delta t}(\mathbf{x}_n) = \mathbf{x}_n + \Delta t \cdot \mathbf{f}(\mathbf{x}_n, t_n)$, which means the next approximation is obtained from the current one by:

$$\mathbf{x}_{n+1} = \mathbf{x}_n + \Delta t \cdot \mathbf{f}(\mathbf{x}_n, t_n). \tag{7.9}$$

This scheme has no practical relevance because it only offers accuracy of order $m = 1$. A higher order can be reached by introducing intermediate points and thus dividing one step into several stages. For example, the famous "RK4" scheme, sometimes also called *the* Runge-Kutta method, has $s = 4$ stages and also order $m = 4$. It is defined as follows:

Table 7.1 Generic Butcher
Tableau with s stages

c_1					
c_2	$a_{2,1}$				
c_3	$a_{3,1}$	$a_{3,2}$			
$\vdots$	$\vdots$		$\ddots$		
c_s	$a_{s,1}$	$a_{s,2}$	$\cdots$	$c_{s,s-1}$	
	b_1	b_2	$\cdots$	b_{s-1}	b_s

Table 7.2 Butcher tableau
with coefficients for the RK4
method

0				
1/2	1/2			
1/2	0	1/2		
1	0	0	1	
	1/6	1/3	1/3	1/6

$$\mathbf{x}_{n+1} = \mathbf{x}_n + \frac{1}{6}\Delta t(\mathbf{k}_1 + 2\mathbf{k}_2 + 2\mathbf{k}_3 + \mathbf{k}_4), \quad \text{with}$$

$$\mathbf{k}_1 = \mathbf{f}(\mathbf{x}_n, t_n),$$

$$\mathbf{k}_2 = \mathbf{f}\left(\mathbf{x}_n + \frac{\Delta t}{2}\mathbf{k}_1, t_n + \frac{\Delta t}{2}\right), \qquad (7.10)$$

$$\mathbf{k}_3 = \mathbf{f}\left(\mathbf{x}_n + \frac{\Delta t}{2}\mathbf{k}_2, t_n + \frac{\Delta t}{2}\right),$$

$$\mathbf{k}_4 = \mathbf{f}(\mathbf{x}_n + \Delta t\,\mathbf{k}_3, t_n + \Delta t).$$

Note, how the subsequent computations of the intermediate results $\mathbf{k}_i$ depend on the results of the previous stages $\mathbf{k}_{j<i}$.

More generally, a Runge-Kutta scheme is defined by its number of stages s and a set of parameters $c_1 \ldots c_s$, $a_{21}, a_{31}, a_{32}, \ldots, a_{ss-1}$ and $b_1 \ldots b_s$. The algorithm to calculate the next approximation x_{n+1} is then given by:

$$x_{n+1} = x_n + \Delta t \sum_{i=1}^{s} b_i k_i, \qquad \text{where}$$

$$k_i = f(x_n + \Delta t \sum_{j=1}^{i-1} a_{ij} k_j, t_n + c_i\,\Delta t). \qquad (7.11)$$

The parameter sets $a_{i,j}$, b_i and c_i define the so-called Butcher tableau (see Tables 7.1 and 7.2) and fully describe the specific Runge-Kutta scheme. The Butcher tableau for the RK4 scheme above is given in Table 7.2. Note, that the above schemes have a lower triangular structure. For tableaus with entries in the upper right region the method becomes an implicit RK-scheme and can not easily be implemented.

Table 7.3 Computational requirements of the Runge-Kutta algorithms

Requirement	Representation in C++	Example
Represent mathematical entities	Template parameter	`vector<`**`double`**`>`, **`double`**
Memory management	Function specialization	`resize<state_type>`
Vector iteration	Template parameter	`container_algebra`
Elementary operations	Template parameter	`default_operations`

7.3 Generic Runge-Kutta Implementation

In this section, we will develop an implementation of the Runge-Kutta schemes described above. The code will be designed in such a way that it separates the algorithm from the underlying computations and thus can be easily ported to GPUs. We will therefore analyze the computational requirements of the Runge-Kutta algorithms and produce a modularized implementation. In this way, we will be able to replace, for example, the memory management and the computational backend with GPU variants and thus obtain a GPU implementation without re-implementing the algorithm itself. This will allow us to easily use the same code with different GPU technologies, i.e. CUDA and OpenCL.

7.3.1 Computational Requirements

To analyze the algorithmic parts involved in a Runge-Kutta scheme, we will start with a straight-forward implementation that does not yet provide any modularization. Listing 7.1 shows such an implementation for the RK4 algorithm as given by Eq. (7.10). It defines a class `runge_kutta4` that provides a member function `do_step` which performs a single RK4 step given a system function `system`, the current state x, the current time t and the time step `dt`. Note how we use a template parameter `System` to specify the system function. This gives us already some flexibility as `do_step` immediately works with function pointers and functor object, but also in more complicated cases like generalized functions objects from `std::function` or `boost::function` [3, 12] or even C++11 lambdas. Basically anything that defines a function call operator with the signature **`operator`**`()(state_type &x, state_type &k, `**`double`**` t)` can be supplied as `system` in `do_step`.

In the following we will extract the computational requirements for the Runge-Kutta algorithms from the simple implementation in Listing 7.1. First, we need to define a representation of the dependent variable **x**. In the `runge_kutta4` class a `vector<`**`double`**`>` from the Standard Template Library [36] is used for that purpose (Line 7). After that, we need to define the type of the independent variable t (called the `time_type` below). In Listing 7.1 (Line 13) we use **`double`** for this purpose. Then we need to introduce variables for temporary results (Line 38) and allocate

Listing 7.1 Simple Runge-Kutta4 implementation   `simple_runge_kutta4.hpp`

```
5  class runge_kutta4 {
6  public:
7    typedef std::vector<double> state_type;
9    runge_kutta4(size_t N)
10     : N(N), x_tmp(N), k1(N), k2(N), k3(N), k4(N) { }
12   template<typename System>
13   void do_step(System system, state_type &x, double t, double dt)
14   {
15     const double dt2 = dt / 2;
16     const double dt3 = dt / 3;
17     const double dt6 = dt / 6;
19     system(x, k1, t);
20     for(size_t i = 0; i < N; ++i)
21       x_tmp[i] = x[i] + dt2 * k1[i];
23     system(x_tmp, k2, t + dt2);
24     for(size_t i = 0 ; i < N; ++i)
25       x_tmp[i] = x[i] + dt2 * k2[i];
27     system(x_tmp, k3, t + dt2);
28     for(size_t i = 0; i < N; ++i)
29       x_tmp[i] = x[i] + dt * k3[i];
31     system(x_tmp, k4, t + dt);
32     for(size_t i = 0; i < N; ++i)
33       x[i] += dt6*k1[i] + dt3*k2[i] + dt3*k3[i] + dt6*k4[i];
34   }
36 private:
37   const size_t N;
38   state_type x_tmp, k1, k2, k3, k4;
39 };
```

Listing 7.2 Runge-Kutta class with templated types

```
1  template<
2     class state_type,
3     class value_type = double,
4     class time_type = value_type
5     >
6  class runge_kutta4 {
7     // ...
8  };
9  typedef runge_kutta4< std::vector<double> > rk_stepper;
```

enough memory for the temporaries, done in the constructor (Line 10). And finally we have to perform the summation and multiplication, in general operations of the form:

$$\mathbf{y} = a_1\mathbf{x}_1 + a_2\mathbf{x}_2 + \cdots + a_s\mathbf{x}_s, \tag{7.12}$$

where $\mathbf{y}$ and $\mathbf{x}_n$ are of state_type and a_s are of floating point type, typically **double**. Hence, from a mathematical view point, these operations are vector-vector addition and scalar-vector multiplication. In the runge_kutta4 class above we specifically perform the iteration over the elements of the state_type and use the intrinsic operators + and * on those elements which are just **double** values here. All the requirements identified above are again listed in Table 7.3. Note how in the runge_kutta4 class in Listing 7.1 the parts to satisfy these requirements are hard-coded into the class. If we want to change, for example, the state_type to some construct that resides on the GPU, we have to completely rewrite the class for a new state_type, but also to change the memory allocation and the vector operations, thus rewriting the whole algorithm, e.g. in terms of a new class runge_kutta4_gpu. In the next section, however, we will present a modularized implementation based on the requirements identified here, which allows to exchange the fundamental types, memory allocation and vector computations so that the code can be ported to GPUs without changing the algorithm itself.

7.3.2 Modularized Design

In the following, we will generalize the basic implementation above by moving the parts addressing the several requirements out of the runge_kutta4 class and keeping only the essential algorithm.

We start with the fundamental types used to represent the mathematical objects in the Runge-Kutta schemes Eq. (7.10). From a computational point of view we identify three different kinds of objects:

1. The state of the solution at some time $\mathbf{x}(t)$, typically more dimensional and represented by a vector<**double**>.
2. The independent variable t, typically the time and represented by a **double**.
3. Parameters of the Runge-Kutta scheme as given in the Butcher Tableau (Table 7.2), usually also represented by **double** values.

The standard way to generalize an algorithm for arbitrary types in C++ is to introduce template parameters. We will also follow this approach and define three class template parameters state_type, value_type and time_type. Listing 7.2 shows the skeleton of the new runge_kutta4 class. Note how we use default template parameters to provide value_type and time_type as **double**, so for the most typical case the user only has to specify the state_type, as shown exemplarily in Line 9. It should be noted that the derivatives might require a representation different from the state, especially if arithmetic types with dimensions are used, for example the ones from Boost.Units [33].

Let us now consider the memory allocation. In the basic implementation in Listing 7.1 this is done in the constructor which therefore requires the system size. This implementation relies on the existence of a constructor accepting the numbers of elements N, which is not generic enough because the state_type does not need to be a vector anymore, or even a container at all. Therefore we will change the implementation and introduce a templated helper function `resize` that takes care of the resizing and can be specialized by the user for any given `state_type`. The result is outlined in Listing 7.3. The `resize` function here adjusts the allocated memory of some object `out` using the size of the given object `in`. This is the most flexible way. With this technique the `runge_kutta4` class takes care of the memory automatically, and it works out-of-the-box for all containers that provide a `size` and `resize` member functions. If some other `state_type` is employed, the user can implement an overload of the `resize` function to tell the `runge_kutta4` how to allocate memory. One example could be fixed-size arrays `boost::array<`**`double`**`,N>`, which live on the stack and do not require manual memory allocation. Hence, the `resize` function would just be empty (and disappear during the optimization step of the compilation), shown in Lines 7–11 in Listing 7.3. Note that this implementation already supports the case when the system size changes during the integration, i.e. if the size of x changes between `do_step` calls. However, checking the system size at each step of the algorithm is not necessary for almost all situations and thus it is a waste of performance. This can be solved by adding a trivial logic that only calls `resize` during the first call of `do_step` (not shown here for clarity).

Now we arrive at the final and most difficult point: the abstraction of the numerical computation. As seen from the mathematical definition of the Runge-Kutta scheme in Eq. (7.11), we need to calculate vector-vector sums and scalar-vector products to perform a Runge-Kutta step. In the simplistic implementation above (Listing 7.1), this is done by explicit **for** loops and arithmetic operators + and *. In our abstraction of this computation, we divide these computations into two distinct parts: *iteration* and *operation*. The first one will be responsible for iterating over the elements of the involved state types, i.e. it addresses the vector character of the computation. The code structure that performs these iterations will be called `Algebra`. The *operation* on the other hand represents the computation that is performed for each element, i.e. within the iteration. The respective code structure will be called `Operation`.

We start with the `Algebra`. For the RK4 algorithm we need to provide two functions that do iteration over three and six container instances. A possible `Algebra` is presented in Listing 7.4, where for the sake of clarity only the `for_each3` method is shown.

The iteration is performed in terms of `for_each` functions that are gathered in a **struct** called `container_algebra`. The `for_each` functions expect a number of containers and an operation object as parameters. They simply perform the iteration over the elements of the containers and execute the given operation on each of the container's elements. Here we use a raw hand written for-loop which requires a `size()` member function and the `[]`-operator for the given container types `s1,s2...`. This loop could easily be generalized to use iterators which is the preferred and recommended way in C++ to iterate over containers. Inside the loop the functors `op` are applied to the elements of the containers. Listing 7.5 shows an exemplary implementation of such operations designed to be used within the `container_algebra` above. It consists of functor types organized in a **struct**

Listing 7.3 Memory allocation

```
1  template<class state_type>
2  void resize(const state_type &in, state_type &out) {
3    // standard implementation works for containers
4    out.resize(in.size());
5  }
7  // specialization for boost::array
8  template<class T, size_t N>
9  void resize(const boost::array<T, N> &, boost::array<T,N>& ) {
10   /* arrays don't need resizing */
11 }
13 template< ... >
14 class runge_kutta4 {
15   // ...
16   template<class Sys>
17   void do_step(Sys sys, state_type &x, time_type t, time_type~dt)
18   {
19     adjust_size(x);
20     // ...
21   }
23   void adjust_size(const state_type &x) {
24     resize(x, x_tmp);
25     resize(x, k1);
26     resize(x, k2);
27     resize(x, k3);
28     resize(x, k4);
29   }
30 }
```

Listing 7.4 Example algebra for the RK4   `container_algebra.hpp`

```
6  struct container_algebra {
7    template<class S1, class S2, class S3, class Op>
8    static void for_each3(S1 &s1, S2 &s2, S3 &s3, Op op) {
9      const size_t dim = s1.size();
10     for(size_t n = 0; n < dim; ++n)
11       op(s1[n], s2[n], s3[n]);
12   }
20 };
```

called `default_operations`. The `scale_sum2` works with the `for_each3` above, while the `scale_sum5` that interacts with `for_each6` is again omitted. Those functors consist of a number of parameters `alpha1, alpha2...` and a function call operator that calculates a simple product-sum (Listing 7.5).

With these abstractions we have moved the computational details away from the algorithm into separate code structures and thus reached a generic implementation of the RK4 algorithm (shown in Listing 7.6). The `runge_kutta4` class got two

Listing 7.5 Example operations for the RK4   `default_operations.hpp`

```
6   struct default_operations {
7     template<class Fac1 = double, class Fac2 = Fac1>
8     struct scale_sum2 {
9       typedef void result_type;
11      const Fac1 alpha1;
12      const Fac2 alpha2;
14      scale_sum2(Fac1 alpha1, Fac2 alpha2)
15          : alpha1(alpha1), alpha2(alpha2) { }
17      template<class T0, class T1, class T2>
18      void operator()(T0 &t0, const T1 &t1, const T2 &t2) const {
19          t0 = alpha1 * t1 + alpha2 * t2;
20      }
21    };
48  };
```

more template parameters specifying the algebra and operations, i.e. the computational backend used for the calculation. We use the `container_algebra` and `default_operations` from Listings 7.4 and 7.5 as the default values that will work for almost all cases. In the `do_step` method we now use the `for_each` functions from the given `Algebra` in combination with the `scale_sum` functors from the given `Operations` to perform the required computations. So the explicit **for**-loops, that were hard-coded into the algorithm in the first implementation (Listing 7.1), have been separated into two parts, an `algebra` and `operations`. Those parts are supplied to the algorithm in terms of template parameters and can thus be easily replaced without changing the algorithm itself. This flexibility now allows us to port the RK4 implementation to GPUs. The idea is to first provide a GPU data structure, e.g. a `gpu_vector` with the respective `resize` functions as required by the algorithm (Listing 7.3). Then we only need a `gpu_algebra` and `gpu_operations` to do the vector computations on the GPU in a parallelized way. Assuming we have implemented those pieces, the following code would give us a RK4 algorithm running on the GPU:

```
typedef runge_kutta4< gpu_vector<double>, double, double,
    gpu_algebra, gpu_operations > gpu_stepper;
```

So with the generalized implementation we have greatly simplified the problem of implementing a Runge-Kutta scheme on the GPU. Instead of having to start from scratch, we now only have to implement a basic data structure for the GPU (`gpu_vector`), provide low-level functions for memory allocation (`resize`), iteration (`algebra`) and fundamental calculations (`operations`). But the real strength of this approach is that these remaining problems are so fundamental that they are already solved for GPUs. Of course, there are libraries that provide data

Listing 7.6 Generic RK4 implementation $\qquad$ runge_kutta4.hpp

```
10  template<class state_type, class value_type = double,
12        class time_type = value_type,
13        class algebra = container_algebra,
14        class operations = default_operations>
15  class runge_kutta4 {
16  public:
17      template<typename System>
18      void do_step(System &system, state_type &x,
19                   time_type t, time_type dt)
20      {
21          adjust_size( x );
22          const value_type one = 1;
23          const time_type dt2 = dt/2, dt3 = dt/3, dt6 = dt/6;
25          typedef typename operations::template scale_sum2<
26                  value_type, time_type> scale_sum2;
28          typedef typename operations::template scale_sum5<
29                  value_type, time_type, time_type,
30                  time_type, time_type> scale_sum5;
32          system(x, k1, t);
33          algebra::for_each3(x_tmp, x, k1, scale_sum2(one, dt2));
35          system(x_tmp, k2, t + dt2);
36          algebra::for_each3(x_tmp, x, k2, scale_sum2(one, dt2));
38          system(x_tmp, k3, t + dt2);
39          algebra::for_each3(x_tmp, x, k3, scale_sum2(one, dt));
41          system(x_tmp, k4, t + dt);
42          algebra::for_each6(x, x, k1, k2, k3, k4,
43                          scale_sum5(one, dt6, dt3, dt3, dt6));
44      }
45  private:
46      state_type x_tmp, k1, k2, k3, k4;
48      void adjust_size(const state_type &x) {
49          resize(x, x_tmp);
50          resize(x, k1); resize(x, k2);
51          resize(x, k3); resize(x, k4);
52      }
53  };
```

structures and memory management for the GPU, as well as parallelized iteration and element-wise computations. In the following sections we will introduce two such libraries and show how they are combined with the RK4 implementation from Listing 7.6 to produce a GPU-version.

It should be noted that this approach of separating the algorithm from the computations is not only valuable when aiming at GPU computations. With the implementation above we can, for example, also easily create a RK4 algorithm that works with arbitrary precision types instead of the usual **double**. Another example would be an ODE solver based on interval arithmetic [22], also easily implementable by providing some `interval_operations`.

7.3.3 Lorenz Attractor Example

Before considering the GPU backends we want to show how to use the codes above to compute a trajectory of the famous Lorenz system (7.5) introduced earlier in Sect. 7.2.1. Listing 7.7 shows the implementation of a simulation of a trajectory for this system based on the `runge_kutta4` class developed above. As seen there, all that is left to do is to define the `state_type`, implement the RHS of the ODE, here done in terms of a functor `lorenz`, and define the initial conditions (Line 30). Now we can use the Runge-Kutta algorithm implemented above (Listing 7.6) to iterate along the trajectory using a step size of $\Delta t = 0.1$.

7.4 GPU Backends

Having reduced the problem of running the ODE solver on GPUs to memory management and some basic algebra operations, we finally come to the point of implementing those necessities. Instead of relying on low-level GPU programming and thus essentially reinventing the wheel, we will use existing high-level libraries that offer GPU data structures as well as routines for algebraic operations. To cover all available GPU technologies we will develop two GPU backends, the first one based on the NVIDIA CUDA technology, the second one for the OpenCL framework. For the CUDA environments, we will employ the Thrust library [4], which is part of the NVIDIA CUDA SDK [24]. In the case of OpenCL, we will rely on the VexCL library [7], an open source library developed at the Supercomputer Center of Russian Academy of Sciences.

7.4.1 Thrust Backend

The Thrust library is a C++ template library that provides containers and algorithms similar to the Standard Template Library (STL) [36], but capable of running parallel on a CUDA GPU. Besides the CUDA backend, Thrust also supports CPU parallelization via OpenMP [25] and Intel's Thread Building Block (TBB) [31], configurable at compile time by preprocessor variables. As said above, Thrust is part of the NVIDIA CUDA framework and thus requires the use of the nvcc compiler to generate code that can be executed on GPUs. For a thorough introduction into CUDA programming and Thrust in particular, we refer to the respective documentation [4, 24].

Listing 7.7 Computing a trajectory of the Lorenz system `lorenz_single.cpp`

```cpp
#include <iostream>
#include <vector>
#include "runge_kutta4.hpp"
using namespace std;
typedef std::vector<double> state_type;
typedef ncwg::runge_kutta4< state_type > rk4_type;
struct lorenz {
    const double sigma, R, b;
    lorenz(const double sigma, const double R, const double b)
        : sigma(sigma), R(R), b(b) { }
    void operator()(const state_type &x,state_type &dxdt,double t)
    {
        dxdt[0] = sigma * ( x[1] - x[0] );
        dxdt[1] = R * x[0] - x[1] - x[0] * x[2];
        dxdt[2] = -b * x[2] + x[0] * x[1];
    }
};
int main() {
    const int steps = 5000;
    const double dt = 0.01;
    rk4_type stepper;
    lorenz system(10.0, 28.0, 8.0/3.0);
    state_type x(3, 1.0);
    x[0] = 10.0;
    for( size_t n=0 ; n<steps ; ++n ) {
        stepper.do_step(system, x, n*dt, dt);
        cout << n*dt << ' ';
        cout << x[0] << ' ' << x[1] << ' ' << x[2] << endl;
    }
}
```

To handle the memory on the GPU, Thrust provides a `thrust::device_vector` template class similar to `std::vector` from the STL. This will be our basic `state_type` representing the state **x** of the dynamical system. As Thrust mimics the STL, the `thrust::device_vector` also has `size` and `resize` member functions, which means that the memory management for `std::vectors` given in Listing 7.3 also works nicely with `thrust::device_vectors`—no specialization is required. This is a nice example of how well-designed libraries, such as Thrust, decrease the required programming effort by increasing the re-usability of your code.

To ensure that the vector computations are executed in parallel on the GPU, we introduce a `thrust_algebra` as a replacement of the `container_algebra` (see Listing 7.4) above. To implement the `for_each3` and `for_each6` functions required

Listing 7.8 The Thrust algebra   `thrust_algebra.hpp`

```
6   struct thrust_algebra {
7     template<class S1, class S2, class S3, class Op>
8     static void for_each3(S1 &s1, S2 &s2, S3 &s3, Op op) {
9       thrust::for_each(
10          thrust::make_zip_iterator( thrust::make_tuple(
11              s1.begin(), s2.begin(), s3.begin() ) ),
12          thrust::make_zip_iterator( thrust::make_tuple(
13              s1.end(), s2.end(), s3.end() ) ),
14          op);
15    }
39  };
```

in the algebra, we will employ Thrust's `thrust::for_each` routine. This routine has the following signature:

```
thust::for_each(Iterator begin, Iterator end, UnaryOperator op)
```

where the iterators `begin` and `end` define a range of data in a `device_vector` and `op` defines the operation performed for each element of the sequence. As seen from the signature above, `thrust::for_each` iterates only over a single range from `begin` to `end`, but for our `for_each3` and `for_each6` we need to iterate over several device vectors at once. Fortunately, this can be easily achieved by using `zip_iterators` that combine an arbitrary number of iterators into a single iterator and thus allows us to use `thrust::for_each` for iterating over several ranges at once. The implementation of the `thrust_algebra` based on `thrust::for_each` and `make_zip_iterator` in combination with `make_tuple` is shown in Listing 7.8. The usage of `make_zip_iterator` and `make_tuple` is almost self-explanatory: `make_tuple` combines the given parameters (iterators in this case) into a single tuple, and `make_zip_iterator` then converts this tuple of iterators into a single `zip_iterator` that can then be passed to the `for_each` algorithm. Note that the implementation of the `for_each6` algorithm is omitted here for clarity.

Of course, we also need to replace the `default_operations`, containing the `scale_sum` functors (see Listing 7.5), by a CUDA-compatible implementation. These functions contain the code that in the end will run in parallel on the GPU, which means that they will be compiled into so-called *kernels*. Therefore, they need to be decorated by specific compiler instruction to make the nvcc compiler generate specific GPU code for those functions. For this purpose, CUDA provides the keywords `__device__` and `__host__`. The former indicates that a function will run on a GPU, and the latter assures that the compiler will also generate a CPU version. Listing 7.9 shows the implementation of the `thrust_operations`. The keywords are used before the function definition in Line 19.

Furthermore, we have to bear in mind that since we used `zip_iterators` in the `for_each`, the `scale_sum` functors also get the elements from several ranges packed in a single tuple. To access the individual elements, we have to unpack the tuple,

Listing 7.9 The Thrust operations   `thrust_operations.hpp`

```
 9  struct thrust_operations {
10    template<class Fac1 = double, class Fac2 = Fac1>
11    struct scale_sum2 {
12      const Fac1 m_alpha1;
13      const Fac2 m_alpha2;
15      scale_sum2(const Fac1 alpha1, const Fac2 alpha2)
16        : m_alpha1(alpha1), m_alpha2(alpha2) { }
18      template< class Tuple >
19      __host__ __device__ void operator()(Tuple t) const {
20        thrust::get<0>(t) = m_alpha1 * thrust::get<1>(t) +
21                            m_alpha2 * thrust::get<2>(t);
22      }
23    };
48  };
```

which can be done by the Thrust's `get<N>(tuple)` function that simply returns the
N-th entry of the given `tuple`. Together with the `thrust_algebra` (see Listing 7.8)
this completes the CUDA backend for the RK4 scheme. The following code defines
a `gpu_stepper` class that computes an approximate trajectory using the GPU:

```
typedef thrust::device_vector<double> state_type;
typedef runge_kutta4< state_type, double, double,
    thrust_algebra, thrust_operations > gpu_stepper_type;
```

With this, we have successfully ported the RK4 scheme to GPUs using functionality from the Thrust library. However, for a complete simulation we also have
to implement the RHS function such that it is also computed on the GPU. This is
highly non-trivial and will be discussed in detail for several examples in Sect. 7.6.

7.4.2 VexCL Backend

The Thrust backend above allows to run ODE integration on NVIDIA GPUs only
as it is based on the CUDA technology. To address a wider range of hardware, we
will now present a computational backend based on OpenCL (Open Computing
Language) [23]. OpenCL supports NVIDIA as well as AMD/ATI GPUs, but can
also be used for parallel runs on multi-core CPUs.

As above, we will not start from scratch but rather employ the modern, well-
designed GPGPU library VexCL [7]. The library does not only provide the required
data structures, but also covers the vector operations which makes our work even
simpler than with Thrust. As the data structure for representing a `state_type` we
will use a `vex::vector`, which is again similar to a `std::vector`. Listing 7.10
shows the `resize` function specialized for the `vex::vector<T>`. Note how we have

Listing 7.10 Memory allocation for VexCL   `vexcl_resize.hpp`

```
11  template<class T>
12  void resize(const vex::vector<T> &in, vex::vector<T> &out) {
13      out.resize(in.queue_list(), in.size());
14  }
16  template<class T, size_t N>
17  void resize(const vex::multivector<T,N> &in,
18          vex::multivector<T,N> &out)
19  {
20      out.resize(in.queue_list(), in.size());
21  }
```

Listing 7.11 Vector space algebra   `vector_space_algebra.hpp`

```
6   struct vector_space_algebra {
7       template<class S1, class S2, class S3, class Op>
8       static void for_each3(S1 &s1, S2 &s2, S3 &s3, Op op) {
9           op(s1, s2, s3);
10      }
15  };
```

to pass on the list of OpenCL command queues that contains crucial information about where the data will reside (i.e. which compute device) to the vector's `resize` function. Just like the required size, we extract this information from the given `vex::vector` instance `in`. Additionally to the usual vectors, VexCL also provides a `vex::multivector<T,N>`, which is basically a group of N instances of `vex::vector<T>` and can be quite handy for some problems. Hence, we also provide the resize functionality for `vex::multivector<T,N>` in Listing 7.10.

We are left with the vector operations, but as mentioned above this is very simple with VexCL. Being a library designed specifically for linear algebra, VexCL natively supports vector-vector addition and scalar-vector multiplication. Assuming x, y and z are of type `vex::vector<double>` and a and b are `double` values, the following code performs the element-wise summation and scalar multiplication of the vectors:

```
z = a * x + b * y;
```

That means that the VexCL library intrinsically performs the iteration over the elements of the vector in parallel on an OpenCL compute device (i.e. a GPU). Mathematically, one can say that the `vex::vector` together with the standard + and * operators form a *vector space*. Hence, it is not required for us to implement a parallelized iteration ourselves and the existence of an `algebra` is not necessarily required, in contrast to the Thrust backend above (c.f. Listing 7.8). But as the `algebra` is part of the structure of our ODE solver and can not be neglected, we provide a trivial `vector_space_algebra` that simply forwards the operation directly to the vectors without performing an iteration. This is shown in Listing 7.11.

This implementation is not only useful for VexCL and its `vex::vector`, but also for any other vector library that provides vector operations in terms of + and * operators, e.g. MTL4 [11] or Boost.uBLAS [39]. To account for this generality we call this trivial algebra a `vector_space_algebra`, as it works with any type that forms a vector space. From the above it is also clear that for VexCL we do not need to take special care of the operations. As VexCL redefines the operators + and * itself, we can simply plug in the `default_operations` from the beginning (Listing 7.5). Therefore, the computational backend for OpenCL based on VexCL is finished and we can construct an algorithm that is capable of running on a GPU device with the following code:

```
typedef vex::vector<double> state_type;
typedef runge_kutta4< state_type, double, double,
    vector_space_algebra, default_operations > gpu_stepper_type;
```

7.5   The Boost.odeint Library

Above, we have shown how to implement the RK4 scheme in a generic way such that it can be easily ported to GPUs. We have demonstrated the strengths of this approach by providing two backends that address CUDA and OpenCL devices respectively. However, there is a vast potential for improvement and extension of this code. Although this goes well beyond the scope of the present text, we want to mention that a highly sophisticated implementation of the ideas and techniques above exists in the Boost.odeint library [1,2]. Boost.odeint also separates memory allocation, iteration and fundamental operations from the actual algorithm in the same way as described above in Sect. 7.3.2. But in contrast to the ad hoc implementation presented here, Boost.odeint is a fully grown library consisting of about 25,000 lines of C++ code. It includes a vastly larger functionality and we shortly list the most important points below:

- Arbitrary explicit Runge-Kutta schemes, predefined schemes: Dormand-Prince 5, Cash-Karp, Runge-Kutta78.
- Symplectic Runge-Kutta-Nyström schemes.
- Variable order method: Bulirsch-Stoer.
- Multistep methods: Adams-Bashforth, Adams-Bashforth-Moulton.
- Implicit routines: Rosenbrock method, implicit Euler.
- Step-size control and dense output.
- Integrate routines with observer support.
- Iterator and range interfaces.
- Support of arbitrary precision arithmetic with Boost.Multiprecision.
- Support of additional backends: eigen [13], GSL vectors [10], Math Kernel Library [17], Matrix Template Library [11], ViennaCL [32].

Listing 7.12   `lorenz_thrust_v1.hpp`

```
30  typedef thrust::device_vector<double> state_type;
31  struct lorenz_system {
40      struct lorenz_functor {
41          double sigma, b;
42          lorenz_functor(double sigma, double b)
43              : sigma(sigma), b(b) {}
45          template<class T>
46          __host__ __device__ void operator()(T t) const {
47              double x = thrust::get<0>( t );
48              double y = thrust::get<1>( t );
49              double z = thrust::get<2>( t );
50              double R = thrust::get<3>( t );
52              thrust::get<4>( t ) = sigma * ( y - x );
53              thrust::get<5>( t ) = R * x - y - x * z;
54              thrust::get<6>( t ) = -b * z + x * y;
55          }
56      };
58      template<class State, class Deriv>
59      void operator()(const State &x, Deriv &dxdt, double t) const {
60          BOOST_AUTO(start,
61                  thrust::make_zip_iterator( thrust::make_tuple(
62                      x.begin(),
63                      x.begin() + n,
64                      x.begin() + 2 * n,
65                      R.begin(),
66                      dxdt.begin(),
67                      dxdt.begin() + n,
68                      dxdt.begin() + 2 * n
69                      ) )
70                  );
72          thrust::for_each(start, start+n, lorenz_functor(sigma, b));
73      }
74  };
```

If Boost.odeint provides the necessary algorithms and functionality to solve a problem, we strongly advise to use this library. However, some problems require specialized schemes or additional computations. In this case the code developed in the previous pages should represent a good starting point to develop a specific algorithm in a generalized way that is easily portable to GPUs.

Listing 7.13   `lorenz_vexcl_v1.cpp`

```
28  typedef vex::multivector<double, 3> state_type;
29  struct lorenz_system {
36    void operator()(const state_type &x, state_type &dxdt,
37                    double t) const
38    {
39      dxdt = std::tie(
40            sigma * (x(1) - x(0)),
41            R * x(0) - x(1) - x(0) * x(2),
42            x(0) * x(1) - b * x(2) );
43    }
44  };
```

7.6 Example Problems

7.6.1 Lorenz Attractor Ensemble

In the first example we consider the Lorenz system (7.5). Solutions of the Lorenz system usually furnish very interesting behavior in dependence on one of its parameters. For example, one might want to study the chaoticity in dependence on the parameter R. Therefore, one would create a large set of Lorenz systems (each with a different parameter R), pack them all into one system and solve them simultaneously. In a real study of chaoticity one may also calculate the Lyapunov exponents [26], which requires to solve the Lorenz system and their linear perturbations.

In the Thrust version of the example we define the state type as `device_vector` of size $3n$, where n is the system size. The X, Y, and Z components of the state are held in the continuous partitions of the vector. The system functor holds the model parameters and provides a function call operator with the necessary signature. Here we use the standard Thrust technique of packing the state components into a zip iterator which is then passed to a `thrust::for_each` algorithm (Listing 7.12).

The system function object for the VexCL version of the Lorenz attractor example is more compact than the Thrust variant because VexCL supports a rich set of vector expressions. We represent the three components of attractor trajectory as a `multivector<double,3>`. Since VexCL provides all necessary overloads for the `multivector` type, we are able to use the `vector_space_algebra` in this case (Listing 7.13).

Figure 7.1 shows performance results for the Thrust, VexCL, and CPU versions of the Lorenz attractor example. Time in seconds required to make a 1,000 of RK4 iterations is plotted against the ensemble size N. Lines denoted "Thrust v1" and "VexCL v1" correspond to the versions presented above. "CPU v1" is the Thrust version compiled for the OpenMP backend. Times for the Thrust and the

VexCL versions of the code are given for the NVIDIA Tesla K20c GPU. Times for the CPU runs are given for the Intel Core i7 920 CPU (all four cores of which were used through OpenMP technology). It is clear from the figure that the initial implementations for the Thrust and the VexCL libraries perform equally well for large problem sizes and are about 14 times faster than the CPU version. VexCL has higher initialization costs and hence is a bit slower than Thrust for smaller problems. However, the distinction seems not as important once we note that both the Thrust and the VexCL versions loose to the CPU version for $N \lesssim 10^4$.

Note that both the Thrust and the VexCL versions above have the same drawback. Namely, both of them use device vectors as state type. Hence, intermediate state variables used in the steppers are stored in the global GPU memory. Moreover, each operation results in a launch of a separate compute kernel. A kernel launch has nonzero overhead both in CUDA and in OpenCL, but more importantly, each kernel needs to both read and write intermediate states from/to the global GPU memory. Since the problem is memory bound, this leads to a severe drop in performance.

We could overcome the above problem by providing a monolithic kernel which would encode the stepper logic and provide the complete solution in a single launch. However, the use of such kernel would also mean the loss of the flexibility we achieved so far by separation of algorithm and the underlying computations: one would have to completely re-implement the kernel for each new problem. Luckily, VexCL library allows us to generate such a fused kernel automatically by providing the `vex::symbolic<T>` class template. Instances of the type dump to the specified output stream any arithmetic operations they are being subjected to. For example, in the following code snippet two symbolic variables are declared and participate in an arithmetic expression:

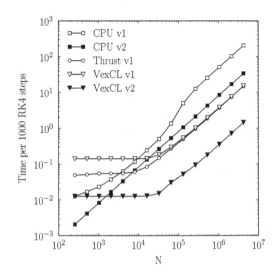

Fig. 7.1 Performance results for the Lorenz attractor example

```
vex::generator::set_recorder(std::cout);
vex::symbolic<double> x = 6, y = 7;
x = sin( x * y );
```

This generates the following output:

```
double var1 = 6;
double var2 = 7;
var1 = sin( (var1 * var2) );
```

This is implemented by overloading arithmetic operators and mathematical functions for the symbolic classes. So when two symbolic variables are being added, the overloaded addition operator just outputs names of the variables divided by symbol "+" to the specified output stream. By defining the state type to be `boost::array< vex::symbolic<double>, 3>`, and using the same algebra and the system function as in Listing 7.7, we are able to record the sequence of arithmetic operations made by a Runge-Kutta stepper. This gives us a fused kernel which is as effective as a manually written one (Listing 7.14).

This approach has some obvious restrictions: namely, it only supports embarrassingly parallel problems (no data dependencies between threads of execution), and it does not allow conditional statements or loops with non-constant number of iterations. But when the method works, it works very well. This version of the code is denoted "VexCL v2" in Fig. 7.1 and is about ten times faster than the initial VexCL implementation.

We use a similar approach in order to accelerate the CPU version of the example. Namely, we create a Boost.odeint stepper for a single Lorenz attractor (state type is `boost::array<double,3>`), and then we use an outer loop which iterates over the complete ensemble (Listing 7.15). This version of the code ("CPU v2") uses less memory and is more cache-friendly. As a result, it is about 6 times faster than the Thrust example with the OpenMP backend. Unfortunately, the Thrust library does not allow the same type of optimization. We could in principle create a device function that would operate on a single attractor (by calling `runge_kutta4<...>::do_step` from inside the function), and apply the function to the complete ensemble with the help of the `thrust::for_each` algorithm. But CUDA requires all device functions to be decorated with __device__ keyword, and the Boost.odeint functions are not marked as such.

7.6.2 Chain of Coupled Phase Oscillators

As a second example we consider a chain of coupled phase oscillators. A phase oscillator describes the dynamics of an autonomous oscillator [18]. Its evolution is governed by the phase φ, which is a 2π-periodic variable growing linearly in time, i.e. $\dot{\varphi} = \omega$, where ω is the phase velocity. The amplitude of the oscillator does not occur in this equation, so interesting behavior can only be observed if many of

Listing 7.14   `lorenz_vexcl_v2.cpp`

```cpp
34  typedef vex::symbolic<double> sym_vector;
35  typedef boost::array<sym_vector, 3> sym_state;
64  // Custom kernel body will be recorded here
65  std::ostringstream body;
66  vex::generator::set_recorder(body);
68  // State types that would become kernel parameters
69  sym_state sym_S = {{
70      sym_vector(sym_vector::VectorParameter),
71      sym_vector(sym_vector::VectorParameter),
72      sym_vector(sym_vector::VectorParameter)
73  }};
75  sym_vector sym_R(sym_vector::VectorParameter, sym_vector::Const);
77  // Stepper type
78  odeint::runge_kutta4_classic<
79      sym_state, double, sym_state, double,
80      odeint::container_algebra, odeint::default_operations
81      > stepper;
83  // Record single RK4 step
84  lorenz_system sys(sym_R);
85  stepper.do_step(sys, sym_S, 0, dt);
87  // Generate the kernel from the recorded sequence
88  auto kernel = vex::generator::build_kernel(ctx, "lorenz",
89          body.str(), sym_S[0], sym_S[1], sym_S[2], sym_R);
91  // Real state initialization
92  vex::vector<double> X(ctx, n), Y(ctx, n), Z(ctx, n), R(ctx, n);
93  X = Y = Z = 10.0;
94  R = Rmin + dR * vex::element_index();
99  // Integration loop
100 for(double t = 0; t < t_max; t += dt)
101     kernel(X, Y, Z, R);
```

such oscillators are coupled. In fact, such a system can be used to study phenomena like synchronization, wave and pattern formation, phase chaos, or oscillation death [20, 27]. It is a prominent example of an emergent system where the coupled system shows a more complex behavior than its constituents.

The concrete example we analyze here is a chain of nearest-neighbor coupled phase oscillators [5]:

$$\dot{\varphi}_i = \omega_i + \sin(\varphi_{i+1} - \varphi_i) + \sin(\varphi_i - \varphi_{i-1}). \qquad (7.13)$$

The index i denotes here the i-th phase in the chain. Note, that the phase velocity is different for each oscillator.

From the implementation point of view, the main difference between the phase oscillator chain and the Lorenz attractor examples is that in the former example the values of neighboring vector elements are needed in order to compute the system function. In the Thrust version this is implemented with help of fancy iterators.

Listing 7.15 `lorenz_cpu_v2.cpp`

```
67  #pragma omp parallel for
68  for(size_t i = 0; i < n; ++i) {
69      odeint::runge_kutta4_classic<
70          state_type, double, state_type, double,
71          odeint::container_algebra, odeint::default_operations
72          > stepper;
74      lorenz_system sys(R[i]);
75      for(double t = 0; t < t_max; t += dt)
76          stepper.do_step(sys, x[i], t, dt);
77  }
```

First, we define device functors `left_nbr` and `right_nbr` returning left and right neighbor positions for the i-th element. Then we create a couple of permutation iterators from transformed counting iterators (with `left_nbr` and `right_nbr` used as transformation functors), pack the resulting iterators together with iterators `x`, `omega`, and `dxdt` into a `zip_iterator`. Finally we call the `thrust::for_each` algorithm with the accordingly defined system functor (Listing 7.16).

We use a similar technique for the VexCL version of the example. VexCL provides the `vex::permutation` function that allows to permute arbitrary expressions (Listing 7.17). Note how the use of C++11 `auto` keyword in Lines 38–40 allows us to conveniently capture intermediate expressions and thus simplify the code in Line 42.

Fig. 7.2 Performance results for the chain of coupled phase oscillators example

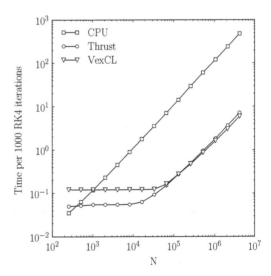

Listing 7.16   `po_thrust.cpp`

```
25  typedef thrust::device_vector< double > state_type;
26  struct phase_oscillators {
33      struct left_nbr : thrust::unary_function<size_t, size_t> {
34          __host__ __device__ size_t operator()(size_t i) const {
35              return (i > 0) ? i - 1 : 0;
36          }
37      };
47      struct sys_functor {
48          template< class Tuple >
49          __host__ __device__ void operator()( Tuple t ) {
50              double phi_c = thrust::get<0>(t);
51              double phi_l = thrust::get<1>(t);
52              double phi_r = thrust::get<2>(t);
53              double omega = thrust::get<3>(t);
55              thrust::get<4>(t) = omega +
56                  sin(phi_r - phi_c) + sin(phi_c - phi_l);
57          }
58      };
60      void operator() (const state_type &x, state_type &dxdt,
61          double dt)
62      {
63          BOOST_AUTO(start, thrust::make_zip_iterator(
64                  thrust::make_tuple(
65                      x.begin(),
66                      thrust::make_permutation_iterator(
67                          x.begin(),
68                          thrust::make_transform_iterator(
69                              thrust::counting_iterator<size_t>(0),
70                              left_nbr()
71                          )
72                      ),
73                      thrust::make_permutation_iterator(
74                          x.begin(),
75                          thrust::make_transform_iterator(
76                              thrust::counting_iterator<size_t>(0),
77                              right_nbr(n - 1)
78                          )
79                      ),
80                      omega.begin(),
81                      dxdt.begin()
82                  )
83              )
84          );
86          thrust::for_each(start, start + n, sys_functor());
87      }
88  };
```

Listing 7.17 `po_vexcl.cpp`

```
25  typedef vex::vector<double> state_type;
26  struct phase_oscillators {
31     void operator()(const state_type &phi, state_type &dxdt,
32        double t) const
33     {
34        VEX_FUNCTION(left, size_t(size_t),
35           "return (prm1 > 0) ? prm1 - 1 : 0;");
36        VEX_FUNCTION(right, size_t(size_t, size_t),
37           "return (prm1 >= prm2) ? prm2 : prm1 + 1;");
39        auto idx = vex::element_index();
40        auto phi_l=vex::permutation(left(idx))(phi);
41        auto phi_r=vex::permutation(right(idx,phi.size()-1))(phi);
43        dxdt = omega + sin(phi_r - phi) + sin(phi - phi_l);
44     }
45  };
```

The performance results for the chain of coupled phase oscillators are presented in Fig. 7.2. Again, the Thrust and the VexCL versions show similar results for large problems (with VexCL being faster by about 20 %). The GPU versions are 70–80 times faster than the CPU version (which is the Thrust version compiled for the OpenMP backend). The higher acceleration w.r.t. the Lorenz attractor example is explained by the higher FLOP/byte ratio of the problem.

7.6.3 Molecular Dynamics

Molecular dynamics (MD) are a simulation technique for a large number of small interacting particles, typically with local interaction forces. Examples are systems of molecules [9], granular systems [28], or coarse-grained models of fluid molecules.

Here, we study a two dimensional MD simulation described by the following equations of motion for particle i

$$m_i \ddot{x}_i = f_{loc}(x_i) + f_{fric}(\dot{x}_i) + \sum_{j \in S_i} f_{int}(x_i, x_j) . \qquad (7.14)$$

m_i is the mass of the particle, f_{loc} is a local external force, for example the gravity. $f_{int}(x_i, x_j)$ is the (low-range) interaction between the particles i and j and the sum goes over all particles in an appropriate surrounding S_i of particle i. The second term is the friction which usually is only velocity dependent. Of course, other terms might also be included here, but for our purposes the above equation is generic enough to explain most details of implementing a molecular dynamics simulation. The restriction to two dimensions is easily generalizable to three dimensions. In fact, most of the following code is already independent of the concrete dimension.

For the interaction we use the Lennard-Jones potential [19]

$$f_{int}(x_i, x_j) = -\frac{r}{|r|}\frac{dV}{dr} \quad \text{with} \quad r = x_i - x_j \qquad (7.15)$$

with

$$V(r) = 4\varepsilon \left(\left(\frac{\sigma}{r}\right)^{12} - \left(\frac{\sigma}{r}\right)^{6} \right). \qquad (7.16)$$

It is used to describe the interaction of chemically unbounded atoms and molecules. Here ε is the strength of the interaction and σ denotes the interaction radius. The interaction decreases very fast with increasing distance of the particles $f \sim r^{-7}$. So, to speed up the simulations one usually restricts the interactions for particle i to particles withing its surrounding $S_i = \{j : |x_i - x_j| < 4\sigma\}$. Of course, this means that mathematically the Lennard-Jones is not continuous anymore, but this is only of minor importance for our sample application. In practice several possibilities to overcome this discontinuity exist.

How can one implement such rather complicated systems of ODEs in a high-performance way on GPUs? The obvious idea would be to discard the locality of the potential and calculate all pairwise interaction for all particles. Unfortunately, this brute-force solution is far from being optimal. The computational complexity is $O(n^2)$ since all possible pairwise interactions are calculated. As explained above the interaction decreases very fast with increasing particle distance, so one should only take neighboring particles into account. In the following we present an algorithm for this problem and its GPU-implementation.

The basic idea is to assign particles to a regular grid of relatively large cells and calculate the interaction of particle i only with the particles located in neighboring cells, see Listing 7.18. This method is also known as cell list algorithm. Another popular ansatz for the interaction computation,—the neighbor list—takes only the neighbors of particle i into account [38]. In the following we will only concentrate on the first method.

In the two-dimensional case each cell can be identified either by a two dimensional index (j_x, j_y) or by a one dimensional index $j = i_x + n_x j_y$ where n_x is the number of cells in x-direction. The ordering of the particles is done in two steps. First, the cell index j of each particle is calculated and stored in a vector cell_idx, lines 213–220. Secondly, the particles are sorted in ascending order according to the cell index. Of course, the vector of particles is not ordered itself. Instead, a vector with indices is created and sorted according to the cell indices. This is done by the sort_by_key algorithm from Thrust which sorts the first container and reorders the second container according to the order of the first one. The part_ord vector is then used as the index to refer to the original element in the particles vector. This kind of sort algorithm is also know as bucket sort [6].

The cell_idx vector now consists of a sorted array of the cell indices for each particles. Next we find the range (begin and end) for each cell in cell_idx which

Listing 7.18   `mdc_thrust_v2.cu`

```
71   template< typename LocalForce , typename Interaction >
72   struct md_system_bs {
204    void operator() (point_vector const &x, point_vector const &v,
205        point_vector &a, double t) const
206    {
207      typedef thrust::counting_iterator< size_t > ci;
209      // Reset the ordering.
210      thrust::copy(ci(0), ci(prm.n), part_ord.begin());
212      // Assign each particle to a cell.
213      thrust::for_each(
214        thrust::make_zip_iterator ( thrust::make_tuple(
215          x.begin(), cell_coo.begin(), cell_idx.begin()
216          ) ) ,
217        thrust::make_zip_iterator ( thrust::make_tuple(
218          x.end(), cell_coo.end(), cell_idx.end()
219          ) ) ,
220        fill_index_n_hash ( prm ));
222      // Sort particle numbers in part_ord by cell numbers.
223      thrust::sort_by_key(cell_idx.begin(), cell_idx.end(),
224          part_ord.begin());
226      // Find range of each cell in cell_idx array.
227      thrust::lower_bound(cell_idx.begin(), cell_idx.end(),
228          ci(0), ci(prm.n_cells), cells_begin.begin());
230      thrust::upper_bound(cell_idx.begin(), cell_idx.end(),
231          ci(0), ci(prm.n_cells), cells_end.begin());
233      // Handle boundary conditions
234      thrust::transform(x.begin(), x.end(), x_bc.begin(),
235          bc_functor(prm));
237      // Calculate the local and interacttion forces.
238      thrust::for_each(
239        thrust::make_zip_iterator ( thrust::make_tuple(
240          x_bc.begin(), v.begin(), cell_coo.begin(),
241          ci(0), a.begin()
242          ) ) ,
243        thrust::make_zip_iterator ( thrust::make_tuple(
244          x_bc.end(), v.end(), cell_coo.end(),
245          ci(prm.n), a.end()
246          ) ) ,
247        interaction_functor(cells_begin, cells_end, part_ord,
248          x, v, prm)
249        ) ;
250    }
288   };
```

corresponds to particles located in each of the cells (Lines 227–231). The range limits are stored in the `cells_begin` and `cells_end` arrays.

The final step is to compute the local forces and interactions for all particles, see Lines 238–249. Here we loop over all particles and velocities. The result is

Listing 7.19   `mdc_thrust_v2.cu`

```
71  template< typename LocalForce , typename Interaction >
72  struct md_system_bs {
139    struct interaction_functor {
169      template< typename Tuple >
170      __host__ __device__ void operator()( Tuple const &t ) const {
171        point_type X = thrust::get<0>( t );
172        point_type V = thrust::get<1>( t );
173        index_type index = thrust::get<2>( t );
174        size_t cell_idx = thrust::get<3>( t );
176        point_type A = local_force(X, V);
178        for(int i = -1; i <= 1; ++i) {
179          for(int j = -1; j <= 1; ++j) {
180            index_type cell_index = index + index_type(i, j);
181            size_t cell_hash = get_cell_idx(cell_index, nx, ny);
182            for(size_t ii = cells_begin[cell_hash],
183                      ee = cells_end[cell_hash]; ii < ee; ++ii)
184            {
185              size_t jj = order[ii];
187              if( jj == cell_idx ) continue;
188              point_type Y = x[jj];
190              if( cell_index[0] >= nx ) Y[0] += xmax;
191              if( cell_index[0] < 0 ) Y[0] -= xmax;
192              if( cell_index[1] >= ny ) Y[1] += ymax;
193              if( cell_index[1] < 0 ) Y[1] -= ymax;
195              A += interaction(X, Y);
196            }
197          }
198        }
200        thrust::get<4>( t ) = A;
201      }
202    };
288  };
```

the acceleration which is stored in the vector a. The vector `cell_coo` contains the index of the cell in which the current particle is located. The interaction functor is shown in Listing 7.19. First, the local force is calculated in Line 176. Then two loops iterate over all neighboring cells of the current particle. Inside that loop the interaction between all particles in this cell and the particle is calculated. Lines 190–193 perform checks and corrections if particles are out of boundaries or are located on the opposite side of the considered domain.

At this point we only need to define the concrete solver type. A classical solver for molecular dynamic simulation is the Velocity-Verlet algorithm [9], which is used for second order ODEs and makes single RHS evaluation during one step. Here we use the implementation of the method from Boost.odeint.

The VexCL implementation follows the Thrust variant closely, so we omit the code for the sake of conciseness. VexCL provides `sort_by_key` primitive, and we

Fig. 7.3 Performance results for the molecular dynamics example

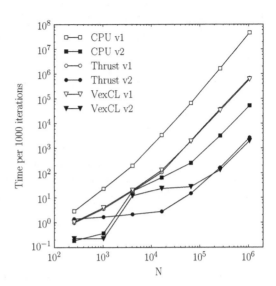

had to implement `lower_bound` and `upper_bound` algorithms in form of custom VexCL functions. We also had to use custom kernel in order to compute the interaction force. The kernel source is very similar to the Thrust interaction functor (Listing 7.19). See `md_vexcl_v2.cpp` file for the complete VexCL solution.

Figure 7.3 shows performance results for the different versions of the molecular dynamics example. Versions denoted by "v1" implement the straight-forward algorithm with $O(n^2)$ complexity. "v2" versions employ the bucket sort optimization. Both of the CPU versions use separate code which was again omitted from the text. The versions that use bucket sort optimization are expectedly faster than the "v1" algorithm. The Thrust and the VexCL versions show similar performance for large enough problems on the same hardware (with VexCL by 10–30 % faster than Thrust). For both versions the GPU implementations are orders of magnitude faster than the CPU implementation (factor 75 for "v1" and 25 for "v2"). But the biggest performance boost comes from the algorithmic complexity reduction: e.g. the optimized VexCL version runs 300 times faster than the straight-forward one.

7.7 Summary and Conclusions

We have presented a high-level approach to compute numerical solutions of ODEs by developing a generic implementation of common ODE solvers. The proposed framework is very flexible and is able to adapt several CPU and GPU backends. The Thrust and the VexCL backends considered here are very different with respect to their interface design, but nevertheless are easily incorporated with our approach to generic algorithms. The proposed ideas and techniques are already implemented in the Boost.odeint library, which offers a vastly larger functionality, including more steppers and more backends.

Regarding the backend choice, it seems that the use of VexCL results in generally shorter and cleaner code for the kind of problems we considered here. Admittedly, for the more advanced molecular dynamics example we had to implement a custom OpenCL kernel, although the implementation was very similar to the corresponding Thrust functor. Performance-wise, VexCL showed slightly better results for the larger problems, but due to OpenCL initialization cost was slower for the smaller problem sizes. The main advantage of VexCL (and of OpenCL libraries in general) seems to be the larger set of supported hardware. It should be noted that Boost.odeint supports many other backends, which allows the user to choose the one best suited for the problem at hand, or the one they feel most comfortable with. This freedom is the great advantage of the modularized, generic design that we presented here for ODE solvers. It is clear that this technique can be applied to other numerical algorithms as well.

Acknowledgements This work has been partially supported by the Russian Foundation for Basic Research (RFBR) grants No. 12-07-0007, 12-01-00333a, and by the Russian Government Program of Competitive Growth of Kazan Federal University. M. M. thankfully acknowledges financial support through the visitors program of the MPIPKS, Dresden (Germany).

References

1. Ahnert, K.: Odeint v2 - Solving Ordinary Differential Equations in C++. http://www. codeproject.com/Articles/268589/odeint-v2-Solving-ordinary-differential-equations (Oct 2011)
2. Ahnert, K., Mulansky, M.: Odeint — solving ordinary differential equations in C++. AIP Conf. Proc. **1389**, 1586–1589 (2011)
3. Alexandrescu, A.: Modern C++ Design: Generic Programming and Design Patterns Applied. Addison-Wesley Longman Publishing, Boston (2001)
4. Bell, N., Hoberock, J.: Thrust: A Productivity-Oriented Library for CUDA, Chap. 26, pp. 359–371. Elsevier, USA (2011)
5. Cohen, A.H., Holmes, P.J., Rand, R.H.: The nature of the coupling between segmental oscillators of the lamprey spinal generator for locomotion: a mathematical model. J. Math. Biol. **13**, 345–369 (1982)
6. Cormen, T.H., Leiserson, C.E., Rivest, R.L., Stein, C.: Introduction to Algorithms. MIT Press, Cambridge (2001)
7. Demidov, D.: VexCL: Vector Expression Template Library for OpenCL. http://www. codeproject.com/Articles/415058/VexCL-Vector-expression-template-library-for-OpenC (July 2012)
8. Demidov, D., Ahnert, K., Rupp, K., Gottschling, P.: Programming CUDA and OpenCL: a case study using modern C++ libraries. SIAM J. Sci. Comput. **35**(5), C453–C472 (2013)
9. Frenkel, D., Smit, B.: Understanding Molecular Simulation from Algorithms to Applications. Academic, San Diego (2002)
10. Galassi, M., Davies, J., Theiler, J., Gough, B., Jungman, G., Alken, P., Booth, M., Rossi, F.: GNU Scientific Library. Network Theory, Bristol (2007)
11. Gottschling, P., Wise, D.S., Adams, M.D.: Representation-transparent matrix algorithms with scalable performance. In: Proceedings of the 21st Annual International Conference on Supercomputing, pp. 116–125. ACM, New York (2007)

12. Gregor, D.: The Boost Function Library. http://www.boost.org/doc/libs/release/libs/function (2001)
13. Guennebaud, G., Jacob, B., et al.: Eigen v3. http://eigen.tuxfamily.org (2010)
14. Hairer, E., Nørsett, S.P., Wanner, G.: Solving Ordinary Differential Equations I: Nonstiff Problems, 2nd edn. Springer, Berlin (1993) (corr. 3rd printing 1993; corr. 3rd edn. 2009)
15. Hairer, E., Wanner, G.: Solving Ordinary Differential Equations II: Stiff and Differential-Algebraic Problems, 2nd edn. Springer, Berlin (1996) (2nd printing edn. 2010)
16. Hindmarsh, A.C., Brown, P.N., Grant, K.E., Lee, S.L., Serban, R., Shumaker, D.E., Woodward, C.S.: SUNDIALS: suite of nonlinear and differential/algebraic equation solvers. ACM Trans. Math. Softw. (TOMS) 31(3), 363–396 (2005)
17. Intel Math Kernel Library Reference Manual, Intel, Version 11.1 (2013)
18. Izhikevich, E.M., Ermentrout, B.: Phase model. Scholarpedia 3(10), 1487 (2008)
19. Jones, J.E.: On the determination of molecular fields. II. From the equation of state of a gas. R. Soc. Lond. Proc. Ser. A 106, 463–477 (1924)
20. Kuramoto, Y.: Chemical Oscillations, Waves, and Turbulence. Springer, New York (1984)
21. Leimkuhler, B., Reich, S.: Simulating Hamiltonian Dynamics, vol. 14. Cambridge University Press, Cambridge (2004)
22. Moore, R.E., Kearfott, R.B., Cloud, M.J.: Methods and Applications of Interval Analysis, vol. 2. SIAM, Philadelphia (1979)
23. Munshi, A., et al.: The OpenCL Specification. Khronos OpenCL Working Group 1, pp. 11–15 (2009)
24. NVIDIA CUDA Programming Guide. NVIDIA Corporation, Version 5.5 (2013)
25. OpenMP Architecture Review Board: OpenMP Application Program Interface Version 3.0. http://www.openmp.org/mp-documents/spec30.pdf (May 2008)
26. Ott, E.: Chaos in Dynamical Systems. Cambridge University Press, Cambridge (2002)
27. Pikovsky, A., Rosenblum, M., Kurths, J.: Synchronization: A Universal Concept in Nonlinear Sciences. Cambridge University Press, Cambridge (2001)
28. Poschel, T., Schwager, T.: Computational Granular Dynamics: Models and Algorithms. Springer, Berlin/Heidelberg [u.a.] (2010)
29. Press, W.H., Teukolsky, S.A., Vetterling, W.T., Flannery, B.P.: Numerical Recipes: The Art of Scientific Computing, 3rd edn. Cambridge University Press, Cambridge (2007)
30. R Core Team: R: A Language and Environment for Statistical Computing. R Foundation for Statistical Computing, Vienna (2013)
31. Reinders, J.: Intel Threading Building Blocks: Outfitting C++ for Multi-core Processor Parallelism. O'Reilly Media, Inc., Sebastopol (2010)
32. Rupp, K., Rudolf, F., Weinbub, J.: ViennaCL - A High Level Linear Algebra Library for GPUs and Multi-Core CPUs. In: International Workshop on GPUs and Scientific Applications, pp. 51–56 (2010)
33. Schabel, M.C., Watanabe, S.: The Boost Units Library. http://www.boost.org/doc/libs/release/libs/units (2010)
34. Soetaert, K., Cash, J.R., Mazzia, F.: Solving Differential Equations in R. Springer, Berlin/New York (2012)
35. Sparrow, C.: The Lorenz Equations: Bifurcations, Chaos, and Strange Attractors, vol. 41. Springer, New York (1982)
36. Stepanov, A., Lee, M: The Standard Template Library, vol. 1501. Hewlett Packard Laboratories, Palo Alto (1995)
37. Teschl, G.: Ordinary Differential Equations and Dynamical Systems, vol. 140. AMS Bookstore, Providence (2012)
38. Verlet, L.: Computer "experiments" on classical fluids. I. Thermodynamical properties of Lennard-Jones molecules. Phys. Rev. 159(1), 98–103 (1967)
39. Walter, J., Koch, M.: The Boost uBLAS Library. http://www.boost.org/doc/libs/release/libs/numeric/ublas/doc/index.htm (2002)

Chapter 8
GPU-Based Parallel Integration of Large Numbers of Independent ODE Systems

Kyle E. Niemeyer and Chih-Jen Sung

8.1 Introduction

In a number of science and engineering applications, researchers are faced with the task of solving large numbers of independent systems of ordinary differential equations (ODEs). One prominent example is the simulation of reactive flows for modeling combustion [5, 15, 25, 27, 28], atmospheric chemistry [1, 13], and groundwater transport [2,3], where operator splitting [31,33] decouples the solution of the fluid transport (e.g., advection, diffusion) and chemical kinetics terms. This results in large numbers of independent ODEs for the conservation of chemical species masses, with one system for each spatial location. The solution of the aggregate of these ODEs consumes most of the total wall-clock time of such simulations, >90 % in some cases. Simulations of electrical behavior in cardiac tissue use a similar operator splitting technique, which results in ODE systems for cell membrane dynamics [23, 34]. Other areas that deal with solving many independent systems of ODEs include systems biology [6, 40] and Monte Carlo methods for sensitivity analysis of ODEs [10, 16, 18].

In such problems, large numbers of the same governing ODEs with different initial conditions and/or input parameters must be integrated; since each system is independent, the entire set of ODEs can be integrated concurrently. While performance can be improved by using parallel central processing unit (CPU) methods, this embarrassingly parallel problem is especially well-suited to acceleration with the thread-based parallelism of graphics processing units (GPUs), as demonstrated

K.E. Niemeyer (✉)
School of Mechanical, Industrial, & Manufacturing Engineering, Oregon State University,
Corvallis, OR, USA
e-mail: Kyle.Niemeyer@oregonstate.edu

C.-J. Sung
Department of Mechanical Engineering, University of Connecticut, Storrs, CT, USA
e-mail: cjsung@engr.uconn.edu

V. Kindratenko (ed.), *Numerical Computations with GPUs*,
DOI 10.1007/978-3-319-06548-9__8, © Springer International Publishing Switzerland 2014

for reactive-flow simulations [21, 22, 29, 32]. In particular, Niemeyer and Sung [21] recently developed a GPU-based approach for the integration of moderately stiff chemical kinetics ODEs using explicit integration algorithms, using an adaptive fifth-order Runge–Kutta–Cash–Karp (RKCK) method for nonstiff cases and a stabilized second-order Runge–Kutta–Chebyshev (RKC) method for problems with greater stiffness. For large numbers of ODEs, they demonstrated that the GPU-based RKCK and RKC algorithms performed up to 126 and 65 times faster, respectively, than CPU versions of the same algorithm on a single CPU core. Furthermore, with moderate levels of stiffness, the GPU-based RKC offered a speedup factor of 57 compared to a conventional implicit algorithm executed in parallel on a six-core CPU. The specific acceleration factor demonstrated depended on the problem studied (e.g., the kinetic mechanism) and number of ODEs considered. Due to the favorable performance of these methods, in this chapter we present the integration algorithms, associated GPU source code, and implementation details so that interested readers can apply them to more general applications (e.g., the areas described above).

8.2 Mathematical Background

In this chapter, we represent a generic system of ODEs using

$$\frac{d\mathbf{y}}{dt} = \mathbf{f}(t, \mathbf{y}(t), \mathbf{g}), \qquad (8.1)$$

where $\mathbf{y}(t)$ is the vector of unknown dependent variables at some time t, $\mathbf{f}$ is the right-hand-side vector function, and $\mathbf{g}$ is a vector of constant parameters (e.g., pressure or temperature for chemical kinetics). The size of $\mathbf{y}$ (i.e., the number of equations/unknowns) is N. For the types of problems with which we are concerned here, a large number of ODE systems, N_{ode}, each given by Eq. (8.1) must be integrated independently from some time $t = t_n$ to t_{n+1}, with different initial conditions $\mathbf{y}(t_n)$ and constant parameters $\mathbf{g}$ for each system. The numerical approximation to the exact solution $\mathbf{y}(t_n)$ is $\mathbf{y}_n$, and the step size for a given step is $\delta t_n = t_{n+1} - t_n$.

Nonstiff ODEs, or those with little stiffness, can be solved using explicit integration methods. Many such methods exist, and algorithms can be classified in general into Runge–Kutta and linear multistep methods, and also into explicit or implicit methods; see Hairer and Wanner [9] for more details. Stiffness, a concept somewhat difficult to quantify, refers to the quality of an ODE making standard explicit methods perform poorly due to the requirement for unreasonably small time-step sizes—otherwise the solutions become unstable and oscillate wildly [8].

Traditionally, implicit integration algorithms such as those based on backwards difference formulas have been used to handle stiff equations, but these require expensive linear algebra operations on the Jacobian matrix (e.g., LU decomposition, matrix inversion). The complex logical flow of such operations results in highly divergent instructions for different initial conditions, making implicit algorithms unsuitable for operation on GPUs. In fact, Stone and Davis [32] implemented a traditional high-order implicit algorithm on GPUs, and found that it performed only slightly better than a multi-core CPU version of the same algorithm would. While implicit algorithms may be required for ODE systems with extreme levels of stiffness (suggesting that new solutions need to be found for GPU acceleration of such problems), other options can be used for cases of little-to-moderate stiffness. Here, we describe two integration algorithms suitable for use solving many independent systems of ODEs on GPUs.

8.2.1 Runge–Kutta–Cash–Karp

For nonstiff ODEs, an appropriate explicit algorithm is the fifth-order Runge–Kutta method developed by Cash and Karp [4]: the RKCK method. The RKCK method estimates the local truncation error using an embedded fourth-order method, by taking the difference between the fourth- and fifth-order solutions. It then uses this estimate to adaptively select the step size [26].

Using the terminology established above, the RKCK formulas—which also apply to any general fifth-order Runge–Kutta method—are

$$\mathbf{k}_1 = \delta t_n \, \mathbf{f}\left(t_n, \mathbf{y}_n, \mathbf{g}\right) , \tag{8.2}$$

$$\mathbf{k}_2 = \delta t_n \, \mathbf{f}\left(t_n + a_2 \, \delta t_n, \mathbf{y}_n + b_{21}\mathbf{k}_1, \mathbf{g}\right) , \tag{8.3}$$

$$\mathbf{k}_3 = \delta t_n \, \mathbf{f}\left(t_n + a_3 \, \delta t_n, \mathbf{y}_n + b_{31}\mathbf{k}_1 + b_{32}\mathbf{k}_2, \mathbf{g}\right) , \tag{8.4}$$

$$\mathbf{k}_4 = \delta t_n \, \mathbf{f}\left(t_n + a_4 \, \delta t_n, \mathbf{y}_n + b_{41}\mathbf{k}_1 + b_{42}\mathbf{k}_2 + b_{43}\mathbf{k}_3, \mathbf{g}\right) , \tag{8.5}$$

$$\mathbf{k}_5 = \delta t_n \, \mathbf{f}\left(t_n + a_5 \, \delta t_n, \mathbf{y}_n + b_{51}\mathbf{k}_1 + b_{52}\mathbf{k}_2 + b_{53}\mathbf{k}_3 + b_{54}\mathbf{k}_4, \mathbf{g}\right) , \tag{8.6}$$

$$\mathbf{k}_6 = \delta t_n \, \mathbf{f}\left(t_n + a_6 \, \delta t_n, \mathbf{y}_n + b_{61}\mathbf{k}_1 + b_{62}\mathbf{k}_2 + b_{63}\mathbf{k}_3 + b_{64}\mathbf{k}_4 + b_{65}\mathbf{k}_5, \mathbf{g}\right) , \tag{8.7}$$

$$\mathbf{y}_{n+1} = \mathbf{y}_n + c_1\mathbf{k}_1 + c_2\mathbf{k}_2 + c_3\mathbf{k}_3 + c_4\mathbf{k}_4 + c_5\mathbf{k}_5 + c_6\mathbf{k}_6 , \tag{8.8}$$

$$\mathbf{y}_{n+1}^* = \mathbf{y}_n + c_1^*\mathbf{k}_1 + c_2^*\mathbf{k}_2 + c_3^*\mathbf{k}_3 + c_4^*\mathbf{k}_4 + c_5^*\mathbf{k}_5 + c_6^*\mathbf{k}_6 , \tag{8.9}$$

where $\mathbf{y}_{n+1}$ is the fifth-order solution and $\mathbf{y}_{n+1}^*$ is the solution of the embedded fourth-order method. The RKCK coefficients are given in Table 8.1. The local error $\mathbf{\Delta}_{n+1}$ is estimated using the difference between the fourth- and fifth-order solutions:

$$\mathbf{\Delta}_{n+1} = \mathbf{y}_{n+1} - \mathbf{y}_{n+1}^* = \sum_{i=1}^{6} \left(c_i - c_i^*\right) \mathbf{k}_i . \tag{8.10}$$

Then, this error is compared against a desired accuracy, $\mathbf{\Delta}_0$, defined by

$$\mathbf{\Delta}_0 = \varepsilon \left(|\mathbf{y}_n| + |\delta t_n \, \mathbf{f}\left(t_n, \mathbf{y}_n, \mathbf{g}\right)| + \delta\right) , \tag{8.11}$$

Table 8.1 Coefficients for the fifth-order Runge–Kutta–Cash–Karp method, adopted from Press et al. [26]

i	a_i	b_{ij}					c_i	c_i^*
1							$\frac{37}{378}$	$\frac{2825}{27648}$
2	$\frac{1}{5}$	$\frac{1}{5}$					0	0
3	$\frac{3}{10}$	$\frac{3}{40}$	$\frac{9}{40}$				$\frac{250}{621}$	$\frac{18575}{48384}$
4	$\frac{3}{5}$	$\frac{3}{10}$	$-\frac{9}{10}$	$\frac{6}{5}$			$\frac{125}{594}$	$\frac{13525}{55296}$
5	1	$-\frac{11}{54}$	$\frac{5}{2}$	$-\frac{70}{27}$	$\frac{35}{27}$		0	$\frac{277}{14336}$
6	$\frac{7}{8}$	$\frac{1631}{55296}$	$\frac{175}{512}$	$\frac{575}{13824}$	$\frac{44275}{110592}$	$\frac{253}{4096}$	$\frac{512}{1771}$	$\frac{1}{4}$
j		1	2	3	4	5		

where ε is a tolerance level and δ represents a small value (e.g., 10^{-30}). When the estimated error rises above the desired accuracy ($\mathbf{\Delta}_{n+1} > \mathbf{\Delta}_0$), the algorithm rejects the current step and calculates a smaller step size. Correspondingly, the algorithm accepts a step with error at or below the desired accuracy ($\mathbf{\Delta}_{n+1} \leq \mathbf{\Delta}_0$) and calculates a new step size for the next step using

$$
\delta t_{\text{new}} =
\begin{cases}
S\,\delta t_n\,\max_i\left(\left|\frac{\Delta_{0,i}}{\Delta_{n+1,i}}\right|\right)^{1/5} & \text{if}\,\mathbf{\Delta}_{n+1} \leq \mathbf{\Delta}_0\,,\text{ or} \\
S\,\delta t_n\,\max_i\left(\left|\frac{\Delta_{0,i}}{\Delta_{n+1,i}}\right|\right)^{1/4} & \text{if}\,\mathbf{\Delta}_{n+1} > \mathbf{\Delta}_0\,.
\end{cases}
\tag{8.12}
$$

Here, i represents the ith element of the related vector and S denotes a safety factor slightly smaller than unity (e.g., 0.9). Equation (8.12) is used to calculate the next time step size both for an accepted step and also for a new, smaller step size in the case of a step rejection. In practice, step size decreases and increases are limited to factors of 10 and 5, respectively [26].

8.2.2 Runge–Kutta–Chebyshev

For ODEs with moderate levels of stiffness, one alternative to implicit algorithms is a stabilized explicit scheme such as the RKC method [30, 35–39]. While the RKC method is explicit, it handles stiffness through additional stages—past the first two required for second-order accuracy—that extend its stability domain along the negative real axis of eigenvalues.

Our RKC implementation is taken from Sommeijer et al. [30] and Verwer et al. [39]. Following the same terminology as above, the formulas for the second-order RKC are

$$
\mathbf{w}_0 = \mathbf{y}_n\,,
\tag{8.13}
$$

$$
\mathbf{w}_1 = \mathbf{w}_0 + \tilde{\mu}_1\,\delta t_n\,\mathbf{f}_0\,,
\tag{8.14}
$$

$$\mathbf{w}_j = (1 - \mu_j - \nu_j)\mathbf{w}_0 + \mu_j \mathbf{w}_{j-1}$$

$$+ \nu_j \mathbf{w}_{j-2} + \tilde{\mu}_j \, \delta t_n \, \mathbf{f}_{j-1} + \tilde{\gamma}_j \, \delta t_n \, \mathbf{f}_0, \quad j = 2, \ldots, s \,, \tag{8.15}$$

$$\mathbf{y}_{n+1} = \mathbf{w}_s \,, \tag{8.16}$$

where s is the total number of stages and $\mathbf{w}_j$ are internal vectors for the stages. The right-hand-side derivative vector function, $\mathbf{f}_j$, is evaluated at each stage, such that $\mathbf{f}_j = \mathbf{f}(t_n + c_j \, \delta t_n, \mathbf{w}_j, \mathbf{g})$. The recursive nature of $\mathbf{w}_j$ allows the use of only five arrays for storage. The coefficients used in Eqs. (8.14) and (8.15) can be obtained analytically for any number of stages $s \geq 2$:

$$\tilde{\mu}_1 = b_1 \omega_1, \quad \mu_j = \frac{2b_j \omega_0}{b_{j-1}}, \quad \nu_j = \frac{-b_j}{b_{j-2}}, \quad \tilde{\mu}_j = \frac{2b_j \omega_1}{b_{j-1}} \,, \tag{8.17}$$

$$\tilde{\gamma}_j = -a_{j-1}\tilde{\mu}_j, \quad b_0 = b_2, \quad b_1 = \frac{1}{\omega_0}, \quad b_j = \frac{T_j''(\omega_0)}{\left(T_j'(\omega_0)\right)^2} \,, \tag{8.18}$$

$$w_0 = 1 + \frac{\kappa}{s^2}, \quad \omega_1 = \frac{T_s'(\omega_0)}{T_s''(\omega_0)}, \quad a_j = 1 - b_j T_j(\omega_0) \,, \tag{8.19}$$

for $j = 2, \ldots, s$, where $\kappa \geq 0$ is the damping parameter (e.g., $\kappa = 2/13$ [30, 39]). The Chebyshev polynomials of the first kind, $T_j(x)$, with first and second derivatives $T_j'(x)$ and $T_j''(x)$, respectively, are defined recursively as

$$T_j(x) = 2x T_{j-1}(x) - T_{j-2}(x), \quad j = 2, \ldots, s \,, \tag{8.20}$$

where $T_0(x) = 1$ and $T_1(x) = x$. The c_j used in the function evaluations are

$$c_1 = \frac{c_2}{T_2'(\omega_0)} \approx \frac{c_2}{4} \,, \tag{8.21}$$

$$c_j = \frac{T_s'(\omega_0)}{T_s''(\omega_0)} \frac{T_j''(\omega_0)}{T_j'(\omega_0)} \approx \frac{j^2 - 1}{s^2 - 1}, \quad 2 \leq j \leq s - 1 \,, \tag{8.22}$$

$$c_s = 1 \,. \tag{8.23}$$

As with the RKCK method in Sect. 8.2.1, the RKC method can also be used with an adaptive time stepping method for error control, as given by Sommeijer et al. [30]. The error accrued in taking the step $t_{n+1} = t_n + \delta t_n$ and obtaining $\mathbf{y}_{n+1}$ is estimated using

$$\mathbf{\Delta}_{n+1} = \frac{4}{5}(\mathbf{y}_n - \mathbf{y}_{n+1}) + \frac{2}{5}\delta t_n (\mathbf{f}_n + \mathbf{f}_{n+1}) \,. \tag{8.24}$$

We then obtain the weighted RMS norm of error using this error estimate with absolute and relative tolerances:

$$\|\mathbf{\Delta}_{n+1}\|_{\text{RMS}} = \left\|\frac{\mathbf{\Delta}_{n+1}}{\mathbf{T}\sqrt{N}}\right\|_2 , \tag{8.25}$$

$$\mathbf{T} = \mathbf{A} + R \cdot \max(|\mathbf{y}_n|, |\mathbf{y}_{n+1}|) , \tag{8.26}$$

where N represents the number of unknown variables as defined previously, $\mathbf{A}$ is the vector of absolute tolerances, and R is the relative tolerance. The norm $\|\cdot\|_2$ indicates the Euclidean or L_2 norm. If $\|\mathbf{\Delta}_{n+1}\|_{\text{RMS}} \leq 1$, the step is accepted; otherwise, it is rejected and repeated using a smaller step size. Finally, a new step size is calculated using the weighted RMS norm of error for the current and prior steps, as well as the associated step sizes, via

$$\delta t_{n+1} = \min(10, \max(0.1, f)) \delta t_n , \tag{8.27}$$

$$f = 0.8 \left(\frac{\|\mathbf{\Delta}_n\|_{\text{RMS}}^{1/(p+1)}}{\|\mathbf{\Delta}_{n+1}\|_{\text{RMS}}^{1/(p+1)}} \frac{\delta t_n}{\delta t_{n-1}}\right) \frac{1}{\|\mathbf{\Delta}_n\|_{\text{RMS}}^{1/(p+1)}} , \tag{8.28}$$

where $p = 2$, the order of the algorithm. We use Eq. (8.27) with a modified relation to calculate a new step size for a step rejection:

$$f = \frac{0.8}{\|\mathbf{\Delta}_n\|_{\text{RMS}}^{1/(p+1)}} . \tag{8.29}$$

Determining the initial time step size requires special consideration. First, the algorithm takes a tentative integration step, using the inverse of the spectral radius σ—the magnitude of the largest eigenvalue—of the Jacobian matrix as the step size. Then, after estimating the error associated with this tentative step, it calculates a new step size following a similar procedure to that given in Eqs. (8.27) and (8.28):

$$\delta t_0 = \frac{1}{\sigma} , \tag{8.30}$$

$$\mathbf{\Delta}_0 = \delta t_0 \left(\mathbf{f}(t_0 + \delta t_0, \mathbf{y}_0 + \delta t_0 \mathbf{f}(t_0, \mathbf{y}_0)) - \mathbf{f}(t_0, \mathbf{y}_0)\right) , \tag{8.31}$$

$$\delta t_1 = 0.1 \frac{\delta t_0}{\|\mathbf{\Delta}_0\|_{\text{RMS}}^{1/2}} , \tag{8.32}$$

where $\|\mathbf{\Delta}_0\|_{\text{RMS}}$ is evaluated in the same manner as $\|\mathbf{\Delta}_{n+1}\|_{\text{RMS}}$ using Eq. (8.25).

After selecting the optimal time step size to control local error, the algorithm next determines the optimal number of RKC stages in order to remain stable. Due to stiffness, too few stages could lead to instability; in contrast, using more stages than required would add unnecessary computational effort. The number of stages is determined using the spectral radius and time step size:

$$s = 1 + \sqrt{1 + 1.54 \delta t_n \sigma} , \tag{8.33}$$

as suggested by Sommeijer et al. [30], where the value 1.54 is related to the stability boundary of the algorithm. Note that s may vary between time steps due to a changing spectral radius and time step size. We recommend using a nonlinear power method [30] to calculate the spectral radius with our RKC implementation; this choice costs an additional vector to store the computed eigenvector, but avoids storing or calculating the Jacobian matrix directly. Following Sommeijer et al. [30], our RKC implementation estimates the spectral radius every 25 (internal) steps or after a step rejection.

8.3 Source Code

Next, we provide implementation details and source code for the GPU versions of the algorithms described above. The number of unknowns (and corresponding equations) N is represented with the variable NEQN, and the number of ODE systems N_{ode} is defined as numODE in the following code. In order for the GPU algorithms to offer a performance increase over CPU algorithms, N_{ode} should be relatively large. Although the exact number of ODEs where the GPU algorithm becomes faster than its CPU equivalent is problem dependent, Niemeijer and Sung [21] showed that a GPU implementation of the RKCK algorithm for chemical kinetics outperforms an equivalent serial CPU version for as little as 128 ODE systems. All operations are given here in double precision, although depending on the particular needs of the specific application single-precision calculations may be preferable to reduce the computational expense.

In order to take advantage of global memory coalescing on the GPU, we recommend storing the set of dependent variable vectors y_i, where $i = 1, \ldots, N_{ode}$, in a single one-dimensional array, where variables corresponding to the same unknown for consecutive systems sit adjacent in memory. In other words, if **Y** is a two-dimensional matrix with N_{ode} rows and N columns, where the ith row contains the unknown vector y_i, then **Y** should be stored in memory as a one-dimensional array in column-major ordering. This ensures that adjacent GPU threads in the same warp access adjacent global memory locations when reading or writing equivalent array elements. See Kirk and Hwu [14] or Jang et al. [12] for more details and examples on global memory coalescing.

The following code snippet shows the proper loading of a host array yHost from an arbitrary array y that contains initial conditions for all ODEs:

```
1 double *yHost;
2 yHost = (double *) malloc (numODE * NEQN * sizeof(double));
3
4 for (int i = 0; i < numODE; ++i) {
5     for (int j = 0; j < NEQN; ++j) {
6         yHost[i + numODE * j] = y[i][j];
7     }
8 }
```

A similar procedure should be used for the constant parameter vector **g** if needed.

Next, the GPU global memory arrays should be declared and initialized, and the block/grid dimensions set up using

```
double *yDevice;
cudaMalloc ((void**) &yDevice, numODE * NEQN * sizeof(double));

int blockSize;
if (numODE < 4194304) {
  blockSize = 64;
} else if (numODE < 8388608) {
  blockSize = 128;
} else if (numODE < 16777216) {
  blockSize = 256;
} else {
  blockSize = 512;
}
dim3 dimBlock (blockSize, 1);
dim3 dimGrid (numODE / dimBlock.x, 1);
```

Here, we use simple one-dimensional block and grid dimensions; reshaping the grid should not affect performance, but can be done for convenience. We chose 64 as the block size for problems with less than 4,194,304 ODEs based on experience for chemical kinetics problems [21]. The size should remain a power of two, but different block sizes may be optimal for other problems.

The final step is to set up the ODE integration loop and kernel function execution. The integration driver kernel, to be described shortly, will perform internal substepping as necessary to reach the specified end time. Depending on the objectives, there are various ways to approach this:

- If only the final integrated results are needed, then a single GPU kernel can be invoked.
- If intermediate integration results are needed, then an acceptable outer step size over which results will be spaced should be chosen, and the GPU kernel should be invoked inside a loop.

We will leave the code as general as possible by following the second approach, although modifications should be made depending on the desired functionality. In both cases, the global memory array holding the variables to be integrated needs to be transferred to the GPU before, and from the GPU after, each kernel invocation.

```
// set initial time
double t = t0;
double tNext = t + h;

while (t < tEnd) {
  // transfer memory to GPU
```

```
7   cudaMemcpy (yDevice, yHost, numODE*NEQN*sizeof(double),
        cudaMemcpyHostToDevice);
8   intDriver <<<dimGrid, dimBlock>>> (t, tNext, numODE, gDevice,
        yDevice);
9
10  // transfer memory back to CPU
11  cudaMemcpy (yHost, yDevice, numODE*NEQN*sizeof(double),
        cudaMemcpyDeviceToHost);
12
13  t = tNext;
14  tNext += h;
15  }
16
17  cudaFree (gDevice);
18  cudaFree (yDevice);
```

Here, t0 refers to the initial time, tEnd the desired final time, and h the outer
step size. In the current form, each outer integration step performed by the GPU
will be a "restart" integration, meaning no information about previous steps (e.g.,
error estimates, step sizes) will be used to assist the startup. This is necessary in
certain applications such as reactive-flow simulations (and other simulation methods
that use operator splitting), where, after each outer step, integration results are
combined with changes due to fluid transport, thereby invalidating stored integration
information. However, where possible, better performance may be obtained by
transferring the appropriate data from the GPU and using it in the next overall
integration step.

The next code snippet contains the general integration driver kernel, suitable for
either algorithm:

```
1   __global__ void
2   intDriver (const double t, const double tEnd, const int numODE,
3              const double* gGlobal, double* yGlobal) {
4     // unique thread ID, based on local ID in block and block ID
5     int tid = threadIdx.x + (blockDim.x * blockIdx.x);
6
7     // ensure thread within limit
8     if (tid < numODE) {
9
10      // local array with initial values
11      double yLocal[NEQN];
12
13      // constant parameter(s)
14      double gLocal = gGlobal[tid];
15
16      // load local array with initial values from global array
17      for (int i = 0; i < NEQN; ++i) {
18        yLocal[i] = yGlobal[tid + numODE * i];
19      }
20
21      // call integrator for one time step
22      integrator (t, tEnd, yLocal, gGlobal);
23
```

```
24    // update global array with integrated values
25    for (int i = 0; i < NEQN; ++i) {
26      yGlobal[tid + numODE * i] = yLocal[i];
27    }
28    }
29  }
```

The function `integrator` should be replaced with `rkckDriver` or `rkcDriver` (given below) depending on the desired integration algorithm.

8.3.1 RKCK Code

In the following, the source code for the RKCK driver device function is given in functional snippets. First, the minimum and maximum allowable time step sizes are defined, and the initial step size is set as half the integration time.

```
1   __device__ void
2   rkckDriver (double t, const double tEnd, const double g,
3               double* y) {
4
5     // maximum and minimum allowable step sizes
6     const double hMax = fabs(tEnd - t);
7     const double hMin = 1.0e-20;
8
9     // initial step size
10    double h = 0.5 * fabs(tEnd - t);
```

Then, inside the time integration loop, the algorithm takes a trial integration step and estimates the error of that step.

```
1     // integrate until specified end time
2     while (t < tEnd) {
3
4       // limit step size based on remaining time
5       h = fmin(tEnd - t, h);
6
7       // y and error vectors temporary until error determined
8       double yTemp[NEQN], yErr[NEQN];
9
10      // evaluate derivative
11      double F[NEQN];
12      dydt (t, y, g, F);
13
14      // take a trial step
15      rkckStep (t, y, g, F, h, yTemp, yErr);
16
17      // calculate error
18      double err = 0.0;
19      int nanFlag = 0;
20      for (int i = 0; i < NEQN; ++i) {
21        if (isnan(yErr[i])) nanFlag = 1;
```

```
22
23        err = fmax(err, fabs(yErr[i] / (fabs(y[i]) + fabs(h * F[i])
              + TINY)));
24      }
25    err /= eps;
```

If the error is too large, the step size is decreased and the step retaken; otherwise, the algorithm accepts the step and calculates the next step size, then repeats the process.

```
1      // check if error too large
2      if ((err > 1.0) || isnan(err) || (nanFlag == 1)) {
3        // step failed, error too large
4        if (isnan(err) || (nanFlag == 1)) {
5          h *= P1;
6        } else {
7          h = fmax(SAFETY * h * pow(err, PSHRNK), P1 * h);
8        }
9
10     } else {
11       // step accepted
12       t += h;
13
14       if (err > ERRCON) {
15         h = SAFETY * h * pow(err, PGROW);
16       } else {
17         h *= 5.0;
18       }
19
20       // ensure step size is bounded
21       h = fmax(hMin, fmin(hMax, h));
22
23       for (int i = 0; i < NEQN; ++i)
24         y[i] = yTemp[i];
25     }
26   }
27 }
```

The device function dydt evaluates the derivative function F for the particular problem as in Eq. (8.1) using the input time t, vector of dependent variables y, and constant parameter(s) g. The device function rkcStep, not given here, takes a single integration step using Eqs. (8.2)–(8.9), returning the vector of integrated values yTemp as well as the error vector yErr. A number of constants were used in this function, given here:

```
1 #define UROUND (2.22e-16)
2 #define SAFETY 0.9
3 #define PGROW (-0.2)
4 #define PSHRNK (-0.25)
5 #define ERRCON (1.89e-4)
6 #define TINY (1.0e-30)
7 const double eps = 1.0e-10;
```

8.3.2 RKC Code

The RKC driver algorithm is next given. For this algorithm, the number of stages must be determined at each step to handle local stiffness; to avoid excess computation, a maximum number of stages is first set. In addition, minimum and maximum allowable time step sizes are defined.

```
1  __device__ void
2  rkcDriver(double t, const double tEnd, const double g, double* y)
       {
3    // number of steps
4    int numStep = 0;
5
6    // maximum allowable number of RKC stages
7    int mMax = (int)(round(sqrt(relTol / (10.0 * UROUND))));
8
9    // RKC needs at least two stages for second-order accuracy
10   if (mMax < 2) mMax = 2;
11
12   // maximum allowable step size
13   const double stepSizeMax = fabs(tEnd - t);
14
15   // minimum allowable step size
16   double stepSizeMin = 10.0*UROUND*fmax(fabs(t), stepSizeMax);
```

Then, the algorithm evaluates the derivative using the initial conditions for use as the initial eigenvector estimate for the spectral radius calculation. The calculated eigenvectors are stored and used as initial guesses in later steps.

```
1    // internal y vector
2    double y_n[NEQN];
3    for (int i = 0; i < NEQN; ++i)
4      y_n[i] = y[i];
5
6    // calculate F_n for initial y
7    double F_n[NEQN];
8    dydt (t, y_n, g, F_n);
9
10   // internal work vector
11   double work[4 + NEQN];
12
13   // load initial estimate for eigenvector
14   if (work[2] < UROUND) {
15     for (int i = 0; i < NEQN; ++i) {
16       work[4 + i] = F_n[i];
17     }
18   }
```

Inside the time integration loop, the algorithm calculates the spectral radius for the first step—which it next uses to determine the initial step size—and every 25 steps thereafter.

```
1    // perform internal sub-stepping
```

```
2    while (t < tEnd) {
3      double tempArr[NEQN], tempArr2[NEQN], err;
4
5      // if last step, limit step size
6      if ((1.1 * work[2]) >= fabs(tEnd - t)) work[2] = fabs(tEnd -
         t);
7
8      // estimate Jacobian spectral radius if 25 steps passed
9      if ((numStep % 25) == 0) {
10       work[3] = rkcSpecRad (t, y_n, g, F_n, stepSizeMax, &work
           [4], tempArr2);
11     }
```

For the initial step, a trial step is taken using the inverse of the spectral radius as the step size; the resulting error is used to determine an appropriate step size that satisfies error control.

```
1      // if this is initial step
2      if (work[2] < UROUND) {
3        // estimate first time step
4        work[2] = stepSizeMax;
5
6        if ((work[3] * work[2]) > 1.0) work[2] = 1.0 / work[3];
7        work[2] = fmax(work[2], stepSizeMin);
8
9        for (int i = 0; i < NEQN; ++i) {
10         temp_arr[i] = y_n[i] + (work[2] * F_n[i]);
11       }
12       dydt (t + work[2], tempArr, g, tempArr2);
13
14       err = 0.0;
15       for (int i = 0; i < NEQN; ++i) {
16         double est = (tempArr2[i] - F_n[i]) / (absTol + relTol *
             fabs(y_n[i]));
17         err += est * est;
18       }
19       err = work[2] * sqrt(err / NEQN);
20
21       if ((P1 * work[2]) < (stepSizeMax * sqrt(err))) {
22         work[2] = fmax(P1 * work[2] / sqrt(err), stepSizeMin);
23       } else {
24         work[2] = stepSizeMax;
25       }
26     }
```

For all steps following the first, the value stored in work[2] is used for the time step size.

Next, the number of stages is determined using the spectral radius and current time step size, and a tentative integration step performed.

```
1      // calculate number of steps
2      int m = 1 + (int)(sqrt(1.54 * work[2] * work[3] + 1.0));
3
4      // modify step size based on stages
```

```
5    if (m > mMax) {
6      m = mMax;
7      work[2] = ((double)(m * m - 1)) / (1.54 * work[3]);
8    }
9
10   // perform tentative time step
11   rkcStep (t, y_n, g, F_n, work[2], m, y);
```

The algorithm then estimates the error of that step.

```
1    // calculate derivative F_np1 with tentative y_np1
2    dydt (t + work[2], y, g, tempArr);
3
4    // estimate error
5    err = 0.0;
6    for (int i = 0; i < NEQN; ++i) {
7      double est = 0.8 * (y_n[i] - y[i]) + 0.4 * work[2] * (F_n[i
         ] + tempArr[i]);
8      est /= (absTol + relTol * fmax(fabs(y[i]), fabs(y_n[i])));
9      err += est * est;
10   }
11   err = sqrt(err / ((double)N));
```

Based on the error magnitude, the algorithm determines whether to accept the step and proceed to the next step or to decrease the step size and repeat the current step.

```
1    // check value of error
2    if (err > 1.0) {
3      // error too large, step is rejected
4      // select smaller step size
5      work[2] = 0.8 * work[2] / (pow(err, (1.0 / 3.0)));
6
7      // reevaluate spectral radius
8      work[3] = rkcSpecRad (t, y_n, g, F_n, stepSizeMax, &work
         [4], tempArr2);
9    } else {
10     // step accepted
11     t += work[2];
12     numStep++;
```

Finally, for an accepted step, the current step size and error are stored and the next step size is calculated.

```
1    double fac = 10.0;
2    double temp1, temp2;
3
4    if (work[1] < UROUND) {
5      temp2 = pow(err, (1.0 / 3.0));
6      if (0.8 < (fac * temp2)) fac = 0.8 / temp2;
7    } else {
8      temp1 = 0.8 * work[2] * pow(work[0], (1.0 / 3.0));
9      temp2 = work[1] * pow(err, (2.0 / 3.0));
10     if (temp1 < (fac * temp2)) fac = temp1 / temp2;
11   }
```

```
12
13     // set "old" values to those for current time step
14     work[0] = err;
15     work[1] = work[2];
16
17     for (int i = 0; i < NEQN; ++i) {
18       y_n[i] = y[i];
19       F_n[i] = tempArr[i];
20     }
21
22     work[2] *= fmax(P1, fac);
23     work[2] = fmax(stepSizeMin, fmin(stepSizeMax, work[2]));
24   }
25   }
26 }
```

As before, we do not provide the RKC integration step device function `rkcStep`, which evaluates Eqs. (8.13)–(8.16). The absolute and relative tolerances `absTol` and `relTol` are set as defined constants, e.g.,:

```
1 const double abs_tol = 1.0e-10;
2 const double rel_tol = 1.0e-6;
```

Note that these may be modified to more stringent tolerances if desired. The constant UROUND is defined the same as in the RKCK code above. The local work array `work` contains, in element order:

0 the previous step error estimation,
1 previous time step,
2 current time step,
3 spectral radius, and
4 vector of eigenvalues (of size N).

The device function `rkcSpecRad` returns the spectral radius, the largest magnitude eigenvalue; various methods may be used for this purpose depending on the case. We provide GPU source code for a nonlinear power method adopted from Sommeijer et al. [30] that may be used for general applications in the Appendix.

8.4   Performance Results

We tested the performance of the GPU-based RKCK and RKC integration algorithms using two ODE test cases, ranging the number of ODE systems from 10^1 to 10^5. For both cases, all calculations were performed in double precision using a single GPU and single CPU; we compared the performance of the GPU algorithm against serial CPU calculations as well as parallelized CPU performance—via OpenMP [24]—on four cores. We performed the GPU calculations using an NVIDIA Tesla c2075 GPU with 6 GB of global memory, and an Intel Xeon X5650 CPU, running at 2.67 GHz with 256 kB of L2 cache memory per core and 12 MB of L3 cache memory, served as the host processor both for the GPU calculations and the CPU single- and four-core OpenMP calculations. We used the GNU Compiler Collection (gcc) version 4.6.2 (with the compiler options "-O3 -ffast-math -std=c99 -m64") to compile the CPU programs and the CUDA 5.5 compiler nvcc version 5.5.0 ("-O3 -arch=sm_20 -m64") to compile the GPU versions. We set the GPU to persistence mode, but also used the cudaSetDevice() to hide any further device initialization delay in the CUDA implementations prior to the timing.

The integration algorithms take as input initial conditions and a global time step, performing internal sub-stepping as necessary. The computational wall-clock times reported represent the average over ten global time steps, which for the GPU versions includes the overhead required for transmitting data between the CPU and GPU before and after each global step. The integrator restarts at each global time step, not storing any data from the previous step—although any sub-stepping performed by the algorithm within these larger steps does benefit from retained information from prior sub-steps. Interested readers should refer to Niemeyer and Sung [21] for more detailed performance evaluations of these algorithms in the context of chemical kinetics problems.

8.4.1 RKCK Results

We used the Pleiades ODE test problem (PLEI) of Hairer et al. [9, 20] to test the GPU- and CPU-based versions of the RKCK integrator. This nonstiff test case originates from a celestial mechanics problem tracking the coordinates of seven stars; it consists of a set of 14 second-order ODEs based on Newtonian gravitational forces, in the form

$$z'' = \begin{pmatrix} x \\ y \end{pmatrix}'' = \begin{pmatrix} f^{(1)}(x, y) \\ f^{(2)}(x, y) \end{pmatrix}, \quad z \in \Re^{14}, \tag{8.34}$$

$$x_i'' = f_i^{(1)}(x, y) = \sum_{j \neq i} m_j (x_j - x_i)/r_{ij}, \tag{8.35}$$

$$y_i'' = f_i^{(2)}(x, y) = \sum_{j \neq i} m_j (y_j - y_i)/r_{ij}, \tag{8.36}$$

$$r_{ij} = \left((x_i - x_j)^2 + (y_i - y_j)\right)^{3/2}, \quad i, j = 1, \dots, 7, \tag{8.37}$$

where (x_i, y_i) and $m_i = i$ are the coordinates and mass of the ith star, respectively. This second-order system can be converted into a system of 28 first-order ODEs of the form by defining $w = z'$, such that

$$\begin{pmatrix} z \\ w \end{pmatrix}' = \begin{pmatrix} w \\ f(z) \end{pmatrix}, \quad \begin{pmatrix} z \\ w \end{pmatrix} \in \Re^{28}. \tag{8.38}$$

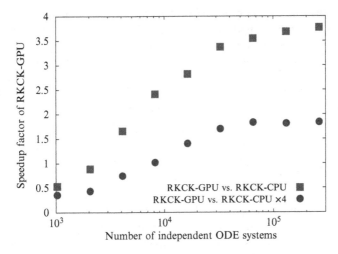

Fig. 8.1 Speedup factors offered by GPU-based explicit RKCK integration algorithm over single-
and four-core CPU-based versions for Pleiades ODE problem. Note that the horizontal axis is
displayed in logarithmic scale

While the original problem offers specific initial conditions for a single ODE
system, here we consider a large number of ODEs with the initial conditions
randomly perturbed by a small factor to emulate a sensitivity analysis. We integrated
the ODE systems from $t = 0$ to $1.0\,\mathrm{s}$ using 1.0×10^{-1} s as the global time step size.
We set the RKCK tolerance ε (eps in the code) to 1.0×10^{-10}.

Figure 8.1 shows the speedup factors, measured as the ratio of computational
times per step, offered by the GPU-based RKCK algorithm over the baseline
CPU version for both serial and four-core parallel operation, for numbers of ODE
systems ranging from 1,024 to 262,144. The GPU-based algorithm ran faster
than the serial and parallel CPU-based algorithms for N_{ode} larger than 4,096 and
8,192, respectively. For the current problem, at best the GPU offered speedup
factors of nearly four and two over the serial and four-core CPU implementations,
respectively. The non-smooth performance scaling resulted from the randomly
perturbed initial conditions.

Note that since each ODE system used randomly perturbed initial conditions,
adjacent threads in the GPU implementation handled potentially extremely different
initial condition values, resulting in thread divergence due to varying internal time
step sizes. Therefore, the results shown here represent a worst-case GPU algorithm performance, particularly compared to applications involving operator-split
reactive-flow codes where adjacent threads/ODE systems correspond to neighboring
spatial locations. In such situations, initial conditions would be more similar
and therefore follow similar instruction pathways. In either case, GPU-based
integration algorithms offer performance benefits over the baseline CPU versions.
See Niemeyer and Sung [21] for more discussion on this topic.

Furthermore, the current problem involved a relatively simple system of ODEs, limiting the calculations performed on the GPU between the memory transfers before and after each integration step. ODE systems with more complex derivative functions would saturate the GPU with operations, increasing performance. For example, the RKCK algorithm demonstrated by Niemeyer and Sung [21] performed up to 126 times faster on a GPU than on a serial CPU, integrating a chemical kinetics ODE system with nine species participating in 38 reaction steps—requiring significantly more floating-point calculations than the case studied here.

8.4.2 RKC Results

To demonstrate the performance of the GPU-based RKC algorithm, we used a chemical kinetics problem: the ODE system describing the constant-volume autoignition of ethanol (C_2H_5OH). We implemented the reaction mechanism of Marinov [19] to describe the oxidation of ethanol, with 57 species participating in 766 irreversible reaction steps. The governing ODE system contained 58 equations: one for temperature T and the rest for species mass fractions $\mathbf{Y}$:

$$\frac{d\mathbf{y}}{dt} = \left(\frac{dT}{dt}, \frac{dY_1}{dt}, \dots, \frac{dY_{N_{sp}}}{dt} \right)^{\mathsf{T}} , \tag{8.39}$$

$$\frac{dT}{dt} = -\frac{1}{\rho c_v} \sum_{i=1}^{N_{sp}} e_i \omega_i W_i , \tag{8.40}$$

$$\frac{dY_i}{dt} = \frac{W_i \omega_i}{\rho}, \quad i = 1, \dots, N_{sp} , \tag{8.41}$$

$$\omega_i = \sum_{j=1}^{N_{reac}} \left(v_{ij}'' - v_{ij}' \right) k_j \prod_{k=1}^{N_{sp}} C_k^{v_{kj}'} , \tag{8.42}$$

where ρ indicates the density, c_v the mass-averaged constant-volume specific heat, e_i the internal energy of the ith species, W_i the molecular weight of the ith species, v_{ij}'' and v_{ij}' the forward and reverse stoichiometric coefficients for the ith species in reaction j, C_k the molar concentration of the kth species, and N_{sp} and N_{reac} are the numbers of species and reactions, respectively. For a reaction j without pressure dependence, the rate coefficient k_j is given in Arrhenius form by

$$k_j = A_j T^{\beta_j} \exp\left(\frac{-E_j}{\mathscr{R}T} \right) , \tag{8.43}$$

where $\mathscr{R}$ is the universal gas constant, A_j the pre-exponential coefficient, β_j the temperature exponent, and E_j the activation energy. Note that reactions can be pressure-dependent (see, e.g., Law [17] for examples of various

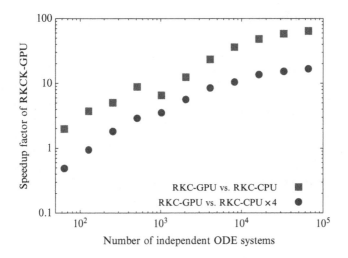

Fig. 8.2 Speedup factors offered by GPU-based explicit RKC integration algorithm over single- and four-core CPU-based versions for chemical kinetics ODE problem. Note that both axes are displayed in logarithmic scale

pressure-dependence formulations); these were also considered in the current implementation.

This problem is moderately stiff using a time step size of $\delta t = 1.0 \times 10^{-6}$ s for ten global time steps. In this case, we generated initial conditions for the set of ODE systems by sampling the solutions obtained from constant-pressure homogeneous ignition simulations, initiated at 1,600 K, 1 atm, and an equivalence ratio of one.[1] We assigned these initial conditions sequentially, such that adjacent threads in the GPU implementation contained data from consecutive time steps in the sample— and therefore such threads handled the integration of similar conditions, emulating adjacent spatial locations in an operator-split reactive-flow simulation.

Figure 8.2 shows the speedup factors offered by the GPU-based RKC algorithm over the baseline CPU version for both serial and four-core parallel operation, for numbers of ODE systems ranging from 64 to 16,384. In this case, the GPU-accelerated code ran faster than the serial CPU version for the entire range of ODE system sizes considered, while it offered better performance than the four-core parallel CPU version for 256 ODEs and higher. At best, the GPU-based RKC algorithm ran nearly 64 and 17 times faster than the serial and four-core CPU implementations, respectively. The discontinuity in speedup seen in Fig. 8.2 corresponded to the inclusion of initial conditions leading to greater stiffness.

---

[1]An equivalence ratio of one indicates the mixture of fuel and oxidizer set to an appropriate ratio for complete combustion.

8.5   Conclusions

In this chapter, we presented two explicit algorithms appropriate for integrating large numbers of independent ODE systems on GPUs. Specifically, we proposed the fifth-order adaptive Runge–Kutta–Cash–Karp (RKCK) method for nonstiff problems and the stabilized second-order adaptive Runge–Kutta–Chebyshev (RKC) method for problems with moderate levels of stiffness. Source code and implementation details were presented to ease the adoption of such methods, and performance comparison results were presented for each method. The examples shown here served to demonstrate the potential of GPU acceleration where many independent systems of ODEs need to be integrated; in the case of the RKC algorithm, we demonstrated more than an order of magnitude performance increase over an equivalent parallel CPU code running on four cores. The types of scientific and engineering problems dealing with large numbers of ODEs—in particular, reactive-flow models that rely on operator splitting—can benefit significantly from GPU acceleration; interested readers can directly implement the algorithms presented here to such ends, or use them as the beginnings for their own solution.

Acknowledgements This work was supported by the US Department of Defense through the National Defense Science and Engineering Graduate Fellowship program, the National Science Foundation Graduate Research Fellowship under grant number DGE-0951783, and the Combustion Energy Frontier Research Center—an Energy Frontier Research Center funded by the US Department of Energy, Office of Science, Office of Basic Energy Sciences under award number DE-SC0001198.

Appendix

Various methods may be used to calculate the spectral radius, including the Gershgorin circle theorem [7, 11] that provides an upper-bound estimate. Here, we provide a function based on a nonlinear power method [30].

```
1  __device__ double
2  rkcSpecRad (const double t, const double* y, const double g,
        const double* F, const double hMax, double* v, double* Fv) {
3    // maximum number of iterations
4    const int itmax = 50;
5
6    double small = 1.0 / hmax;
7
8    double nrm1 = 0.0;
9    double nrm2 = 0.0;
10   for (int i = 0; i < NEQN; ++i) {
11     nrm1 += (y[i] * y[i]);
12     nrm2 += (v[i] * v[i]);
13   }
14   nrm1 = sqrt (nrm1);
```

```
15    nrm2 = sqrt(nrm2);
16
17    double dynrm;
18    if ((nrm1 != 0.0) && (nrm2 != 0.0)) {
19      dynrm = nrm1 * sqrt(UROUND);
20      for (int i = 0; i < NEQN; ++i) {
21        v[i] = y[i] + v[i] * (dynrm / nrm2);
22      }
23    } else if (nrm1 != 0.0) {
24      dynrm = nrm1 * sqrt(UROUND);
25      for (int i = 0; i < NEQN; ++i) {
26        v[i] = y[i] * (1.0 + sqrt(UROUND));
27      }
28    } else if (nrm2 != 0.0) {
29      dynrm = UROUND;
30      for (int i = 0; i < NEQN; ++i) {
31        v[i] *= (dynrm / nrm2);
32      }
33    } else {
34      dynrm = UROUND;
35      for (int i = 0; i < NEQN; ++i) {
36        v[i] = UROUND;
37      }
38    }
39
40    // now iterate using nonlinear power method
41    double sigma = 0.0;
42    for (int iter = 1; iter <= itmax; ++iter) {
43
44      dydt (t, pr, v, Fv);
45
46      nrm1 = 0.0;
47      for (int i = 0; i < NEQN; ++i) {
48        nrm1 += ((Fv[i] - F[i]) * (Fv[i] - F[i]));
49      }
50      nrm1 = sqrt(nrm1);
51      nrm2 = sigma;
52      sigma = nrm1 / dynrm;
53
54      nrm2 = fabs(sigma - nrm2) / sigma;
55      if ((iter >= 2) && (fabs(sigma - nrm2) <= (fmax(sigma, small)
             * 0.01))) {
56        for (int i = 0; i < NEQN; ++i) {
57          v[i] -= y[i];
58        }
59        return (1.2 * sigma);
60      }
61
62      if (nrm1 != 0.0) {
63        for (int i = 0; i < NEQN; ++i) {
64          v[i] = y[i] + ((Fv[i] - F[i]) * (dynrm / nrm1));
65        }
66      } else {
```

```
67        int ind = (iter % NEQN);
68        v[ind] = y[ind] - (v[ind] - y[ind]);
69      }
70    }
71    return (1.2 * sigma);
72  }
```

References

1. Alexandrov, V., Sameh, A., Siddique, Y., Zlatev, Z.: Numerical integration of chemical ODE problems arising in air pollution models. Environ. Monit. Assess. **2**(4), 365–377 (1997). doi:10.1023/A:1019086016734
2. Barry, D., Miller, C., Culligan, P., Bajracharya, K.: Analysis of split operator methods for nonlinear and multispecies groundwater chemical transport models. Math. Comput. Simul. **43**(3–6), 331–341 (1997). doi:10.1016/S0378-4754(97)00017-7
3. Barry, D., Bajracharya, K., Crapper, M., Prommer, H., Cunningham, C.: Comparison of split-operator methods for solving coupled chemical non-equilibrium reaction/groundwater transport models. Math. Comput. Simul. **53**(1–2), 113–127 (2000). doi:10.1016/S0378-4754(00)00182-8
4. Cash, J.R., Karp, A.H.: A variable order Runge–Kutta method for initial value problems with rapidly varying right-hand sides. ACM Trans. Math. Softw. **16**(3), 201–222 (1990)
5. Day, M.S., Bell, J.B.: Numerical simulation of laminar reacting flows with complex chemistry. Combust. Theory Model. **4**(4), 535–556 (2000). doi:10.1088/1364-7830/4/4/309
6. Dematte, L., Prandi, D.: GPU computing for systems biology. Brief. Bioinform. **11**(3), 323–333 (2010). doi:10.1093/bib/bbq006
7. Geršgorin, S.: Über die abgrenzung der eigenwerte einer matrix. Bulletin de l'Académie des Sciences de l'URSS. Classe des sciences mathématiques et na (6), 749–754 (1931)
8. Hairer, E., Wanner, G.: Solving Ordinary Differential Equations II: Stiff and Differential-Algebraic Problems, 2nd edn. Springer Series in Computational Mathematics, vol. 14. Springer, Berlin/Heidelberg (1996)
9. Hairer, E., Wanner, G., Nørsett, S.P.: Solving Ordinary Differential Equations I: Nonstiff Problems, 2nd edn. Springer Series in Computational Mathematics, vol. 8. Springer, Berlin/Heidelberg (1993). doi:10.1007/978-3-540-78862-1
10. Helton, J., Davis, F.: Latin hypercube sampling and the propagation of uncertainty in analyses of complex systems. Reliab. Eng. Syst. Saf. **81**(1), 23–69 (2003). doi:10.1016/S0951-8320(03)00058-9
11. Horn, R.A., Johnson, C.R.: Matrix Analysis. Cambridge University Press, Cambridge (1990)
12. Jang, B., Schaa, D., Mistry, P., Kaeli, D.: Exploiting memory access patterns to improve memory performance in data-parallel architectures. IEEE Trans. Parallel Distrib. Syst. **22**(1), 105–118 (2011). doi:10.1109/TPDS.2010.107
13. Kim, J., Cho, S.Y.: Computation accuracy and efficiency of the time-splitting method in solving atmospheric transport/chemistry equations. Atmos. Environ. **31**(15), 2215–2224 (1997)
14. Kirk, D.B., Hwu, W.W.: Programming Massively Parallel Processors: A Hands-on Approach. Morgan Kaufmann, Burlington (2010)
15. Knio, O.M., Najm, H.N., Wyckoff, P.S.: A semi-implicit numerical scheme for reacting flow II. Stiff, operator-split formulation. J. Comput. Phys. **154**, 428–467 (1999). doi:10.1006/jcph.1999.6322

16. Kühn, C., Wierling, C., Kühn, A., Klipp, E., Panopoulou, G., Lehrach, H., Poustka, A.: Monte Carlo analysis of an ODE model of the sea urchin endomesoderm network. BMC Syst. Biol. **3**, 83 (2009). doi:10.1186/1752-0509-3-83
17. Law, C.K.: Combustion Physics. Cambridge University Press, New York (2006)
18. Marino, S., Hogue, I.B., Ray, C.J., Kirschner, D.E.: A methodology for performing global uncertainty and sensitivity analysis in systems biology. J. Theor. Biol. **254**(1), 178–19 (2008). doi:10.1016/j.jtbi.2008.04.011
19. Marinov, N.M.: A detailed chemical kinetic model for high temperature ethanol oxidation. Int. J. Chem. Kinet. **31**(3), 183–220 (1999)
20. Mazzia, F., Magherini, C.: Test Set for Initial Value Problem Solvers, Release 2.4. Department of Mathematics, University of Bari and INdAM, Research Unit of Bari (2008). Available at http://www.dm.uniba.it/~testset
21. Niemeyer, K.E., Sung, C.J.: Accelerating moderately stiff chemical kinetics in reactive-flow simulations using GPUs. J. Comput. Phys. **256**, 854–871 (2014). doi:10.1016/j.jcp.2013.09.025
22. Niemeyer, K.E., Sung, C.J., Fotache, C.G., Lee, J.C.: Turbulence-chemistry closure method using graphics processing units: a preliminary test. In: 7th Fall Technical Meeting of the Eastern States Section of the Combustion Institute, Storrs (2011)
23. Nimmagadda, V.K., Akoglu, A., Hariri, S., Moukabary, T.: Cardiac simulation on multi-GPU platform. J. Supercomput. **59**(3), 1360–1378 (2011). doi:10.1007/s11227-010-0540-x
24. OpenMP Architecture Review Board: OpenMP Application Program Interface Version 3.0. http://www.openmp.org/mp-documents/spec30.pdf (2008)
25. Oran, E.S., Boris, J.P.: Numerical Simulation of Reactive Flow, 2nd edn. Cambridge University Press, Cambridge (2001)
26. Press, W.H., Teukolsky, S.A., Vetterling, W.T., Flannery, B.P.: Numerical Recipes in Fortran 77: The Art of Scientific Computing, 2nd edn. Cambridge University Press, Cambridge (1992)
27. Ren, Z., Pope, S.B.: Second-order splitting schemes for a class of reactive systems. J. Comput. Phys. **227**(17), 8165–8176 (2008). doi:10.1016/j.jcp.2008.05.019
28. Schwer, D., Lu, P., Green, W.H., Semiao, V.: A consistent-splitting approach to computing stiff steady-state reacting flows with adaptive chemistry. Combust. Theory Model. **7**(2), 383–399 (2003). doi:10.1088/1364-7830/7/2/310
29. Shi, Y., Green, W.H., Wong, H., Oluwole, O.O.: Accelerating multi-dimensional combustion simulations using hybrid CPU-based implicit/GPU-based explicit ODE integration. Combust. Flame **159**(7), 2388–2397 (2012). doi:10.1016/j.combustflame.2012.02.016
30. Sommeijer, B.P., Shampine, L.F., Verwer, J.G.: RKC: an explicit solver for parabolic PDEs. J. Comput. Appl. Math. **88**(2), 315–326 (1997)
31. Sportisse, B.: An analysis of operator splitting techniques in the stiff case. J. Comput. Phys. **161**(1), 140–168 (2000)
32. Stone, C.P., Davis, R.L.: Techniques for solving stiff chemical kinetics on graphical processing units. J. Propulsion Power **29**(4), 764–773 (2013). doi:10.2514/1.B34874
33. Strang, G.: On the construction and comparison of difference schemes. SIAM J. Numer. Anal. **5**(3), 506–517 (1968)
34. Sundnes, J., Nielsen, B.F., Mardal, K., Cai, X., Lines, G., Tveito, A.: On the computational complexity of the bidomain and the monodomain models of electrophysiology. Ann. Biomed. Eng. **34**(7), 1088–1097 (2006). doi:10.1007/s10439-006-9082-z
35. van der Houwen, P.J.: The development of Runge–Kutta methods for partial differential equations. Appl. Numer. Math. **20**, 261–272 (1996)
36. van der Houwen, P.J., Sommeijer, B.P.: On the internal stability of explicit, m-stage Runge-Kutta methods for large m-values. Z. Angew. Math. Mech. **60**(10), 479–485 (1980)
37. Verwer, J.G.: Explicit Runge–Kutta methods for parabolic partial differential equations. Appl. Numer. Math. **22**, 359–379 (1996)

38. Verwer, J.G., Hundsdorfer, W., Sommeijer, B.P.: Convergence properties of the Runge–Kutta–Chebyshev method. Numer. Math. **57**, 57–178 (1990)
39. Verwer, J.G., Sommeijer, B.P., Hundsdorfer, W.: RKC time-stepping for advection–diffusion–reaction problems. J. Comput. Phys. **201**(1), 61–79 (2004). doi:10.1016/j.jcp.2004.05.002
40. Zhou, Y., Liepe, J., Sheng, X., Stumpf, M.P.H., Barnes, C.: GPU accelerated biochemical network simulation. Bioinformatics **27**(6), 874–876 (2011). doi:10.1093/bioinformatics/btr015

Chapter 9
Finite and Spectral Element Methods on Unstructured Grids for Flow and Wave Propagation Problems

Dominik Göddeke, Dimitri Komatitsch, and Matthias Möller

9.1 Introduction

Many relevant processes and phenomena from a wide range of scientific areas and application domains can be described by mathematical models comprising (a system of) partial differential equations (PDEs). A simple example is the Poisson equation

$$- \Delta u = f, \tag{9.1}$$

which is fulfilled by the scalar quantity u that represents the state of minimal energy subject to load f and appropriate boundary conditions. As an illustration, consider the deformation due to loading of an elastic membrane that is fixed on a frame.

A large class of model problems can be written as first-order systems of the form

$$\partial_t U + \nabla \cdot \mathbf{F}(U) = 0, \tag{9.2}$$

where $\mathbf{F} = [F^1, \ldots, F^n]$ represents an n-dimensional flux function that depends on the solution U but not on its derivatives. As an example, consider the flow of air around an aeroplane at high speeds, which can be modelled by the equations of gas dynamics.

D. Göddeke (✉)
Department of Mathematics, Institute of Applied Mathematics, TU Dortmund,
Dortmund, Germany
e-mail: dominik.goeddeke@math.tu-dortmund.de

D. Komatitsch
Laboratory of Mechanics and Acoustics, CNRS UPR 7051, Aix-Marseille University,
Centrale Marseille, France
e-mail: komatitsch@lma.cnrs-mrs.fr

M. Möller
Delft Institute of Applied Mathematics, Delft University of Technology, Delft, The Netherlands
e-mail: m.moller@tudelft.nl

V. Kindratenko (ed.), *Numerical Computations with GPUs*,
DOI 10.1007/978-3-319-06548-9_9, © Springer International Publishing Switzerland 2014

A third example of an important real-world phenomenon that is modelled by a time-dependent PDE is the propagation of waves, for instance the propagation of seismic waves in the Earth to calculate the effects of earthquakes, or the propagation of ultrasonic acoustic waves in ocean acoustics or in non-destructive testing. The appropriate mathematical model is the elastodynamics equation

$$\varrho \partial_t^2 \mathbf{u} - \nabla \cdot \mathbf{T}(\mathbf{u}, \nabla \mathbf{u}) = \mathbf{f}, \tag{9.3}$$

where the stress tensor $\mathbf{T}$ depends on the multi-dimensional displacement field $\mathbf{u}$ and/or its (transposed) gradient, which yields a second-order system.

The numerical treatment of such PDE problems typically involves two aspects, the *discretisation* that maps the continuous model to a formulation suitable for computers, and the *solver* that computes actual solution approximations. Both from an engineering and mathematical point of view, the finite element method (FEM) is often well suited for the discretisation: Finite elements offer a high degree of geometric flexibility since they can be naturally formulated on unstructured grids. Furthermore, they can deliver high (guaranteed) accuracy and robustness when enhanced with h/hp-adaptation and a-posteriori error estimation techniques, for which solid theoretical foundations exist in the FEM framework. In combination with powerful and robust iterative solvers for the resulting linear or non-linear systems of equations, finite elements form the underlying fabric of many modern simulation tools.

9.2   Finite Element Analysis in a Nutshell

It is far beyond the scope of this chapter to give a comprehensive introduction to the finite element method including all its variants and theoretical aspects. Thus, we only outline the basic concepts of the *continuous* Galerkin finite element method and refer the interested reader to, e.g., [8, 10] for an introduction to practical finite element analysis, and to the more theoretical textbooks [2, 4]. A good overview of *discontinuous* Galerkin methods, which are not considered in this chapter but feature amenable properties that can be helpful for designing highly efficient parallelised codes, can be found for instance in [14] and the references therein.

9.2.1   Variational Formulation

Let the scalar quantity u be governed by the generic PDE model problem

$$Lu = f$$

within the n-dimensional domain $\Omega \subset \mathbb{R}^n$, $d = 1, 2, 3$, where L represents a linear spatial differential operator. Moreover, u has to fulfil certain conditions at the boundary $\Gamma := \partial \Omega$, e.g., a combination of Dirichlet and Neumann conditions:

$$u = u_D \quad \text{on} \quad \Gamma_D \subseteq \Gamma,$$

$$\partial_n u = g \quad \text{on} \quad \Gamma_N = \Gamma \setminus \Gamma_D.$$

The first step in deriving the finite element method is to translate the problem at hand into its variational form, which amounts to integrating the weighted residual over the domain Ω and forcing the result to vanish. Thus, the solution u is sought in some suitable function space V, referred to as trial space, such that

$$\int_\Omega w(Lu - f)\, d\mathbf{x} = 0 \qquad \forall w \in W, \tag{9.4}$$

where W denotes the space of test functions w. Both spaces have to comply with the demands of the differential operator L and the Dirichlet boundary conditions.

In the second step, the infinite dimensional function spaces V and W are approximated by finite dimensional ones, denoted by V_h and W_h.

9.2.2 Galerkin Discretisation

To simplify the presentation, let us adopt the same set of basis functions $\{\varphi_i\}_{i=1}^N$ for the discrete test and trial spaces W_h and V_h, respectively. The approximate solution and its derivatives can then be represented as:

$$u(\mathbf{x}) \approx u_h(\mathbf{x}) = \sum_{j=1}^N \varphi_j(\mathbf{x}) u_j, \qquad Lu(\mathbf{x}) \approx Lu_h(\mathbf{x}) = \sum_{j=1}^N L\varphi_j(\mathbf{x}) u_j$$

Substituting them into the weak form (9.4) and replacing the weighting function w by all possible φ_i yields a linear system of equations for the vector of unknowns $u = [u_1, \ldots u_N]^T$:

$$\sum_{j=1}^N \left[\int_\Omega \varphi_i L\varphi_j \, d\mathbf{x} \right] u_j = \int_\Omega \varphi_i f \, d\mathbf{x}, \qquad i = 1, 2, \ldots, N$$

As an example, consider Poisson's equation (9.1), which corresponds to defining the differential operator according to $L[\cdot] := -\Delta[\cdot] = -\nabla \cdot \nabla[\cdot]$. Performing integration by parts results in seeking $u \in V := \{u \in \mathscr{H}^1(\Omega) : u = u_D \text{ on } \Gamma_D\}$ such that

$$\int_\Omega \nabla w \cdot \nabla u \, d\mathbf{x} = \int_{\Gamma_N} w g \, ds + \int_\Omega w f \, d\mathbf{x}$$

for all test functions $w \in W := \{w \in \mathcal{H}^1(\Omega) \,:\, w = 0 \text{ on } \Gamma_D\}$ that vanish on the Dirichlet boundary part. Here and below $\mathcal{H}^1(\Omega)$ denotes the space of square integrable functions with square integrable first weak derivatives. The discrete counterpart of the problem at hand can be expressed in compact matrix form as

$$Su = b \quad \text{with} \quad s_{ij} = \int_{\Omega} \nabla\varphi_i \cdot \nabla\varphi_j \, dx = s_{ji} \quad \text{and} \quad b_i = \int_{\Gamma_N} \varphi_i g \, ds + \int_{\Omega} \varphi_i f \, dx.$$
(9.5)

Note that $s_{ij} \neq 0$ if and only if φ_i and φ_j have overlapping supports, and that basis functions are typically constructed so that they fulfil a local support property. For practical applications, the system (9.5) may thus be very large but will remain sparse.

9.2.3 Element-Based Assembly

In all finite element methods, the domain Ω is covered by non-overlapping simple geometric objects, e.g., triangles and/or quadrilaterals in two space dimensions, and tetrahedra, hexahedra and/or prisms in three space dimensions. This (fully unstructured) partition $\mathcal{T}_h = \{T_1, \ldots, T_M\}$ is referred to as a mesh, or triangulation, of Ω. It is common practice to associate the degrees of freedom with entities of this mesh such as element vertices or midpoints of edges/faces. Depending on the choice of basis functions, the unknowns u_j may represent nodal solution values, i.e. $u_j = u_h(\mathbf{x}_j)$, integral mean values of the solution, or they can be related to solution derivatives.

The global integral terms are then assembled by summing over individual element contributions that may either be computed exactly or approximated by some cubature rule (quadrature formula), e.g., Gaussian quadrature. The weighting coefficients $\hat{\omega}_k$ and cubature points $\hat{\mathbf{x}}_k$ are typically tabulated for some reference element $\hat{T}$ with a regular shape. That is,

$$\int_{\hat{T}} I(\hat{\mathbf{x}}) \, d\hat{\mathbf{x}} \approx \sum_{k=1}^{N_{\text{cub}}} \hat{\omega}_k I(\hat{\mathbf{x}}_k)$$
(9.6)

for a generic integrand $I(\cdot)$. As a result the numerical evaluation procedure reads

$$\int_{\Phi_T(\hat{T})} \varphi_i(\mathbf{x}) L\varphi_j(\mathbf{x}) \, dx = \int_{\hat{T}} \varphi_i(\Phi_T(\hat{\mathbf{x}})) L\varphi_j(\Phi_T(\hat{\mathbf{x}})) \, |\det J(\hat{\mathbf{x}})| \, d\hat{\mathbf{x}}.$$
(9.7)

That is, it involves a change of variables, where $\Phi_T : \hat{T} \mapsto T$ is the mapping from the reference mesh element to the physical one so that coordinates are related by $\mathbf{x} = \Phi_T(\hat{\mathbf{x}})$, and $J = D\Phi_T$ denotes the Jacobian matrix of the transformation.

Let the basis $\{\varphi_i\}_{i=1}^N$ be given by the definition of shape functions on individual elements. For instance, using linear polynomials that equal unity at one vertex of a triangle/tetrahedron and vanish at all other vertices leads to Lagrange finite elements of degree $p = 1$. The choice of shape/basis functions determines the order of the final approximation. In so-called p-adaptive schemes the shape functions, and hence, the approximation order may therefore differ from one element to the other.

In parametric finite elements, the reference element is also used to define the shape functions in terms of referential coordinates, i.e., $\hat{\varphi}_i(\hat{\mathbf{x}})$. Substituting the relation $\varphi_i(\Phi_T(\hat{\mathbf{x}})) = \hat{\varphi}_i(\hat{\mathbf{x}})$ into expression (9.7) yields the final integration formula (9.6) to be implemented into a finite element code. This approach makes it possible to, say, adopt higher-order basis functions but still use a mapping of low order, called sub-parametric approach. On the other hand, using the same order for both components (iso-parametric) or even increasing the order of the mapping (super-parametric) may be beneficial if curved boundaries need to be approximated with high accuracy.

In summary, the assembly procedure for the global stiffness matrix S of the Poisson problem reduces to evaluating all non-vanishing matrix coefficients

$$s_{ij} = \sum_{T \in \mathcal{T}_h} \int_{\hat{T}} \underbrace{\left(J^{-T}(\hat{\mathbf{x}})\hat{\nabla}\hat{\varphi}_i(\hat{\mathbf{x}})\right) \cdot \left(J^{-T}(\hat{\mathbf{x}})\hat{\nabla}\hat{\varphi}_j(\hat{\mathbf{x}})\right) |\det J(\hat{\mathbf{x}})|}_{=:I(\hat{\mathbf{x}},i,j)} \, d\hat{\mathbf{x}} \qquad (9.8)$$

after applying the chain rule and the theorem of local inverses.

The assembly procedure of the volumetric part of the right-hand side vector

$$b_i = \sum_{T \in \mathcal{T}_h} \int_{\hat{T}} \underbrace{\hat{\varphi}_i(\hat{\mathbf{x}}) f(\Phi_T(\hat{\mathbf{x}})) |\det J(\hat{\mathbf{x}})|}_{=:I(\hat{\mathbf{x}},i)} \, d\hat{\mathbf{x}} \qquad (9.9)$$

is sketched in Fig. 9.1. Note that numerical cubature rules of higher order are typically adopted to integrate the function f with sufficient accuracy.

Fig. 9.1 Assembly of (9.9): the local element integrals for all five elements contributing to node i are computed independently and assembled into the global vector

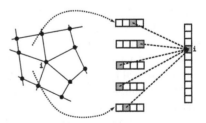

The same approach can be applied to the entries of the bilinear form if, say, the differential operator L is replaced by a non-linear one as it is the case for the elastodynamics equation (9.3). The above strategy is not restricted to linear and bilinear forms, it can be naturally extended to multi-linear forms such as the non-linear convection term $\mathbf{u} \cdot \nabla \mathbf{u}$ that plays a central role in the Navier–Stokes equations. The tensor contraction approach [13] is an alternative concept for the assembly of arbitrary multi-linear forms not addressed here due to space constraints. For the extension of the tensor contraction approach to GPUs, the interested reader is referred to [15].

9.2.4 Group Finite Element Formulation

An alternative approach that is commonly used in finite elements for conservation laws such as the first-order system (9.2) is the *group finite element formulation* developed by Fletcher [9]. Instead of evaluating the non-linear flux function $\mathbf{F}(U) \approx \mathbf{F}(U_h)$ based on the interpolated solution $U_h(\mathbf{x}, t) = \sum_{j=1}^{N} \varphi_j(\mathbf{x}) U_j(t)$, the same basis is adopted to interpolate the fluxes

$$\mathbf{F}(U) \approx \mathbf{F}_h(\mathbf{x}, t) = \sum_{j=1}^{N} \varphi_j(\mathbf{x}) \mathbf{F}_j(t), \qquad \mathbf{F}_j(t) = \mathbf{F}(U_{j(t)}), \qquad U_j(t) = U(\mathbf{x}_{j,t}) .$$

The resulting semi-discretised variational formulation of system (9.2) reads

$$\sum_{j=1}^{N} \left[m_{ij} \frac{dU_{j(t)}}{dt} + \mathbf{c}_{ij} \cdot \mathbf{F}_j(t) \right] = 0, \qquad (9.10)$$

where $M = \{m_{ij}\}$ is the consistent mass matrix with $N_\mathrm{D} \times N_\mathrm{D}$ blocks defined by $m_{ij} = m_{ij} \mathbb{I}$, where $\mathbb{I}$ stands for the identity tensor. The coefficients of the mass matrix $M = \{m_{ij}\}$ and those of the discrete gradient operator $\mathbf{C} = \{\mathbf{c}_{ij}\}$ are given by

$$m_{ij} = \int_{\Omega} \varphi_i(\mathbf{x}) \varphi_j(\mathbf{x}) \, d\mathbf{x}, \qquad \mathbf{c}_{ij} = \int_{\Omega} \varphi_i(\mathbf{x}) \nabla \varphi_j(\mathbf{x}) \, d\mathbf{x}. \qquad (9.11)$$

These coefficients remain constant unless the mesh is changed, and therefore, they can be computed a priori and stored (maybe adopting an optimized renumbering strategy) for later use. It goes without saying that the element-based assembly is readily applicable to assemble M and $\mathbf{C}$ in the preprocessing step.

As an example, consider the linear advection equation $\partial_t u + \nabla \cdot \mathbf{f}(u) = 0$ with flux function $\mathbf{f}(u) = \mathbf{v}u$, whereby the externally given velocity field $\mathbf{v} = \mathbf{v}(\mathbf{x}, t)$ may vary both in time and space. Adopting the group finite element formulation, the convective operator $K = \{k_{ij}\}$ can be updated very efficiently letting $k_{ij} = \mathbf{c}_{ij} \cdot \mathbf{v}_j(t)$ without the need to perform costly numerical integration in every time step.

9.2.5 Edge-Based Assembly

For finite elements featuring the partition-of-unity property $\sum_{j=1}^{N} \varphi_j(\mathbf{x}) = 1$, which holds for instance for Lagrangian or B-spline basis functions, the matrix of auxiliary coefficients has zero row sums, i.e., $\sum_{j=1}^{N} \mathbf{c}_{ij} = 0 \ \forall i$. This property makes it possible to cast (9.10) into the following edge-based form [20]

$$\sum_{j=1}^{N} m_{ij} \frac{du_j(t)}{dt} + \sum_{j \in \mathscr{S}_i} \mathbf{c}_{ij} \cdot [f_j(t) - f_i(t)] = 0, \tag{9.12}$$

where the index set $\mathscr{S}_i = \{1 \leq i \neq j \leq N : \text{supp } \varphi_i \cap \text{supp } \varphi_j \neq \emptyset\}$ extends over all neighbouring degrees of freedom j that share a common edge with i. In [19], an alternative flux decomposition has been developed that amounts to performing integration by parts in the spatial discretisation of the divergence terms, yielding

$$\mathbf{c}_{ij} = -\mathbf{c}_{ji} + \mathbf{s}_{ij}, \qquad \mathbf{s}_{ij} = \int_{\Gamma} \varphi_i(\mathbf{x}) \varphi_j(\mathbf{x}) \mathbf{n} \, ds,$$

whereby the symmetric boundary term $\mathbf{s}_{ij}$ vanishes in the interior. Finally, (9.12) can be recast into the equivalent edge-based formulation [19]

$$\sum_{j=1}^{N} \left[M_{ij} \frac{du_j(t)}{dt} + \mathbf{s}_{ij} \cdot F_j(t) \right] + \sum_{j \in \mathscr{S}_i} G_{ij}(t) = 0 \tag{9.13}$$

with skew-symmetric Galerkin fluxes $G_{ij}(t) = \mathbf{c}_{ij} \cdot F_i(t) - \mathbf{c}_{ji} \cdot F_j(t) = -G_{ji}(t)$.

9.3 Implementation and Parallelisation Strategies

At a sufficiently high level of abstraction, element-based FEM as depicted, e.g., in (9.8), reduces to a double-nested loop over matrix positions (i, j) and elements $T \in \mathscr{T}_h$, respectively. Within each element, another triple-nested loop over all cubature points for all pairs of test and trial functions is executed to compute the integral(s). The edge-based approach shares many features with the element-based one, so we can focus on elements. Depending on how one arranges the loops, one obtains different assembly algorithms. We first discuss advantages and disadvantages of choosing the ij- or T-loop as the outermost one, and then consider the order of the per-element loops. We emphasise two aspects that are clear from Sect. 9.2: First, only the resulting matrix entries s_{ij} are written to memory, all other operations either read data from memory or perform arithmetics. This simple observation is crucial for extracting parallelism in finite element methods. Second, due to the unstructured nature of the mesh $\mathscr{T}_h$, memory indirections cannot be avoided.

9.3.1 Choice of the Outermost Loop

Nonzeros-First. The *nonzeros-first* approach can be seen as the direct implementation of expression (9.8). The outermost loop iterates over all non-vanishing matrix entries s_{ij}, and for each entry, all elemental contributions are *gathered* by computing each integral in the sum for which φ_i and φ_j have overlapping support. In the example depicted in Fig. 9.1 for a right hand side assembly, we need to compute five per-element contributions for node i, corresponding to the vector entry i. Based on the observation above, the method is intrinsically parallel and free of synchronisation requirements because all computations can be performed independently for all non-vanishing matrix entries. This constitutes a major conceptual advantage of this algorithm, but immediately implies its potential disadvantage: All per-element integrations that contribute to the ij sum are performed redundantly, e.g., a given element is computed first for node i and again for node j if both nodes are in its support. This method is called "assembly-by-nonzeros" in [3].

Elements-First. To avoid this redundancy, one can alternatively iterate over all elements in the outermost loop, and *scatter* the contributions stemming from each element to the matrix entries representing the support of the element. We mostly focus on this technique throughout this chapter. Subsets of the generic approach presented here have been considered before, for instance "assembly-by-elements" in [3], our own work (to our knowledge the first high-order FEM-GPU implementation [16]), and the "add-to" algorithm [23]. The generic algorithm proceeds as follows:

Loop over all elements $T \in \mathscr{T}_h$
　　Determine N_{test}, N_{trial} and N_{cub}, $\{\hat{\omega}\}_{\text{cub}}$, $\{\hat{\mathbf{x}}\}_{\text{cub}}$
　　Loop over all $i_{\text{test}} = 1, \ldots, N_{\text{test}}$ test functions
　　　　Loop over all $j_{\text{trial}} = 1, \ldots, N_{\text{trial}}$ trial functions
　　　　　　Loop over all $k_{\text{cub}} = 1, \ldots, N_{\text{cub}}$ cubature points
　　　　　　　　Compute the integrand $I(\hat{\mathbf{x}}_{k_{\text{cub}}}, i_{\text{test}}, j_{\text{trial}})$
　　　　　　　　Scatter the result: $s_{G(i_{\text{test}}, j_{\text{trial}})} \mathrel{+}= \hat{\omega}_{k_{\text{cub}}} I(\hat{\mathbf{x}}_{k_{\text{cub}}}, i_{\text{test}}, j_{\text{trial}})$

Here, $I(\hat{\mathbf{x}}_{k_{\text{cub}}}, i_{\text{test}}, j_{\text{trial}})$ denotes the evaluation of the integrand in (9.8) for the given cubature point and pair of test/trial functions, and $\hat{\omega}_{k_{\text{cub}}}$ is the cubature weight that depends on the cubature formula. G is the mapping from local degrees of freedom i_{test} and j_{trial} to their global counterparts i and j, depicted in Fig. 9.1 for the vector-assembly case. This way of casting the assembly into a loop ordering removes all redundant computations. However, it comes at a cost because the inherent parallelism of the "nonzeros-first" approach is partially lost. To see this, consider again Fig. 9.1, and assume that all five elements are computed simultaneously via some parallelisation of the outermost loop in the algorithm sketched above. Due to the first observation on input and output data on the previous page, this is not an issue for the bulk of computations in the assembly process: As only local computations and read operations for global data are performed per element, the local portion of the computation is still trivially parallel. However, once all five parallel threads reach the accumulation operation (the += statement), a *race condition* occurs because five threads update the i-th memory location in the target array (or matrix) simultaneously and thus there is no guarantee about the final sum: In many cases it will be partial and thus incorrect. In practice, the result can even vary over several executions of the program on the same machine.

To ensure correctness in the accumulation step, the increments need to be made *mutually exclusive*, i.e., one needs to ensure that during the "read-modify-write" sequence performed by one thread, no other threads interfere. There are essentially two different solutions to this problem, leading to different assembly algorithms.

Synchronisation. Most parallel programming environments for multicores and GPUs provide built-in mechanisms to synchronise on certain sequences of operations and/or on memory locations. To use them, code statements that may cause data races must be labelled with specific keywords, or dedicated function calls can be used. The general idea is that the hardware and/or the runtime can serialise the sequence of operations automatically, because they have been made aware of the condition. For instance, both NVIDIA CUDA and OpenMP provide so-called *atomic memory updates* for, e.g., increment operations. Using them ensures that as many operations as possible remain parallel, because only the actual increment operation becomes protected through an automatic serialisation. In longer sequences of operations, exclusive access to resources (e.g., array entries) must be ensured by other synchronisation techniques. In OpenMP for instance, so-called "critical sections" can be used, while in CUDA, cheap barrier synchronisation between blocks of threads is available.

Decoupling. To resolve the conflicting writes without resorting to synchronisation, the computation can also be rearranged so that the race condition can never occur. The basic idea is to take advantage of the fact that the support for each degree of freedom constitutes, in typical large meshes, just a few neighbouring elements. Any two elements may safely increment "their" memory locations if the supports do not overlap, i.e., if the memory locations they write to are independent. In the decoupling technique, entities (e.g., elements or edges) that can lead to problems are a priori partitioned accordingly into disjoint sets. The parallel loop over all entities is then replaced with a sequential loop over all such sets, while all entities within one set can still be treated independently in parallel. Such independent sets are typically computed through some kind of *colouring* algorithm [7].

Matrix-Free Methods. In this chapter, we focus on methods that yield actual matrices and vectors, in some standard format such as CSR. An alternative approach are "matrix-free" methods. One example that has been successfully demonstrated on GPUs in [23] is the "local matrix approach": All local element matrices are computed independently (e.g., as in the elements-first algorithms we describe below) and stored for later use, but the global system is never actually assembled. The complete "proto-assembly" is thus free of both synchronisation and redundancy. The idea is then to modify all operations that would normally make use of the matrix (or vectors), e.g., sparse matrix vector multiplies. This method is most beneficial if the assembly rather than the solver dominates the complete simulation.

9.3.2   Per-Element Loops

Both the nonzeros-first and the elements-first approach assume for correctness that the inner triple-nested loop is executed sequentially. For large meshes with a high number of elements, this is a valid assumption on CPU-type architectures. On GPUs however, it can be advantageous to expose parallelism in these loops as well, in particular in higher-order finite element methods. More parallelism can often increase overall throughput substantially owing to less granularity effects and less resource contention. The corresponding code transformations typically lead to data structures that are also beneficial on CPU-type architectures, e.g., through better exploitation of the vectorisation capabilities (SIMD units). We return to this issue in Sect. 9.4, because it is highly dependent on the actual finite element method.

9.3.3   An Improved Blocked Version

The loop structure given in the previous section reveals that large parts of the algorithm depend on the actual element type and discrete operator(s). Let us emphasise a few important examples:

- The determinant of the Jacobian is computed differently in the transformation to different reference elements (e.g., Cartesian vs. barycentric coordinates), or is even the same for all elements.
- Cubature formulae vary, per se and also depending on the element type.
- Coefficients of the various derivatives of the operator(s) must be evaluated at each cubature point.

The naive solution to this problem is to include large amounts of nested conditional statements and callback functions. In terms of efficiency, it is generally a bad idea to do so in the innermost loops, due to the comparatively high function call overhead, and the resulting branch divergence that prevents SIMD/SIMT execution. C++ template metaprogramming or creative use of the preprocessor can partially alleviate the issue at compile time, although this is no longer possible when runtime decisions are needed, e.g., when mixed-element methods and/or p-adaptivity are employed.

As a remedy, the elements can be reorganised into "distributions"

$$\mathscr{T}_h = D_1 \cup D_2 \cup \ldots \cup D_{N_D}$$

of elements featuring identical or similar properties (e.g., polynomial degree, shape). This additional level of partitioning allows one to "hose" all conditionals out of the innermost loops, and to drastically reduce the amount of callback functions. Note that the colouring approach can be incorporated into this scheme in a natural way via an additional outermost sequential loop, and if element types are not mixed, colour groups coincide with distributions. Finally, the distribution loop may be executed in parallel or sequentially, depending on the (relative) size of the various distributions $D_l \subset \mathscr{T}_h$.

The next pseudocode snippet illustrates the loop ordering that stems from this distribution-based approach, including the mandatory synchronisation points. We introduce an additional blocking layer that can be used to adapt the computation to the hardware at hand. We refer to this algorithmic template as "sets-first".

For all distributions $D_l \subset \mathscr{T}_h$ sequentially or in parallel
    Preallocate work memory for a block of B elements of type D_l
    —Barrier synchronisation if outer loop is parallel—
    Compute static, common data for element type D_l, store in work memory
    —Barrier synchronisation if outer loop is parallel—
    For all sets $S = \{T_{e_1}, \ldots, T_{e_B}\} \subset D_l$ of size B in parallel
        For all elements $T \in S$ in parallel
            Loop over the test and trial functions, compute the integral(s) using the precomputed data in work memory, and scatter the result

9.3.4 Implementation on Multicore CPUs

On CPU-type architectures, it is not necessary to exploit all algorithmically available degrees of parallelism. Instead, the entire body of the distribution loop can be executed in parallel, equidistributed among all available threads. Load balancing is needed if the cardinality of each distribution is not large enough to keep all available threads busy. The additional blocking layer should be chosen with respect to cache sizes to increase locality and improve efficiency. It is sometimes beneficial to split the loop over all elements: The local element matrices in each set can first be computed independently into work memory, followed by a second nested loop that performs the actual scattering, protected by a "critical section".

9.3.5   Implementation on GPUs

Throughout this chapter, we use CUDA terminology, but emphasise that the implementation guidelines are equally valid in other programming environments. One important general recommendation is to implement one kernel for each type of distribution, to facilitate clean and reusable code. The individual kernels can additionally be equipped with C++ template metaprogramming and/or preprocessor statements to reduce boilerplate overhead by, e.g., treating all Lagrangian elements on triangles by one meta-kernel. The loop over all distributions and/or colour groups then naturally translates to separate kernel launches.

Within each kernel, the main concern towards an efficient implementation is to translate algorithmic algorithmic to GPU concepts. Examples include the mapping of loop nesting levels to CUDA entities (blocks, warps, single threads), and the mapping to various memory spaces (global, shared, constant, registers). The following characteristics of the CUDA programming model must be taken into consideration: (1) Global synchronisation is only possible at the kernel launch granularity. (2) Threads in a block may synchronise inexpensively, and have access to small, but fast shared memory. (3) Global memory is orders of magnitude slower than registers or shared memory. (4) Instruction divergence within a warp should be avoided. (5) The latter is particularly true for memory instructions, i.e., addresses touched by a single load or store instruction must meet certain alignment criteria so that the hardware can coalesce memory accesses by a single warp into ideally a single memory transaction. (6) Atomic memory updates are not equally efficient on all hardware generations and for all data types.

9.4   Examples and Applications

In this section, we provide guidelines for devising generally applicable yet competitive (with respect to performance) mappings of algorithms to the hardware, under the constraints outlined in Sect. 9.3.5. Our description is based on representative examples: Low-order Lagrangian discretisations are discussed quite generically, high-order (spectral) elements are presented in the scope of a linear wave propagation application, and the edge-based approach is presented for a gas dynamics application. We do not specifically cover the case of Discontinuous Galerkin methods, although its high-order forms are close to the high-order spectral-element method, and refer to [14] instead.

9.4.1   Low-Order Lagrangian-Type Elements

Low-order methods are widely used in practice. In this section, no explicit assumptions are made on the problem at hand (2D vs. 3D, time-dependent, (non)linear, scalar vs. multiple fields), but we do assume a single distribution D_l in the loop structure on p. 193, i.e., elements of the exact same type. In the following, we focus on summarising the main ideas to explore the optimisation space. We explicitly do not aim at describing implementations that perform optimally for a specific finite element method on a specific hardware generation for a specific problem only. Further information can be found in [3, 11, 23, 25].

A naive implementation of the "sets-first" approach developed in Sect. 9.3.3 is straightforward: We simply use the data structures for the (typically, existing) CPU implementation, associate one fixed-size set of elements with a CUDA thread block, and let each thread compute the triple per-element loop. For the final accumulation into the global matrix, either colouring or atomic memory updates can be used. This implementation generally performs poorly because it does not make good use of the available bandwidth, which, due to the low arithmetic intensity of low-order methods, is the main bottleneck: Memory accesses are unstructured, the effective bandwidth is low because coalescing into a minimal amount of memory transactions per warp may be poor, and the comparatively small cache does not enable automatic reuse of, e.g., nodal data that are shared due to common support.

Improved Data Structures. The first optimisation we highlight aims at improving memory access patterns only. We arrange all input data into a column-major 2D array, indexed by elements. In each row of this structure, we store nodal data associated with one element: The first few entries contain static data such as per-element nodal coordinates, the next few entries are associated with target indices in the matrix/vector data structure to be assembled into (plus eventually some padding), and the final optional entries are associated with dynamic data. It depends on the problem at hand whether dynamic fields evaluations (transport directions depending on coordinates in convection-diffusion-equations, the $\mathbf{u} \cdot \nabla \mathbf{u}$ term in the non-linear Navier–Stokes equations, cf. Sect. 9.1) are best copied here, or instead read from their original locations. This data structure can be built from information readily available in the finite element framework.

For the actual computation, we can now associate each thread with one element. The threads execute "their" triple-nested loop to compute an element matrix each, by iterating over this data structure. All data required by consecutive threads are stored contiguously in memory by construction, and the column-major layout ensures that all memory accesses are fully coalesced, i.e., no bandwidth is wasted. This approach implies certain redundancies, as actual data instead of just indices are stored several times.

The final step is to scatter the element matrices to the target matrix/vector in global memory. There are several ways to implement this: The easiest one is to use atomic memory updates for all data, but such an approach may be slow, especially on older hardware generations. Alternatively, a two-pass strategy can be used. All computed element matrices are first written to global memory in a contiguous way, and a second kernel is invoked that performs the actual assembly. In this kernel, each thread is responsible for one target entry in the matrix, gathers all required data from the global array of element matrices, and performs a sequential reduction to compute the final entry. In this case, there is no need to store target indices in the 2D data structure. Instead, an additional "reduction list" is precomputed, which for each nonzero entry contains the indices in the array of element matrices that need to be gathered. To be more precise, this list typically starts with the target index, and the number of subsequent source indices depends on the connectivity of the mesh. Padding up to a fixed maximum size (amount of elements influencing one node) with "negative indices" is a standard way to ensure coalesced memory accesses into this list and SIMT computation during the reduction. Figure 9.2 (left) illustrates the data flow of such a reduction.

This approach is appealing due to its simplicity and versatility: In essence, all GPU-related challenges (regarding input data) are resolved by introducing redundancy in the underlying data structure, which in turn can be easily precomputed in standard finite element programs. The additional storage is well-invested, in particular for schemes where the assembly is invoked several times. Also, indices require the same storage as actual values in single precision. However, the approach does not exploit the fact that several neighbouring elements share nodal data due to the local support property.

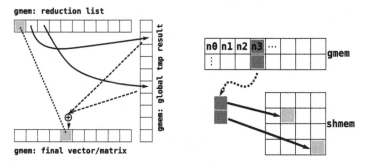

Fig. 9.2 *Left*: Data flow through reduction lists: While index lookups (including the highlighted target index) are perfectly coalesced, the gathering step is not. *Right*: Shared memory implementation, data flow of the lookup of the last node of the first element (2D bilinear quadrilaterals) treated by each warp (only two threads shown). The lookup into the map is perfectly coalesced, only accesses to shared memory are irregular

Improved Data Sharing. The following description of a (more involved) approach that uses shared memory to facilitate data reuse builds upon a proposal by Cecka et al. [3]. The basic aim is to preserve generality while increasing performance.

In a preprocessing phase, we partition the elements into sets of approximately equal size such that (1) the total number of element neighbours is minimised across partitions and (2) all required data (defined below) for one partition fit into shared memory. A tunable parameter is the "size" of shared memory: Data reuse and the number of resident warps to improve latency hiding need to be balanced. A small safety factor should be added to the partition size because CUDA uses a small amount of shared memory for itself, e.g., for kernel parameters. This constrained optimisation problem is commonly encountered in distributed memory parallelisation techniques for unstructured meshes, and standard graph partitioning software such as METIS [12] can be used to compute feasible approximations. Nodal data associated with each partition is then stored contiguously in memory.

We associate one thread block with one partition, and choose a number of threads that evenly divides the number of elements to be computed (associated with the partition), rounded up to the next multiple of the warp size. The first phase of the kernel uses all threads to load all nodal data of the entire partition into shared memory. Since the nodal data are contiguous in global memory, it is easy to find a mapping of threads to memory locations that ensures a fully coalesced access pattern. We now need to find a mapping of global node numbers to their indices in shared memory. To this end, we precompute an auxiliary 2D array per block, quite similar to the data structure described above. In it, we store the local indices for each element in a column-major fashion. This data structure can be stored in global or in shared memory; the former is more advisable because (1) data accesses to it are always fully coalesced and (2) the more actual nodal data we can store in shared memory, the higher the benefits from data sharing.

The second phase of the kernel then iterates over this data structure: Each thread looks up an index, accesses the data in shared memory, and as soon as all nodal data for the computation of one element matrix are available, it is computed (e.g., after every fourth memory access for the Laplace operator and bilinear elements on quadrilaterals). All operations per warp are executed in lock-step, and irregular memory accesses are limited to shared memory (where they constitute less of a problem), as shown in Fig. 9.2 (right).

For the actual assembly into the global matrix/vector, we can again employ the two strategies outlined above, or a third one that avoids the overhead of a two-pass solution: We again exploit the local support property. For all "inner" elements of the partition, we know that no other element from any other partition will influence a nonzero entry. Therefore, we can also perform the reduction directly in shared memory in the same kernel without resorting to a two-pass strategy, and for all other elements, we can use atomic memory updates. It depends on the hardware and on the finite element method which implementation performs best.

We conclude this section by referring to Cecka et al. [3, Sect. 4.3], who propose an approach that uses the nonzeros-first strategy within each set of elements. This technique is quite involved, and it tends to require quite problem-specific implementation adjustments to achieve performance.

9.4.2 High-Order Spectral Element Discretisations for Wave Propagation

Let us now discuss the case of high-order methods by focusing on the spectral element method, which is a variant of the FEM that is well-suited for instance for wave propagation modelling in heterogeneous media [29], e.g., seismic waves in the Earth [24], ultrasonic acoustic waves in ocean acoustics [5] or non-destructive testing [30]. Let us consider a linear anisotropic elastic rheology for a solid model, in which case the differential form of the acoustic wave equation can be written as

$$
\begin{aligned}
\varrho\ddot{\mathbf{u}} &= \nabla \cdot \boldsymbol{\sigma} + \mathbf{f}, \\
\boldsymbol{\sigma} &= \mathbf{C} : \boldsymbol{\varepsilon}, \\
\boldsymbol{\varepsilon} &= \tfrac{1}{2}[\nabla\mathbf{u} + (\nabla\mathbf{u})^{T}],
\end{aligned}
\tag{9.14}
$$

where $\mathbf{u}$ is the displacement vector, $\boldsymbol{\sigma}$ the symmetric, second-order stress tensor, $\boldsymbol{\varepsilon}$ the symmetric, second-order strain tensor, $\mathbf{C}$ the fourth-order stiffness tensor, ϱ the density, and $\mathbf{f}$ an external force representing the acoustic or seismic source. Denoting the physical domain of the model and its boundary by Ω and Γ respectively, we can write the weak form of this equation by dotting it with an arbitrary test function $\mathbf{w}$ and integrating by parts over the whole domain:

$$
\int_{\Omega}\varrho\,\mathbf{w}\cdot\ddot{\mathbf{u}}\,d\Omega + \int_{\Omega}\nabla\mathbf{w} : \mathbf{C} : \nabla\mathbf{u}\,d\Omega = \int_{\Omega}\mathbf{w}\cdot\mathbf{f}\,d\Omega + \int_{\Gamma}(\boldsymbol{\sigma}\cdot\hat{\mathbf{n}})\cdot\mathbf{w}\,d\Gamma
\tag{9.15}
$$

The contour integral of the last term vanishes because of the free surface, i.e., traction-free boundary condition: the traction vector $\boldsymbol{\tau} = \boldsymbol{\sigma}\cdot\hat{\mathbf{n}}$ is zero at the surface.

The physical domain is subdivided into hexahedral mesh cells within which variables are approximated by high order interpolants. The SEM resorts to Lagrange polynomials of degree $n = 4$ to 8 to interpolate functions such as the unknown displacement field [6, 27]. The anchor points are most of the time chosen as the $n + 1$ Gauss–Lobatto–Legendre (GLL) points because the mass matrix then becomes perfectly diagonal, which in turn leads to the use of fully explicit time schemes [29], e.g., a second-order Newmark or a fourth-order Runge–Kutta scheme. Consequently, the method is by design very efficient on large parallel computers [24, 26]. Numerical integration over the elements is performed using a GLL integration rule, and thus each spectral element contains $(n + 1)^3$ such GLL integration points. The final matrix system to solve is

$$
\mathrm{M}\ddot{\mathrm{U}} + \mathrm{KU} = \mathrm{F},
\tag{9.16}
$$

where U is the unknown displacement vector that needs to be computed, M is the diagonal mass matrix, K is the stiffness matrix, and F is the source term, whose detailed expressions can be found for instance in [24, 29].

In almost all wave propagation applications a large number of time steps is performed, and thus in the SEM algorithm the total cost is dominated by the contents of the serial time loop. In addition, since the mesh is static and the algorithm is fully explicit, all the time steps have identical cost, which facilitates optimisation. The computations performed at each time step consist of two very different kinds of operations: global vector updates, whose goal is to march the global vector of unknowns in time, and local matrix–matrix products inside each spectral element followed by an "assembly" phase, whose goal is to perform the local elastic force calculations and sum them into the global elastic force vector to be able to compute the acceleration vector at the next time step.

Operations of the first kind are of the typical type $\mathbf{u}^{\text{new}} = \mathbf{u}^{\text{old}} + \Delta t \, \dot{\mathbf{u}} + \frac{\Delta t}{2} \ddot{\mathbf{u}}$, where $\mathbf{u}$, $\dot{\mathbf{u}}$ and $\ddot{\mathbf{u}}$ are the global displacement, velocity and acceleration vectors, respectively, and Δt is the time step. They are all trivially parallel, and thus a simple CUDA implementation with a thread per degree of freedom to update is sufficient because it contains no dependencies. The second step is by far the more complex one and consists mostly of local matrix products performed inside each element to compute its contribution to the stiffness matrix term KU of (9.16) according to (9.15). The global displacement vector is first copied into each spectral element using a local-to-global mesh numbering mapping that has been precomputed and stored before the beginning of the time loop. Small matrix products are then performed between a derivative matrix, whose components are the derivatives of the Lagrange polynomials at the GLL points, and the displacement vector $\mathbf{u}$ in 2D cut planes along the three local topological directions (i, j, k) of the spectral element. The computed local values are then summed at global mesh points using the local-to-global mesh numbering mapping in order to compute the acceleration vector $\ddot{\mathbf{u}}$. This "assembly process" must in principle imply an atomic sum because different elements add to the same memory location of a global array.

This can be analysed more precisely by recalling that in each of the three spatial directions the Lagrange interpolants, defined on $[-1, 1]$, are built from the GLL points, which include the boundary points -1 and $+1$. For polynomial basis functions of degree n, there are $n + 1$ GLL quadrature points, and thus there are $n - 1$ interior points in addition to $+1$ and -1. In three dimensions, out of the $(n + 1)^3$ GLL points that each spectral element comprises, there are thus $(n - 1)^3$ interior points that are not shared with neighbouring elements in the mesh, and $(n + 1)^3 - (n - 1)^3$ that may be shared (and will very often be shared in practice) with neighbouring elements of the unstructured mesh through a common face, edge or corner. In acoustic wave propagation we use a polynomial degree $n = 4$ because it has been shown to provide an optimal trade-off between accuracy and cost [6,27]. Thus, out of the 125 GLL points of each spectral element, only 27 are interior and not shared, and 98, i.e. a vast majority, are shared with other elements.

Since the contributions to the elastic force vector are calculated locally and independently inside each spectral element before being summed at the potentially shared points, we decide to assign a different thread to each of the 125 points of each spectral element. We thus handle a spectral element with a block of 128 threads (4 warps) because using a multiple of the 32-thread warp size is best, and use one thread per GLL quadrature point. Therefore, 125 out of 128 threads perform actual work, while three are purposely unused and idle. We first copy the global displacement vector corresponding to each element into shared memory using the global-to-local mapping. The derivative matrix of the Lagrange polynomials is stored in constant memory, and the CUDA kernel then multiplies it with the coefficients in shared memory, at the GLL points.

The derivative matrices have size $(n + 1) \times (n + 1)$, i.e., 5×5. We inline these small matrix products manually (they are too small to be efficiently handled by BLAS3 calls), and store them in constant memory to take advantage of its faster access times and cache mechanism: It is as fast as registers if all threads access the same item simultaneously.

In order to maximise efficiency, we also apply a number of optimizations that are specific to CUDA: we arrange data so that accesses to local data stored in global memory can be coalesced into large memory transactions, and we try to avoid bank conflicts in shared memory. However, there are two important limitations in this crucial and dominant part of spectral element calculations: first, it is memory-bound because it performs a relatively large number of memory accesses compared to a relatively small number of calculations, owing to the small size of the matrices involved. Second, indirect local-to-global addressing is required because of the unstructured nature of the mesh, which unavoidably leads to some uncoalesced memory access patterns. On recent hardware, we nonetheless measure very good throughput of this calculation kernel because coalesced memory accesses are an issue that is far less critical than in the past. To further improve performance we perform these global accesses through the texture cache, but the gain is small, as expected.

The final key issue is to decide how to best handle the summation of all the elastic forces, local to each spectral element, into the global vector of elastic forces, in which many global points are shared between adjacent elements as seen above. In principle, this sum could simply be atomic, the only requirement in order to get correct results being to ensure that different warps never update the same shared location simultaneously. In this application, mesh colouring (cf. Sect. 9.3.1) has been shown to be more efficient. To do so, we partition the mesh elements into a finite number of disjoint subsets, with the property that any two elements in a given subset do not share any global mesh nodes. Data at these nodes can therefore be added to their corresponding global location without any possibility of access conflict, thus removing the need for an atomic locking mechanism. Mesh colouring is performed once and for all on the host in a preprocessing stage during the meshing step by pre-computing maximally independent sets of mesh elements. Adding an outer serial loop over the mesh colours, each colour is then simply handled through a call to the CUDA calculation kernel, resulting in one kernel call per colour. This is acceptable if (and only if) the total number of colours for a given mesh remains reasonable, which is always the case in practice. Tests not shown here show that for unstructured finite element meshes we typically need 10–30 colours.

The described approach is implemented in the SPECFEM3D software, and full source code is available at http://www.geodynamics.org/cig/software.

9.4.3 Group FEM for Gas Dynamics

The final example deals with the gas dynamic equations, modelled by a first-order hyperbolic system of non-linear coupled equations that can be written in the divergence form (9.2). In particular, it expresses the conservation laws for the mass, momentum, and energy of an inviscid compressible fluid. That is,

$$U = \begin{bmatrix} \varrho \\ \varrho \mathbf{v} \\ \varrho E \end{bmatrix}, \quad \mathbf{F} = \begin{bmatrix} \varrho \mathbf{v} \\ \varrho \mathbf{v} \otimes \mathbf{v} + p\mathbb{I} \\ \varrho E \mathbf{v} + p\mathbf{v} \end{bmatrix} = \mathbf{v}U + \begin{bmatrix} 0 \\ \mathbb{I} \\ \mathbf{v} \end{bmatrix} p, \tag{9.17}$$

where ϱ is the density, $\mathbf{v} = (v^1, v^2, v^3)$ is the three-dimensional velocity vector, E is the total energy and $\mathbb{I}$ is the 3×3 identity tensor. The equation of state

$$p = (\gamma - 1)\left(\varrho E - 0.5\varrho\|\mathbf{v}\|^2\right)$$

for an ideal polytropic gas with, e.g., $\gamma = 1.4$ for air, is used to relate the pressure p to the conserved quantities.

A general class of high-resolution methods for the compressible Euler equations was introduced in [20] and refined in a series of publications. The interested reader is referred to [17,18] and the references therein for a detailed description of the state of the art of so-called *algebraic flux correction* (AFC) schemes for hyperbolic systems. Here, the focus lies on their efficient implementation on GPUs. The following presentation is partly based on algorithms implemented in the open source software package Featflow2 (http://www.featflow.de).

Let system (9.17) be discretised in space by Fletcher's group finite element formulation [9] as outlined in Sect. 9.2.4. The resulting semi-discrete problem reads

$$M\frac{dU}{dt} = R(U) \tag{9.18}$$

where the entries of the right-hand side $R = \{R_i\}$ according to (9.10) are given by

$$R_i = -\sum_{j=1}^{N} \mathbf{c}_{ij} \cdot \mathbf{F}_j. \tag{9.19}$$

It serves as a base scheme for many high-resolution methods, but it gives rise to non-physical undershoots and overshoots in the vicinity of discontinuities such as shock waves, and hence, it is not applicable per se. A common stabilisation strategy consists in adding artificial viscosity to prevent the creation of wiggles, and ideally, to ensure that physical quantities such as the density and pressure variables remain positive. In the framework of AFC-schemes [17,18,20] this is achieved by replacing the consistent mass matrix with its row-sum lumped counterpart

$$M_L = \text{diag}\{m_i\mathbb{I}\}, \quad m_i = \sum_{j=1}^{N} m_{ij}$$

and augmenting the right-hand side by artificial diffusion and limited antidiffusion

$$R_i := R_i + \sum_{j \in \mathscr{S}_i} D_{ij} (U_j - U_i) + \alpha_{ij} F_{ij} . \tag{9.20}$$

The choices for $D = \{D_{ij}\}$ given in [17, 18, 20] differ in the arithmetic intensity but use the same input data. Hence, we consider a generic discrete diffusion operator which is defined as a symmetric matrix with zero row- and column sums [21]

$$D_{ii} := - \sum_{j \in \mathscr{S}_i} D_{ij}, \qquad D_{ij} = D_{ij} (\mathbf{c}_{ij}, \mathbf{c}_{ji}, U_i, U_j) = D_{ji} . \tag{9.21}$$

In the scope of this chapter, it also suffices to consider the skew-symmetric antidiffusive fluxes F_{ij} and the symmetric limiting coefficients $\alpha_{ij} \in [0, 1]$ to be functions that depend on the precomputed coefficients $m_{ij}, \mathbf{c}_{ij}$ and $\mathbf{c}_{ji}$, and on the dynamically changing data U_i, U_j, dU_i/dt and dU_j/dt, and their edge-neighbouring values. All algorithmic details can be found in [17, 18]. The treatment of boundary conditions is a non-trivial task that cannot be addressed here due to space constraints. Thus, one should bear in mind that additional terms need to be computed at the boundary, which, however, consumes only a negligible fraction of the overall computing time even in case the most naive implementation is adopted.

Integrating the semi-discrete form (9.18) in time by the two-level θ-scheme yields

$$M \frac{U^{n+1} - U^n}{\Delta t} = \theta R(U^{n+1}) + (1 - \theta) R(U^n) \tag{9.22}$$

which is non-linear for $\theta \in (0, 1]$, and hence, needs to be computed iteratively, e.g., by successive approximations [17, 20], or linearised based on a Taylor series expansion [18, 28]. In any case, the left-hand side has the form $[M/\Delta t - \theta P]U$, where

$$P = \{P_{ij}\}, \qquad P_{ij} = K_{ij} + D_{ij} + \text{boundary contributions} \tag{9.23}$$

is an approximation such that $P(U)U \approx R(U)$ or to its Jacobian, i.e. $P(U) \approx \frac{\partial R(U)}{\partial U}$. The Galerkin part $K_{ij} = -\mathbf{c}_{ij} \cdot A_j$ exploits the so-called homogeneity property of the Euler equations, i.e. $F_j = A_j U_j$, where $A_j = \frac{\partial \mathbf{F}}{\partial U}(U_j)$ is the nodal value of the flux Jacobians. In summary, the core components to be implemented on GPUs are:

- Vector assembly procedures for $R(U)$ based on (9.19) or (9.20)
- Matrix assembly procedures for operator (9.23) for the Galerkin scheme ($D \neq 0$) and in the presence of artificial viscosities

Vector Assembly. The assembly of the right-hand side vector (9.19) for the Galerkin discretisation is straightforward because implementation techniques from sparse matrix vector multiplication (SpMV, [1]) can be readily adapted. Let the coefficient matrices $\mathbf{C} = (C^1, C^2, C^3)$ be stored in a GPU-friendly matrix format in global memory, then the multiply-add operation in the standard (scalar) SpMV-kernel is replaced by $+{=}\sum_{d=1}^{3} c_{ij}^d F^d(\mathbf{U}_j)$, where $F^d(\cdot)$ stands for arithmetic operations according to (9.17). It should be noted that the computation of a single field variable of the target vector R, say total energy, requires all five field variables from the input vector U. It is therefore advisable to invest some amount of shared memory to store the relevant parts of U. After the synchronisation of all threads of the CUDA thread block, each entry in the destination vector is then computed by one thread.

The right-hand side (9.20) is assembled differently by resorting to one of the edge-based formulations (9.12) or (9.13) to maximise data reuse of the input solution vector. Without loss of generality, let us present an edge-by-edge assembly based on the second variant. In particular, the contribution of edge ij needs to be computed only once and can then be scattered to positions i and j as follows:

$$\mathbf{R}_i := \mathbf{R}_i + \mathbf{G}_{ij} + \mathbf{D}_{ij}(\mathbf{U}_j - \mathbf{U}_i) + \alpha_{ij}\mathbf{F}_{ij}$$

$$\mathbf{R}_j := \mathbf{R}_j - \mathbf{G}_{ij} - \mathbf{D}_{ij}(\mathbf{U}_j - \mathbf{U}_i) - \alpha_{ij}\mathbf{F}_{ij} \qquad (9.24)$$

The strategy from Sect. 9.4.1 to optimise the memory access pattern of the element-based assembly can be easily adapted to the edge-based procedure. The column-major 2D array, which remains unchanged for fixed meshes, is indexed by the edge number ij and contains the precomputed coefficients m_{ij}, $\mathbf{c}_{ij}$ and $\mathbf{c}_{ji}$ and the integer values i and j which serve both as source and target indices in the solution and right-hand side vector, respectively. A fixed-sized set of edges is then associated with a CUDA thread block that implements (9.24) using either atomic memory updates or the suggested two-pass strategy with "edges" instead of "elements". In our current implementation, the solution values are directly read from their original location instead of copying them into the 2D data structure.

As an alternative, the static data structure can be reordered and partitioned based on an edge-colouring algorithm (cf. Sect. 9.3.1), accompanied by a permutation of the edge numbering. This yields a contiguous storage of exactly those edges which have no common start and end points, and that can therefore be processed independently without synchronisation. Following Sects. 9.3.5 and 9.4.1, one sequentially-launched CUDA kernel is used for all edges of the same colour. One potential drawback of this colouring strategy is that data reuse is precluded by construction. Even worse, the jump $|j - i|$ between the two indices of the edge ij but also the jump between the index pairs i and i' as well as j and j' of the succeeding edge $i'j'$ may become large and lead to extremely unstructured memory accesses. A reordering algorithm that reduces the jumps in memory access is developed in [22]. The remaining terms of the right-hand side of (9.22) can be handled efficiently by standard SpMV kernels and its extension to interleaved matrices, respectively.

Matrix Assembly. As for the vector assembly, the assembly of the Galerkin part of the operator (9.23) is straightforward. Moreover, neither atomic memory updates nor colouring is required to augment the off-diagonal blocks with the artificial viscosity tensor D_{ij} if either the static 2D data structure or the reduction list employed in the multi-pass strategy is extended by the target positions (i, j) and (j, i) in the global matrix. It is only the diagonal entries that give rise to concurrent updates, see (9.21).

We therefore propose the following strategy combining the static data structure and the reduction list to maximise data reuse and minimise latency due to atomic memory updates. As before, a fixed-sized set of edges is associated with a CUDA thread block running a single kernel. In the first phase, solution data that is required to process the edges under consideration are gathered into shared memory. Next, *all* local blocks K_{ij} for the Galerkin part and the artificial viscosities D_{ij} with $j \neq i$ are computed and stored in shared memory. In the following phase, $K_{ij} + D_{ij}$ is scattered to the non-contentious off-diagonal position (i, j) in global memory based on the static 2D data structure. An additional "reduction list" is used to calculate the diagonal entries $K_{ij} - \sum_{j \in \mathscr{S}_i} D_{ij}$ from the previously computed local data and store the result in global memory. Between each of the different phases, the threads of the CUDA thread block are synchronised. However, it should be noted that this approach has a relatively high demand on shared memory so that a careful tuning of size-parameters is necessary. This is particularly true for compute intensive artificial viscosities such as Roe's approximate Riemann solver which essentially requires two 5×5 matrix–matrix multiplications per edge. If shared memory becomes the limiting factor then this multi-phase approach can be replaced by a multi-pass strategy similar to the one adopted in Sect. 9.4.1 for the element assembly. That is, the different tasks are implemented in individual kernels launched for the same fixed-sized set of edges, whereby intermediate results are written to global memory.

Acknowledgements This work was supported in part by the German Research Foundation (DFG) through the Priority Programme 1648 "Software for Exascale Computing" (SPPEXA), through DFG SFB 708 "3D Surface Engineering of Tools for the Sheet Metal Forming—Manufacturing, Modelling, Machining—", by the European "Mont-Blanc: European scalable and power efficient HPC platform based on low-power embedded technology" #288777 project of call FP7-ICT-2011-7, and by the G8 and French ANR "Interdisciplinary Program on Application Software towards Exascale Computing for Global Scale Issues" (SEISMIC IMAGING project, ANR-10-G8EX-002). This work was granted access to the high-performance computing resources of the French supercomputing centre CCRT under allocation #2012-046351 awarded by GENCI (Grand Equipement National de Calcul Intensif).

References

1. Bell, N., Garland, M.: Implementing sparse matrix-vector multiplication on throughput-oriented processors. In: SC '09: Proceedings of the 2009 ACM/IEEE Conference on Supercomputing, pp. 18:1–18:11 (2009)
2. Brenner, S., Scott, L.R.: The Mathematical Theory of Finite Element Methods. Texts in Applied Mathematics, vol. 15. Springer, New York (1994)

3. Cecka, C., Lew, A.J., Darve, E.: Assembly of finite element methods on graphics processors. Int. J. Numer. Methods Eng. **85**(5), 640–669 (2011)
4. Ciarlet, P.: The Finite Element Methods for Elliptic Problems. North-Holland, Amsterdam (1978)
5. Cristini, P., Komatitsch, D.: Some illustrative examples of the use of a spectral-element method in ocean acoustics. J. Acoust. Soc. Am. **131**(3), EL229–EL235 (2012)
6. De Basabe, J.D., Sen, M.K.: Grid dispersion and stability criteria of some common finite-element methods for acoustic and elastic wave equations. Geophysics **72**(6), T81–T95 (2007)
7. Diestel, R.: Graph Theory, 4th edn. Graduate Texts in Mathematics, vol. 173. Springer, Heidelberg (2010)
8. Donea, J., Huerta, A.: Finite Element Methods for Flow Problems. Wiley, New York (2003)
9. Fletcher, C.: The group finite element formulation. Comput. Methods Appl. Mech. Eng. **37**, 225–243 (1983)
10. Hughes, T.J.R.: The Finite Element Method, Linear Static and Dynamic Finite Element Analysis. Prentice-Hall, Englewood Cliffs (1987)
11. Huthwaite, P.: Accelerated finite element elastodynamic simulations using the GPU. J. Comput. Phys. **257**, 687–707 (2014)
12. Karypis, G., Kumar, V.: A fast and high-quality multilevel scheme for partitioning irregular graphs. SIAM J. Sci. Comput. **20**(1), 359–392 (1998)
13. Kirby, R.C., Logg, A.: A compiler for variational forms. ACM Trans. Math. Softw. **32**(3), 417–444 (2006)
14. Klöckner, A., Warburton, T., Bridge, J., Hesthaven, J.S.: Nodal discontinuous Galerkin methods on graphics processors. J. Comput. Phys. **228**(21), 7863–7882 (2009)
15. Knepley, M.G., Terrel, A.R.: Finite element integration on GPUs. ACM Trans. Math. Softw. **39**(2), 10 (2013) 10:1–10:13
16. Komatitsch, D., Michéa, D., Erlebacher, G.: Porting a high-order finite-element earthquake modeling application to NVIDIA graphics cards using CUDA. J. Parallel Distrib. Comput. **69**(5), 451–460 (2009)
17. Kuzmin, D., Möller, M.: Algebraic flux correction II: compressible flow problems. Flux-Corrected Transport: Principles, Algorithms, and Applications, 1st edn. Scientific Computation, pp. 207–250. Springer Berlin Heidelberg (2005)
18. Kuzmin, D., Möller, M., Gurris, M.: Algebraic flux Correction II: compressible flow problems. Flux-Corrected Transport: Principles, Algorithms, and Applications, 2nd edn. Scientific Computation, pp. 193–238. Springer Berlin Heidelberg (2012)
19. Kuzmin, D., Möller, M., Turek, S.: Multidimensional FEM-FCT schemes for arbitrary time stepping. Int. J. Numer. Methods Fluids **42**(3), 265–295 (2003)
20. Kuzmin, D., Möller, M., Turek, S.: High-resolution FEM-FCT schemes for multidimensional conservation laws. Comput. Methods Appl. Mech. Eng. **193**, 4915–4946 (2004)
21. Kuzmin, D., Turek, S.: Flux correction tools for finite elements. J. Comput. Phys. **175**, 525–558 (2002)
22. Löhner, R.: Cache-efficient renumbering for vectorization. Int. J. Numer. Methods Biomed. Eng. **26**, 628–636 (2008)
23. Markall, G., Slemmer, A., Ham, D., Kelly, P., Cantwell, C., Sherwin, S.: Finite element assembly strategies on multi-core and many-core architectures. Int. J. Numer. Methods Fluids **71**(1), 80–97 (2013)
24. Peter, D., Komatitsch, D., Luo, Y., Martin, R., Le Goff, N., Casarotti, E., Le Loher, P., Magnoni, F., Liu, Q., Blitz, C., Nissen-Meyer, T., Basini, P., Tromp, J.: Forward and adjoint simulations of seismic wave propagation on fully unstructured hexahedral meshes. Geophys. J. Int. **186**(2), 721–739 (2011)
25. Plaszewski, P., Banas, K., Maciol, P.: Higher order FEM numerical integration on GPUs with OpenCL. In: International Multiconference on Computer Science and Information Technology, pp. 337–342 (2010)

26. Rietmann, M., Messmer, P., Nissen-Meyer, T., Peter, D., Basini, P., Komatitsch, D., Schenk, O., Tromp, J., Boschi, L., Giardini, D.: Forward and adjoint simulations of seismic wave propagation on emerging large-scale GPU architectures. In: Proceedings of the SC'12 ACM/IEEE Conference on Supercomputing, pp. 38:1–38:11 (2012)
27. Seriani, G., Oliveira, S.P.: Dispersion analysis of spectral-element methods for elastic wave propagation. Wave Motion **45**, 729–744 (2008)
28. Trépanier, J.Y., Reggio, M., Ait-Ali-Yahia, D.: An implicit flux-difference splitting method for solving the Euler equations on adaptive triangular grids. Int. J. Numer. Methods Heat Fluid Flows **3**, 63–77 (1993)
29. Tromp, J., Komatitsch, D., Liu, Q.: Spectral-element and adjoint methods in seismology. Commun. Comput. Phys. **3**(1), 1–32 (2008)
30. van Wijk, K., Komatitsch, D., Scales, J.A., Tromp, J.: Analysis of strong scattering at the micro-scale. J. Acoust. Soc. Am. **115**(3), 1006–1011 (2004)

Chapter 10
A GPU Implementation for Solving the Convection Diffusion Equation Using the Local Modified SOR Method

Yiannis Cotronis, Elias Konstantinidis, and Nikolaos M. Missirlis

10.1 Introduction

Commodity GPUs have increased computational power compared to modern CPUs and thus they are proposed as more efficient compute units in solving scientific problems with large computational load. Since the appropriate programming environments (CUDA [29], OpenCL [19]) are getting mature they can be used to develop GPU programs in order to exploit the capabilities of GPUs. In this chapter we use GPUs for the numerical solution of Partial Differential equations (PDEs).

We focus on partial differential equations (PDEs) as they constitute an important sector of the computational science field. In particular, we consider the solution of the second order convection diffusion equation

$$\Delta u - f(x, y)\frac{\partial u}{\partial x} - g(x, y)\frac{\partial u}{\partial y} = 0 \qquad (10.1)$$

on a domain $\Omega = \{(x, y)\}|0 \leq x \leq 1, 0 \leq y \leq 1\}$, where $u = u(x, y)$ is prescribed on the boundary $\partial\Omega$. The solution of PDEs reduces to a system of N linear equations by using the finite difference method through the discretization on a rectangular grid $M_1 \times M_2$.

These systems of equations are sparse and thus iterative methods are preferred in order to solve them. In these methods computations are applied to stencils (Fig. 10.1) where point values are iteratively recomputed till they converge to certain values.

The Successive Overrelaxation (SOR) iterative method is an important solver for large, sparse, linear systems [35, 36]. However, the SOR method is essentially sequential in its original form. In order to use a parallel form of an iterative method

Y. Cotronis • E. Konstantinidis (✉) • N.M. Missirlis
Department of Informatics and Telecommunications, University of Athens,
Panepistimiopolis, 15784 Athens, Greece
e-mail: cotronis@di.uoa.gr; ekondis@di.uoa.gr; nmis@di.uoa.gr

V. Kindratenko (ed.), *Numerical Computations with GPUs*,
DOI 10.1007/978-3-319-06548-9_10, © Springer International Publishing Switzerland 2014

Fig. 10.1 The 5 point stencil computation involves the current value of the element and the values of the four neighbor elements

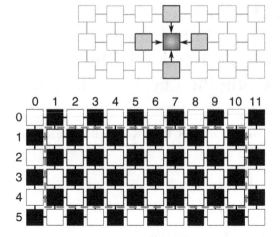

Fig. 10.2 Grid points are colored as *red* and *black* forming a chess board pattern (*red* points are illustrated as *white* due to printing restrictions). This pattern allows the independent computation of all points of the same color

such as SOR with fine grain parallelism we consider the grid points as red-black colored (Fig.10.2). In this case sets of points of the same color can be computed in parallel [32]. The LSOR method was introduced by Ehrlich [12, 13] and Botta and Veldman [4] in an attempt to further increase the rate of convergence of SOR. In [5] it was proved that the local Modified SOR method (LMSOR) possesses a better rate of convergence than LSOR.

Although the focus of this chapter is the Convection Diffusion equation, the particular code should be easily adaptable to other equations as well. We developed three implementations differentiated by the degree of recomputations of read-only matrices thus affecting the flops per element access ratio. Recomputations are applied as an optimization that lowers the memory access requirements of a kernel but raises the instruction workload that has to be executed. Contemporary GPUs feature a quite high $compute/bandwidth$ ratio which allows the instruction overhead to be hidden. In our implementation each GPU program is designed as three different kernel variations each one using either of the three major memory types of GPUs (global memory, texture memory or shared memory) for a particular data array which is subject to data reuse by adjacent threads. In addition, we apply the memory reordering strategy [21, 22] which is proved to benefit this kind of kernels. We assess the performance of the implementations using three NVidia GPUs (GTX480, Tesla S2050 and GTX660) determining which implementation best exploits the capabilities of each GPU.

10.1.1 Related Work

The LMSOR method belongs to the iterative methods family for solving linear systems. Iterative methods in general have already been studied by the research community. The Gauss-Seidel method has been used on GPUs to accelerate fluid dynamic simulation problems [1, 2] and computer vision algorithms [34]. The SOR method has been implemented on GPUs to medical analysis [15], computational fluid dynamics [14, 16, 20, 23, 26, 28] problems, as well as for solving Poisson–Boltzmann equation for electrostatics on molecular systems [7]. Stencil computations in general have been applied on Kepler GPU architecture [27], on AMD APUs [11] which combine a CPU and a GPU on the same silicon area, and even on heterogeneous multi-device implementations [25]. A red-black SOR scheme for solving the steady state heat conduction equation on GPUs has been studied in [17]. Researchers have also studied a variant of the SOR method parallelized on GPUs by employing an unconventional tiling format in order to extract parallelism [10]. Anzt et al. have developed blocked asynchronous kernel solvers [3] that relax the strict order of operations required by classic iterative solvers in order to improve scalability and performance on GPUs. Going further, research efforts have moved on to multi-GPU implementations in order to address the increased computational load of fluid dynamics problems [18, 33, 37]. A hybrid Jacobi implementation has also been studied [9]. This chapter provides insight to the implementation of the red-black LMSOR method on GPUs [8].

10.2 A General Description of the LMSOR Method

The discretization of (10.1) on a rectangular grid $N = M_1 \times M_2$ within Ω leads to a system of linear equations with N unknowns of the form

$$u_{ij} = \ell_{ij} u_{i-1,j} + r_{ij} u_{i+1,j} + t_{ij} u_{i,j+1} + b_{ij} u_{i,j-1}, \tag{10.2}$$

$$i = 1, 2, \ldots, M_1 , \; j = 1, 2, \ldots, M_2$$

with

$$\ell_{ij} = \frac{k^2}{2(k^2 + h^2)} (1 + \frac{1}{2} h f_{ij}) , \; r_{ij} = \frac{k^2}{2(k^2 + h^2)} (1 - \frac{1}{2} h f_{ij})$$

$$\tag{10.3}$$

$$t_{ij} = \frac{h^2}{2(k^2 + h^2)} (1 - \frac{1}{2} k g_{ij}) , \; b_{ij} = \frac{h^2}{2(k^2 + h^2)} (1 + \frac{1}{2} k g_{ij}),$$

where $h = 1/(M_1 + 1), k = 1/(M_2 + 1), f_{ij} = f(ih, jk)$ and $g_{ij} = g(ih, jk)$.

In a red-black colored grid of points we can choose to call a point (i, j) as red when $i + j$ is even and black when $i + j$ is odd. The local Modified SOR (LMSOR) method [5] to (10.2) can be written as follows:

$$u_{ij}^{(n+1)} = (1 - \omega_{1ij}) u_{ij}^{(n)} + \omega_{1ij} J_{ij} u_{ij}^{(n)}, \quad \text{red points} \tag{10.4}$$

$$u_{ij}^{(n+1)} = (1 - \omega_{2ij})u_{ij}^{(n)} + \omega_{2ij}J_{ij}u_{ij}^{(n+1)}, \quad \text{black points} \qquad (10.5)$$

where

$$J_{ij}u_{ij}^{(n)} = l_{ij}u_{i-1,j}^{(n)} + r_{ij}u_{i+1,j}^{(n)} + t_{ij}u_{i,j+1}^{(n)} + b_{ij}u_{i,j-1}^{(n)} \qquad (10.6)$$

and J_{ij} is called the local Jacobi operator. The parameters $\omega_{1ij}, \omega_{2ij}$ are called local relaxation parameters. Note that if $\omega_{ij} = \omega_{1ij} = \omega_{2ij}$, then (10.4), (10.5) reduce to the LSOR method studied in [24]. Moreover, if $\omega = \omega_{1ij} = \omega_{2ij}$ (10.4), (10.5) degenerate into the classical SOR method with red-black ordering. Using Fourier analysis [6], Boukas and Missirlis [5] found the optimum values of the local relaxation parameters ω_{1ij} and ω_{2ij} for the LMSOR method in case the eigenvalues μ_{ij} of the local Jacobi operator J_{ij} are all real or all imaginary. These optimums are expressed in terms of $\bar{\mu}_{ij}$ and $\underline{\mu}_{ij}$, where

$$\bar{\mu}_{ij} = \max_{k_1,k_2} |\mu_{ij}(k_1, k_2)|, \quad \underline{\mu}_{ij} = \min_{k_1,k_2} |\mu_{ij}(k_1, k_2)|, \qquad (10.7)$$

and (see [12])

$$\mu_{ij}(k_1, k_2) = 2\left(\sqrt{l_{ij}r_{ij}}\cos\frac{k_1\pi}{M_1 + 1} + \sqrt{t_{ij}b_{ij}}\cos\frac{k_2\pi}{M_2 + 1} \right), \qquad (10.8)$$

with $k_1 = 1, 2, \ldots, M_1$, $k_2 = 1, 2, \ldots, M_2$, for Dirichlet boundary conditions. More precisely the optimum values of ω_{1ij} and ω_{2ij} are as follows.

Case 1. μ_{ij} are real. This case applies when $l_{ij}r_{ij} \geq 0$ and $t_{ij}b_{ij} \geq 0$. The optimum values of the LMSOR parameters are given by

$$\omega_{1ij}^{opt} = \frac{2}{1 - \bar{\mu}_{ij}\underline{\mu}_{ij} + \sqrt{(1 - \bar{\mu}_{ij}^2)(1 - \underline{\mu}_{ij}^2)}}$$

and

$$\omega_{2ij}^{opt} = \frac{2}{1 + \bar{\mu}_{ij}\underline{\mu}_{ij} + \sqrt{(1 - \bar{\mu}_{ij}^2)(1 - \underline{\mu}_{ij}^2)}} \qquad (10.9)$$

where

$$\bar{\mu}_{ij} = 2\left(\sqrt{l_{ij}r_{ij}}\cos\pi h + \sqrt{t_{ij}b_{ij}}\cos\pi k \right) \qquad (10.10)$$

and

$$\underline{\mu}_{ij} = 2\left(\sqrt{l_{ij}r_{ij}}\cos\frac{\pi(1-h)}{2} + \sqrt{t_{ij}b_{ij}}\cos\frac{\pi(1-k)}{2} \right). \qquad (10.11)$$

Case 2. μ_{ij} are imaginary. This case applies when $\ell_{ij}r_{ij} \leq 0$ and $t_{ij}b_{ij} \leq 0$. The optimum values of the LMSOR parameters are given by

$$\omega_{1ij}^{opt} = \frac{2}{1 - \overline{\mu}_{ij}\underline{\mu}_{ij} + \sqrt{(1 + \overline{\mu}_{ij}^2)(1 + \underline{\mu}_{ij}^2)}}$$

and   (10.12)

$$\omega_{2ij}^{opt} = \frac{2}{1 + \overline{\mu}_{ij}\underline{\mu}_{ij} + \sqrt{(1 + \overline{\mu}_{ij}^2)(1 + \underline{\mu}_{ij}^2)}}$$

where $\overline{\mu}_{ij}$ and $\underline{\mu}_{ij}$ are computed by (10.10) and (10.11), respectively. By using the LMSOR method instead of the red-black SOR we avoid the computation of the optimum value of the parameter ω which increases considerably the computation time [5].

In our code the core computation is implemented as described in (10.4)–(10.6). The values of u_{ij} are computed by using the elements ω_{1ij}, ω_{2ij}, l_{ij}, r_{ij}, t_{ij}, b_{ij} retained in distinct arrays. The values of ω_{1ij} and ω_{2ij} are combined in a single array ω as for each point computation only one of the elements ω_{1ij} or ω_{2ij} is used but not both. This makes a total of six arrays used during the computation for a straightforward implementation through which only the values of u are updated. The rest five arrays (ω, l, r, t, b) have their values calculated once during the initialization as their values are just read and not modified during the computation. Having two kernel invocations per iteration, one for the computation of black elements and one for the red ones, we can reach to a straightforward implementation without dealing with any major issues. However, a straightforward implementation does not exploit the full capabilities of the GPU. In the rest sections we provide more insight on the optimizations and provide the CUDA source code of the best performing kernel developed in this work.

10.3 GPU Implementation

Stencil computations in general are governed by the memory traffic bottleneck. The type of kernels in which performance is determined by the amount and kind of memory access transactions are called memory bound kernels. For instance, the solution of the Laplace equation using the red-black SOR method is considered as memory bound [21]. The cache sizes are limited and cannot persistently accommodate the whole dataset especially on the GPU. As such in each iteration where all array data are traversed they go through the memory bus. Accessing all array data in each iteration is inevitable for non-trivial sized problems and thus the bottleneck of kernel execution is not the amount of instructions to execute but the amount and type of memory accesses instead.

10.3.1 Applied Optimizations

Having characterized the kernel under consideration we move our focus to optimization methods that could mitigate the memory traffic requirements. Since some factors taking part in the equations could be recomputed on the fly, one strategy to follow is to avoid keeping the required values resident in memory space but recomputing them on demand instead. This strategy introduces repetitive computations of the same factors but reduces the access requirements. In this regard the kernel is developed in three variations differentiated by the recomputation factor.

The GPU implementation involves six arrays (u, ω, l, r, t, b) which correspond to the matrices taking part in the core computation equations (10.4)–(10.6). As the computation pattern follows the red-black scheme all matrices are accessed in a red-black fashion. The most significant optimization applied is a memory reorganization policy which is referred as *reordering by color* [21, 22, 30]. In this regard the data of each matrix are split in to two different arrays, one consisting of the red elements and one of the black (Fig. 10.3).

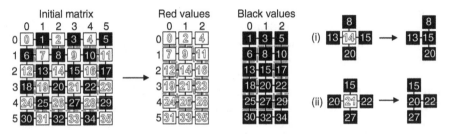

Fig. 10.3 The reordering by color of grid elements. The reordered pattern leads to optimized coalesced accesses

The reordering by color optimization is quite important as it virtually improves the access patterns performed during the core computation. In a straightforward implementation accesses of consecutive threads are forced to be interleaved with unused elements. The ideal pattern would suggest having consecutive threads accessing consecutive memory positions which is referred as *coalescing* [30]. In this regard memory reordering by color assists to achieve the desired access pattern. Coalescing is enforced by using the resulting matrices. This strategy can improve bandwidth utilization by improving locality and coalescing of memory accesses and mostly, by utilizing all values contained in a memory segment, which is not possible with a natural interleaved red-black ordering.

During the core computation accesses on the l, r, t, b, ω arrays are performed sequentially by consecutive threads. No values are shared and the accesses can be perfectly aligned which leads to optimum coalescing. The case for the u array differs. The elements in every array are split in two distinct arrays comprised by the elements of the same color. In case of u the elements of one particular color are computed (red or black) and these elements are accessed exclusively in the same fashion as the rest arrays. However, the elements of the opposite color (black or red) are accessed as neighbor elements and in this case they are not perfectly aligned and there is data sharing evident (e.g. two consecutive threads have one neighbor element in common). This issue of data sharing raises the opportunity to use other memory types (texture memory or shared memory) in order to optimize the local data sharing. In this regard, kernels are developed in three variants able to use either memory type for the accessing of the opposite color elements of the u array.

Additionally, some of the resident read-only matrices used in the computation could be discarded. Having their elements computed during the program initialization, they could be eliminated by replacing accesses to them with recomputations. In each stencil computation the required factors could be recomputed on demand. The GPUs are capable of high instruction throughput and therefore they are tolerable to hiding the additional overhead. Trading memory accesses for computations can be beneficial on such memory bound kernels.

In summary, LMSOR on the GPU was implemented in three variations as three different kernels regarding the amount of recomputation it performs iteratively. Moreover, each kernel was developed in other three variations with regard to the memory type usage (global, texture or shared memory).

GPU Kernel #1—No Recomputations. All values required in (10.4)–(10.6) reside in matrices situated in the GPU device memory (matrices corresponding to elements u_{ij}, ω_{ij}, l_{ij}, r_{ij}, t_{ij} and b_{ij}). This kernel is the natural outcome implementation as no extra computations are performed.

GPU Kernel #2—Recomputations of l_{ij}, r_{ij}, t_{ij}, b_{ij}. The values of the two matrices f_{ij} and g_{ij} multiplied by half the value of h, are precomputed and stored in two matrices ($f'_{ij} = \frac{1}{2}hf_{ij}$, $g'_{ij} = \frac{1}{2}kg_{ij}$) in the device memory. The required terms (i.e. l_{ij}, r_{ij}, t_{ij} and b_{ij}) are generated through f'_{ij}, g'_{ij} values during the LMSOR iterations. Thus, instead of four matrices we just need to access two matrices only in device memory. Memory requirements are lower since only four matrices (u_{ij}, ω_{ij}, f_{ij}, g_{ij}) are accessed from device memory. However, it comes at a cost of extra operations needed to recompute the required terms for the formula on every iteration.

GPU Kernel #3—Recomputations of All Elements (l_{ij}, r_{ij}, t_{ij}, b_{ij}, ω_{ij}). In this implementation, recomputation is additionally applied to the ω_{ij} elements. Note that this recomputation is particularly intensive as it involves more than 20 floating point operations per element which include costly transcendental operations (i.e. sinusoidal, square root, reciprocal operations). During computation u_{ij}, f_{ij}, g_{ij} elements are accessed from memory and all other elements are recomputed on demand.

10.3.2 Kernel's Source Code

In this subsection we provide source code for one kernel variation. The chosen variation is the *GPU Kernel #2* (Recomputations of l, r, t, b) leveraging the texture memory for the neighbor u elements involved in the five point stencil. This particular kernel is the one with the optimum overall performance as it was determined in our experiments. The source code is provided in Listing 10.1.

Listing 10.1 Recompute r,l,t,b with texture memory usage for neighbor elements

```
1   template<int granularity, int calc_red>
2   __global__ void klmsorDP(const unsigned int N, const unsigned int
        pitch, double * __restrict__ dDstU, const double *
        __restrict__ dFFh, const double * __restrict__ dGGh, const
        double * __restrict__ dW, double * __restrict__ sqrerrorD){
3     const texture2Ddouble texSrcU = calc_red ? texUBlack : texURed;
4     int iy = (blockIdx.y*blockDim.y + threadIdx.y)*granularity + 1;
5     const int ix = blockIdx.x * blockDim.x + threadIdx.x;
6     const unsigned int tid = threadIdx.y*blockDim.x + threadIdx.x;
7     double sqrerror = 0.0;
8     int ptrid = ix + iy * pitch; // absolute index to target point
9     // l_i = {0, 1} can be determined during compilation time
10    // in case granularity is a multiple of 2
11    int l_i = granularity%2==0 ? (1-calc_red) : (calc_red+iy)%2;
12    if( ix<N/2 && iy<N-granularity ){ // check if in bounds
13      #pragma unroll
14      for(int i=0; i<granularity; i++){
15        if( ix>=l_i && ix<N/2-(1-l_i) ){ // check if in bounds
16          double oldv = dDstU[ptrid];
17          const double quarter = 1.0/4.0;
18          double r = (1.-dFFh[ptrid])*quarter; // right
19          double l = (1.+dFFh[ptrid])*quarter; // left
20          double t = (1.-dGGh[ptrid])*quarter; // top
21          double b = (1.+dGGh[ptrid])*quarter; // bottom
22          double w = dW[ptrid];                 // omega
23          // compute new element value
24          double newv = ( 1. - w )*oldv + w*(
25            l*tex2Ddouble(texSrcU,       ix, iy-1) +
26            r*tex2Ddouble(texSrcU,       ix, iy+1) +
27            b*tex2Ddouble(texSrcU,   ix-l_i,   iy) +
28            t*tex2Ddouble(texSrcU, ix-l_i+1,   iy) );
29          dDstU[ptrid] = newv;
30          sqrerror += (oldv-newv)*(oldv-newv); // diff^2
31        }
32        l_i = 1-l_i;   // invert l_i (0->1, 1->0)
33        ptrid += pitch; // next row
34        iy++;
35      }
36    }
37    shm_sqrerr[tid] = sqrerror; // store sqrerror in shared memory
38    // perform a reduction of error values of all threads
39    // in the thread block via shared memory
```

```
    __syncthreads();
    shmem_reduction(tid, shm_sqrerr);
    if(tid==0)
        // store accumulated error value of the thread block
        sqrerrorD[ blockIdx.y*gridDim.x+blockIdx.x ] = shm_sqrerr[0];
}
```

The main loop can be unrolled by the compiler as *granularity* is a template
variable and therefore it is known during the compilation time. The template variable
calc_red is set to 1 when computing red elements and 0 otherwise. The *l_i* variable
is used as an offset to the *u* array for accessing the neighbor elements. It is set to 0
when the first element of the current row and the calculated element are opposite-
colored and it is set to 1 when they are same-colored.

There are references to other elements as seen on Listing 10.2. The *shm_sqrerr*
reference is a pointer to the dynamically allocated shared memory address space
that is to be used by the kernel. Since the amount of required shared memory is
determined on runtime it is used as dynamically allocated. The *texture2Ddouble* is a
texture reference defined as texture to *int2* elements. Double precision textures were
not supported in the current version of CUDA. Therefore, an *int2* texture was used
along with the *tex2Ddouble* function which retrieves an *int2* value and converts it to
a 64bit double precision value by combining the two 32bit parts of *int2*.

Listing 10.2 Additional declarations for elements used in the kernel code

```
// shared memory pointer for the reduction of error values
extern __shared__ double shm_sqrerr[];

// texture reference type for U elements
typedef texture<int2,2,cudaReadModeElementType> texture2Ddouble;

// texture references for red and black U elements
texture2Ddouble texURed, texUBlack;

// user defined function for reading double precision values
// from texture memory
__inline__ __device__
double tex2Ddouble(const texture2Ddouble tx, const int x, const
    int y){
    int2 v = tex2D(tx, x, y);
    return __hiloint2double(v.y, v.x);
}
```

The kernel's source code makes use of the *shmem_reduction* function that is
used to reduce the sum of all error values as computed by each individual thread in
a thread block. This function is illustrated in Listing 10.3.

Listing 10.3 shmem_reduction function for reduction of error values in shared memory

```
1   inline __device__
2   void shmem_reduction(unsigned int tid, volatile double *sd){
3     const unsigned int BLOCK_SIZE = blockDim.x*blockDim.y;
4     if(BLOCK_SIZE > 1024){
5       if(tid < min(1024, BLOCK_SIZE-1024)) sd[tid]+=sd[tid+1024];
6       __syncthreads();
7     }
8     if(BLOCK_SIZE >  512){
9       if(tid < min( 512, BLOCK_SIZE -512)) sd[tid]+=sd[tid+ 512];
10      __syncthreads();
11    }
12    if(BLOCK_SIZE >  256){
13      if(tid < min( 256, BLOCK_SIZE -256)) sd[tid]+=sd[tid+ 256];
14      __syncthreads();
15    }
16    if(BLOCK_SIZE >  128){
17      if(tid < min( 128, BLOCK_SIZE -128)) sd[tid]+=sd[tid+ 128];
18      __syncthreads();
19    }
20    if(BLOCK_SIZE >   64){
21      if(tid < min(  64, BLOCK_SIZE  -64)) sd[tid]+=sd[tid+  64];
22      __syncthreads();
23    }
24    if(tid < 32){
25      if(BLOCK_SIZE >  32) sd[tid]+=sd[tid + 32];
26      // Synchronization is not needed at this point due to
27      // implicit synchronization by thread warps
28      sd[tid] += sd[tid + 16];
29      sd[tid] += sd[tid +  8];
30      sd[tid] += sd[tid +  4];
31      sd[tid] += sd[tid +  2];
32      sd[tid] += sd[tid +  1];
33    }
34  }
```

10.4 Performance Results

Performance results are covered in two sections. First, the tuning results are illustrated on Table 10.1 and Fig. 10.4 as the configuration parameters were determined by running multiple experiments on a wide range of different parameter values. The *granularity* is the amount of stencil computations assigned to each thread, the *block size* is the thread block size as determined for the kernel execution and *fast memory config* is the type of memory partitioning to L1 cache and shared memory in the multiprocessors. The problem size under consideration was set to $N = 2,162 \times 2,162$. It is evident that the same kernel type performed best for

Table 10.1 Tuning configuration results for the three GPUs

Device	Recompute factor	Kernel	Granularity	Block size	Fast memory config
GTX480	Recompute r,l,t,b	kernel #2	2	64×2	48 KB/16 KB[a]
S2050	Recompute r,l,t,b	kernel #2	1	64×2	48 KB/16 KB[a]
GTX660	Recompute r,l,t,b	kernel #2	2	64×2	48 KB/16 KB[a]

For each GPU the best performing kernel (recomputation/kernel) is determined along with the
optimum configuration parameters (granularity, thread block size and memory configuration)
[a]L1 Cache/Shared memory configuration

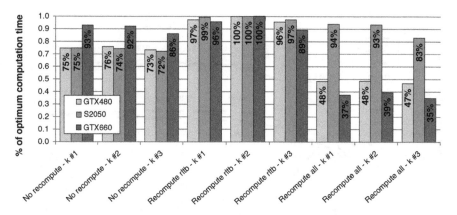

Fig. 10.4 The performance is illustrated as the *optimum time/measured time* ratio in the tuning
procedure. Each kernel (horizontal axis) is characterized by the recomputation factor (no recompute, recompute rltb, recompute all) and the memory type usage (k#1: only global memory,
k#2: texture memory and k#3: shared memory). Problem size in the experiments was $N =
2,162 \times 2,162$ and run for 500 iterations. The optimum times correspond to 0.758, 1.141 and
1.241 s for the GTX480, S2050 and GTX660, respectively

all three devices (*recompute of l,r,t,b with texture memory usage*). The parameter
values that correspond to the minimum execution times were selected as the best for
the next experiments.

As already mentioned the recomputation allows larger problems to solved by the
GPU. GPUs are not equipped with expandable memory and their available memory
is fixed. If more memory is needed then the whole graphic card device has to
be replaced. Let us, for example, consider the theoretical maximum problem size
that can be solved by each kernel and GPU device combination (Table 10.2). The
maximum feasible problem size to be processed by a GPU device can be determined
from formula (10.13). The "*total arrays*" term is divided by eight as double
precision arithmetic is used. For instance, using the "No recompute" kernel (six
arrays) on the GTX480 (1.5 GB device memory) the maximum problem that can be
solved is $N = 33,554,432 \approx 5,792 \times 5,792$. In practice the available memory is
even less and thus the maximum problem size is further decreased.

$$max(N) = \frac{device\ memory}{total\ arrays \times 8} \tag{10.13}$$

Table 10.2 Theoretical maximum applicable problem sizes (N) per kernel

Device name	Device memory (GB)	No recompute (six arrays)	Recompute rltb (four arrays)	Recompute all (three arrays)
GTX480	1.5	5,792 × 5,792	7,094 × 7,094	8,192 × 8,192
S2050	3.0[a]	8,192 × 8,192	10,033 × 10,033	11,585 × 11,585
GTX660	2.0	6,688 × 6,688	8,192 × 8,192	9,459 × 9,459

The amount of required arrays and device memory determines the available memory for each array and the maximum problem size
[a]With ECC disabled

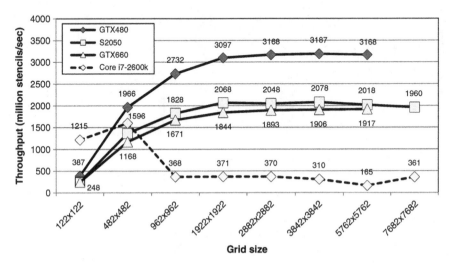

Fig. 10.5 A scaling throughput comparison for the three GPUs and the CPU. Run for problem sizes N = {122 × 122, 482 × 482, 962 × 962, 1,922 × 1,922, 2,882 × 2,882, 3,842 × 3,842, 5,762 × 5,762 and 7,682 × 7,682}

In the next section a scaling experiment was executed. The optimum versions for each GPU was executed on a range of different problem sizes. In addition, they were compared with a CPU (Intel Core i7 2600K). The results are illustrated in Fig. 10.5.

It is worth noting that the OpenMP [31] CPU implementation also employs some of the optimizations applied on the GPU versions. First, the partial recomputation strategy is similarly applied on the CPU. The CPU compute potential of contemporary CPUs is also higher than its memory bandwidth, therefore, through experimentation it was found that the CPU program also benefits by recomputation. In addition the memory reordering strategy is also applied on the CPU and it is essential in order to achieve vectorization through SIMD AVX instructions.

One of the key observations reinforcing the recomputation strategy is the trend of contemporary GPUs and CPUs on increasing their compute potential in a much higher rate than the memory bandwidth does. For instance, the NVidia GTX-285, which is a 5 year old high-end GPU, performs 1,063 GFLOPS peak on single precision and 159 GB/s memory bandwidth. The modern GTX-780 Ti GPU features

5,046 GFLOPS peak on single precision and 336 GB/s memory bandwidth. The *compute/bandwidth* ratio of the former is 6.69 and the latter's ratio is 15.02. In this example this ratio has more than doubled and it is expected to continue increasing in the next years. Therefore, as our endeavour to optimize the particular kernels is focused on memory traffic alleviation, the recomputation policy will probably gain more value on the following generations of CPUs and GPUs.

10.5   Remarks and Conclusions

GPUs have recently gained the interest of the scientific community. They consist a high performance platform for massive parallel computations and the iterative methods for solving PDEs like SOR are pretty well suited. In order to achieve memory coalescing, the locality of accesses must be ensured and one good strategy is to apply the reordering by color optimization.

Recomputation can also be beneficial in cases where memory accesses become a bottleneck for GPUs. Instead of keeping the processing units idle waiting for memory access requests to complete, one strategy is to recompute data in order to lower memory access requirements. This is a tradeoff and in many cases when a kernel is memory bandwidth limited, compute resources can be traded for less memory accesses. In such cases the overall performance can be improved. In addition it potentially allows larger problems to be processed as memory space is released which was initially allocated for the respective arrays. Therefore, the maximum size of the problem to be solved is determined by the type of the applied kernel. Even in cases where recomputation is excessively applied, although performance is worsened on current GPUs, it might prove not to be the case on future GPUs. The increasing trend of *compute/bandwidth* ratio allows to predict that future GPUs will be much more compute capable in comparison to their memory bandwidth and the excessive recomputation strategies are expected to exhibit better performance.

Acknowledgements We would like to acknowledge the kind permission of the Innovative Computing Laboratory at the University of Tennessee to use their NVidia Tesla S2050 installation for the purpose of this work.

References

1. Amador, G., Gomes, A.: A CUDA-based implementation of stable fluids in 3D with internal and moving boundaries. In: 2010 International Conference on Computational Science and Its Applications, pp. 118–128 (2010)
2. Amador, G., Gomes, A.: CUDA-based linear solvers for stable fluids. In: International Conference on Information Science and Applications (ICISA), pp. 1–8 (2010)

3. Anzt, H., Tomov, S., Dongarra, J., Heuveline, V.: Weighted block-asynchronous iteration on GPU-accelerated systems. In: Euro-Par 2012: Parallel Processing Workshops. Lecture Notes in Computer Science, vol. 7640, pp. 145–154 (2013)

4. Botta, E.F., Veldman, A.E.P.: On local relaxation methods and their application to convection-diffusion equations. J. Comput. Phys. **48**, 127–149 (1981)

5. Boukas, L.A., Missirlis, N.M.: The parallel local modified SOR for nonsymmetric linear systems. Int. J. Comput. Math. **68**, 153–174 (1998)

6. Brandt, A.: Multi-level adaptive solutions to boundary-value problems. Math. Comput. **31**(138), 333–390 (1977)

7. Colmenares, J., Ortiz, J., Decherchi, S., Fijany, A., Rocchia, W.: Solving the linearized Poisson-Boltzmann equation on GPUs Using CUDA. In: 21st Euromicro International Conference on Parallel, Distributed and Network-Based Processing (PDP), pp. 420–426 (2013)

8. Cotronis, Y., Konstantinidis, E., Louka, M.A., Missirlis, N.M.: Parallel SOR for solving the convection diffusion equation using GPUs with CUDA. In: EuroPar 2012 Parallel Processing, International European Conference on Parallel and Distributed Computing, Rhodos. Lecture Notes in Computer Science, vol. 7484, pp. 575–586 (2012)

9. Czapiński, M., Thompson, C., Barnes, S.: Reducing communication overhead in multi-GPU hybrid solver for 2D Laplace equation. Int. J. Parallel Program. 1–16 (2013) DOI: 10.1007/s10766-013-0293-2

10. Di, P., Wu, H., Xue, J., Wang, F., Yang, C.: Parallelizing SOR for GPGPUs using alternate loop tiling. Parallel Comput. **38**(6–7), 310–328 (2012)

11. Eberhart, P., Said, I., Fortin, P., Calandra, H.: Hybrid strategy for stencil computations on the APU. In: Proceedings of the 1st International Workshop on High-Performance Stencil Computations, Vienna, pp. 43–49 (2014)

12. Ehrlich, L.W.: An Ad-Hoc SOR method. J. Comput. Phys. **42**, 31–45 (1981)

13. Ehrlich, L.W.: The Ad-Hoc SOR method: a local relaxation scheme. In: Elliptic Problem Solvers II, pp. 257–269. Academic, New York (1984)

14. Gohari, S.M.I., Esfahanian, V., Moqtaderi, H.: Coalesced computations of the incompressible Navier Stokes equations over an airfoil using graphics processing units. Comput. Fluids **80**, 102–115 (2013)

15. Ha, L., Kröger, J., Joshi, S., Silva, C.T.: Multiscale unbiased diffeomorphic atlas construction on multi-GPUs. In: GPU Computing Gems, pp. 771–791. Morgan Kaufmann, Los Altos (2011)

16. Hsieh, C.W., Kuo, S.H., Kuo, F.A., Chou, C.Y.: Solving parabolic problems using multithread and GPU. In: International Symposium on Parallel and Distributed Processing with Applications (ISPA'10), Washington, pp. 75–80 (2010)

17. Itu, L.M., Suciu, C., Moldoveanu, F., Postelnicu, A., Suciu, C.: GPU optimized computation of stencil based algorithms. In: 10th Roedunet International Conference (RoEduNet), pp. 1–6, 23–25 June 2011

18. Khajeh-Saeed, A., Blair Perot, J.: Direct numerical simulation of turbulence using GPU accelerated supercomputers. J. Comput. Phys. **235**, 241–257 (2013)

19. Khronos Group: The OpenCL Specification. Khronos Group, Beaverton (2009) http://www.khronos.org/registry/cl/specs/opencl-1.0.pdf

20. Komatsu, K., Soga, T., Egawa, R., Takizawa, H., Kobayashi, H., Takahashi, S., Sasaki, D., Nakahashi, K.: Parallel processing of the building-cube method on a GPU platform. Comput. Fluids **45**(1), 122–128 (2011)

21. Konstandinidis, E., Cotronis, Y.: Accelerating the red/black SOR method using GPUs with CUDA. In: 9th International Conference on Parallel Processing and Applied Mathematics, Part I, Torun. Lecture Notes in Computer Science, vol. 7203, pp. 589–598 (2012)

22. Konstantinidis, E., Cotronis, Y.: Graphics processing unit acceleration of the red/black SOR method. Concurr. Comput. **25**(8), 1107–1120 (2013)

23. Kosior, A., Kudela, H.: Parallel computations on GPU in 3D using the vortex particle method. Comput. Fluids **80**, 423–428 (2013)

24. Kuo, C.-C.J., Levy, B., Musicus, B.R.: A local relaxation method for solving elliptic PDEs on mesh-connected arrays. SIAM J. Sci. Stat. Comput. **8**(4), 550–573 (1987)

25. Li, P., Brunet, E., Namyst, R.: High performance code generation for stencil computation on heterogeneous multi-device architectures. In: HPCC-15th IEEE International Conference on High Performance Computing and Communications, Zhangjiajie (2013)
26. Liu, J.T., Ma, Z.S., Li,S.H., Zhao, Y.: A GPU accelerated red-black SOR algorithm for computational fluid dynamics problems. Adv. Mater. Res. **320**, 335–340 (2011)
27. Maruyama, N., Aoki, T.: Optimizing stencil computations for NVIDIA Kepler GPUs. In: Proceedings of the 1st International Workshop on High-Performance Stencil Computations, Vienna, pp. 89–95 (2014)
28. Niemeyer, K., Sung, C.: Recent progress and challenges in exploiting graphics processors in computational fluid dynamics. J. Supercomput. **67**(2), 528–564 (2014)
29. NVidia: NVidia CUDA C Programming Guide v.5.0. NVidia (2012)
30. NVidia: NVidia CUDA C Best Practices Guide Version 5.0. NVidia (2012)
31. OpenMP Architecture Review Board: OpenMP Application Program Interface Version 3.0. OpenMP Architecture Review Board (2008)
32. Ortega, J.M., Voight, R.G.: Solution of Partial Differential Equations on Vector and Parallel Computers. SIAM, Philadelphia (1985)
33. Thibault, J., Senocak, I.: Accelerating incompressible flow computations with a Pthreads-CUDA implementation on small-footprint multi-GPU platforms. J. Supercomput. **59**(2), 693–719 (2012)
34. Vandal, N.A., Savvides, M.: CUDA accelerated illumination preprocessing on GPUs. In: 17th International Conference on Digital Signal Processing (DSP), pp. 1–6 (2011)
35. Varga, R.S.: Matrix Iterative Analysis. Prentice-Hall, Englewood Cliffs (1962)
36. Young, D.M.: Iterative Solution of Large Linear Systems. Academic, New York (1971)
37. Zaspel, P., Griebel, M.: Solving incompressible two-phase flows on multi-GPU clusters. Comput. Fluids **80**, 356–364 (2013)

Chapter 11
Finite-Difference in Time-Domain Scalable Implementations on CUDA and OpenCL

Lídia Kuan, Pedro Tomás, and Leonel Sousa

11.1 Finite-Difference in Time-Domain Numerical Method

The FDTD [8, 10] is a popular computational method for solving Maxwell's Equations for electromagnetics [7, 11]. This method was introduced by Kane Yee in [10] and is a time domain solution to the Maxwell's Equations with relatively good accuracy and flexibility. It has become a powerful method for solving a wide variety of different electromagnetics problems.

In [10], Yee presented the Finite-Difference in Time-Domain (FDTD) problem domain as a rectangle composed of cells. Assuming that the dielectric parameters μ, ϵ and σ are independent of time, the following system of scalar equations is equivalent to Maxwell's equations in the rectangular coordinate system (x, y, z):

$$\frac{\partial H_x}{\partial t} = \frac{1}{\mu}\left(\frac{\partial E_y}{\partial z} - \frac{\partial E_z}{\partial y}\right), \tag{11.1a}$$

$$\frac{\partial H_y}{\partial t} = \frac{1}{\mu}\left(\frac{\partial E_z}{\partial x} - \frac{\partial E_x}{\partial z}\right), \tag{11.1b}$$

$$\frac{\partial H_z}{\partial t} = \frac{1}{\mu}\left(\frac{\partial E_x}{\partial y} - \frac{\partial E_y}{\partial x}\right), \tag{11.1c}$$

$$\frac{\partial E_x}{\partial t} = \frac{1}{\epsilon}\left(\frac{\partial H_z}{\partial y} - \frac{\partial H_y}{\partial z} - \sigma E_x\right), \tag{11.1d}$$

$$\frac{\partial E_y}{\partial t} = \frac{1}{\epsilon}\left(\frac{\partial H_x}{\partial y} - \frac{\partial H_z}{\partial x} - \sigma E_y\right), \tag{11.1e}$$

L. Kuan (✉) • P. Tomás • L. Sousa
INESC-ID/IST, Universidade de Lisboa, Rua Alves Redol 9, Lisboa, Portugal
e-mail: lmlk@sips.inesc-id.pt; pfzt@inesc-id.pt; las@inesc-id.pt

V. Kindratenko (ed.), *Numerical Computations with GPUs*,
DOI 10.1007/978-3-319-06548-9_11, © Springer International Publishing Switzerland 2014

$$\frac{\partial E_z}{\partial t} = \frac{1}{\epsilon}\left(\frac{\partial H_y}{\partial x} - \frac{\partial H_x}{\partial y} - \sigma E_z\right). \tag{11.1f}$$

Yee introduced a set of finite-difference equations for the system, where a space cell point is denoted as

$$(x, y, z) = (i\delta x, j\delta y, k\delta z), \tag{11.2}$$

and any function of space and time is denoted as

$$F^n(i, j, k) = F(i\delta, j\delta, k\delta, n\delta t), \tag{11.3}$$

where δ is the space increment, which is considered to be uniform in all x, y and z directions, and δt is the time increment. Finite-difference expressions for the space and time derivatives are used:

$$\frac{\partial F^n(i,j,k)}{\partial x} = \frac{F^n(1+\frac{1}{2},j,k) - F^n(1-\frac{1}{2},j,k)}{\delta} + O(\delta^2), \tag{11.4}$$

$$\frac{\partial F^n(i,j,k)}{\partial t} = \frac{F^{n+\frac{1}{2}}(i,j,k) - F^{n-\frac{1}{2}}(i,j,k)}{\delta t} + O(\delta t^2), \tag{11.5}$$

(11.4) and (11.5) are applied to the E and H field. Therefore, the FDTD algorithm solves Maxwell's equations by first performing the E field equations update for each cell at time-step n, and then performing the H field equations update for each cell at time-step $n + 1/2$. The time resolution of the simulation is determined by the model's spatial resolution, and the number of time-steps is determined by the waveform shape and temporal length of the source being modeled.

There are two modes of electromagnetic waves, namely: transverse electric (TE) wave and transverse magnetic (TM) wave. In the scope of this chapter, the implementation of a wave propagation simulation is addressed by considering that the internal electric field is a uniform, rectangular, space in vacuum, with a rectangular wave source that is assumed to be infinite in the z direction. The incident radiation is assumed to be a $+x$ directed TM wave of frequency f, in this particular example we set f to be $10\,\mathrm{GHz}$. Because there is no variation of either scattered geometry or incident fields in the z direction, this problem may be treated as a two-dimensional scattering of the incident wave, with only E_z, H_x and H_y fields present. Therefore, from this point forward, the focus will be limited to the following finite-difference of Maxwell's equations in a two-dimensional space:

$$H_x^{n+\frac{1}{2}}(i,j+\frac{1}{2}) = H_x^{n-\frac{1}{2}}(i,j+\frac{1}{2}) + \frac{\delta t}{\mu(i,j+\frac{1}{2})\delta} \times$$

$$\left[E_y^n(i,j+\frac{1}{2}) - E_y^n(i,j+\frac{1}{2}) + E_z^n(i,j) - E_z^n(i,j+1)\right], \tag{11.6a}$$

Algorithm 1: High-level overview of the FDTD algorithm.

---

for $n = 0 \rightarrow nsteps$ **do**
  for $j = 0 \rightarrow ydim$ **do**
    for $i = 0 \rightarrow xdim$ **do**
      update $E_x[i][j]$; i=i+1;
    end
    j=j+1;
  end
  for $j = 0 \rightarrow ydim$ **do**
    for $i = 0 \rightarrow xdim$ **do**
      update $H_y[i][j]$;
      update $H_x[i][j]$;
      i=i+1;
    end
    j=j+1;
  end
  n=n+1;
end

$$H_y^{n+\frac{1}{2}}(i+\frac{1}{2},j) = H_y^{n-\frac{1}{2}}(i+\frac{1}{2},j) + \frac{\delta t}{\mu(i+\frac{1}{2},j)\delta} \times$$

$$\left[E_z^n(i+1,j) - E_z^n(i,j) + E_x^n(i+\frac{1}{2},j) - E_x^n(i+\frac{1}{2},j) \right], \qquad (11.6b)$$

$$E_z^{n+1}(i,j) = \left[1 - \frac{\sigma(i,j)\delta t}{\epsilon(i,j)} \right] E_z^n(i,j) + \frac{\delta t}{\epsilon(i,j)\delta} \times \left[H_y^{n+\frac{1}{2}}(i+\frac{1}{2},j) \right.$$

$$\left. - H_y^{n+\frac{1}{2}}(i-\frac{1}{2},j) + H_x^{n+\frac{1}{2}}(i,j-\frac{1}{2}) - H_x^{n+\frac{1}{2}}(i,j+\frac{1}{2}) \right]. \qquad (11.6c)$$

The choice of δ and δt defines accuracy and stability, respectively [8]. To ensure the accuracy of the computed results, δ must be taken as a small fraction of either the minimum wavelength expected in the model or the minimum scatterer dimension. To ensure the stability of the time-stepping algorithm, δt is chosen to satisfy the Courant condition in a two-dimensional space:

$$v_{max}\delta t \leq \sqrt{\frac{1}{\delta x^2} + \frac{1}{\delta y^2}}, \qquad (11.7)$$

where v_{max} is the maximum wave phase velocity expected within the model. The pseudo-code of the sequential FDTD algorithm is shown in Algorithm 1.

In summary, the presented implementation consists of a TM plane wave propagation in a cell grid in vacuum. To avoid wave propagation and/or reflection the E and H fields were set to zero in the left and right grid limits, and a periodic boundary condition was added in the top and bottom grid limits. The source generates a sinusoidal incident wave $E_z = E_o \cos(2\pi f n dt)$ positioned along a column in the grid near the left limit.

While many approaches can be considered to implement the described FDTD algorithm, the objective of this chapter is to show the required optimization steps in order to efficiently explore the GPUs processing power [5]. With this goal in mind, we start in Sect. 11.2 by presenting single GPU CUDA and OpenCL implementations and then address the optimization steps taken to decrease the time required to compute. Section 11.3 presents the algorithm implementation considering larger problem sizes. Computing in multi-GPU systems is addressed in Sect. 11.4. Although, OpenCL framework is target to various devices (i.e. multi-core Central Processing Units (CPUs), AMD and Nvidia GPUs, Altera FPGAs), the focus of the presented work is targeted to Nvidia's GPUs. Therefore, following discussions will focus on that type of GPUs.

11.2 Single GPU Implementation

To implement the FDTD algorithm in a single GPU system, we consider that the problem is defined in a (x, y) grid of size $IE \times JE$. To take advantage of the GPU's architecture, the complete grid is divided in small subsets as shown in Fig. 11.1, where each subset is computed by a different multiprocessor, which in the case of the Nvidia GPU's is designated Streaming Multiprocessor (SM).

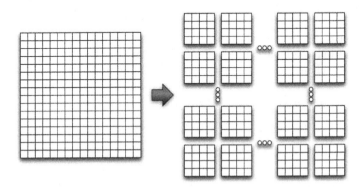

Fig. 11.1 Grid distribution in the GPU

11.2.1 Host Environment Initialization

Although CUDA and OpenCL are similar frameworks, the environment initialization for each of them is different. The CUDA environment initialization can be done by using the cudaGetDeviceCount(), which retrieves the number of CUDA capable GPUs, followed by cudaSetDevice(int *device*), where the argument *device* sets the device on which the active host thread executes the device code.

The OpenCL framework is divided into a platform layer application programming interface (API) and runtime API. The platform API allows an application to query for OpenCL devices and manage them through a context. The runtime API makes use of the context to manage the execution of kernels, which are the programs, on OpenCL devices. To execute an OpenCL program it is necessary to:

1. Query the host system for OpenCL devices.
2. Create a context to associate the OpenCL devices.
3. Create programs (which in OpenCL is a source file with a set of kernels) that will run on one or more associated devices.
4. From the programs objects, select the kernels to execute.
5. Create memory objects on the host or on the device.
6. Copy memory data to the devices as needed.
7. Provide arguments for the kernels.
8. Submit kernels to the command queue for execution.
9. Copy the results from the device to the host.

In OpenCL, when the context is established, command queues can be created that allow commands to be sent to the compute devices associated with this context. In the following sections we mainly focus on items 5–9 from the list above.

11.2.2 Straightforward Implementation

This section describes a straightforward implementation of the parallel algorithm on the GPU, by considering that the problem fits into the GPU's memory. The most basic approach for designing the kernel would be the one presented in Algorithm 1, where the update field for() loops would be turn into kernels and be processed by the GPU [2]. Therefore, there would be two different kernels: one for the E_z update; and another for the H_x and H_y update. As shown in (11.6) E and H fields are evaluated at alternate half-time steps, which implies a calculation dependency between the two fields.

11.2.2.1 General Optimizations

Since the electric and magnetic fields depend on each other, this is equivalent to having two different kernels that cannot be overlapped due to data dependencies, which increases the kernel launch overhead. Thus, in order to reduce the kernel launch overhead these two kernels can be merged into a single kernel, respecting he data dependencies. This new kernel will be the one that we will use form now on and the whole problem size is defined in a space of size $IE \times JE$.

In the presented implementation, the number of threads per block was chosen in order to maximize the targeted GPU resource utilization. Thus, in the following example, for a grid of size $IE \times JE$, if the number of threads is 256, a two dimensional 16×16 kernel environment is set, as it is shown in Fig. 11.2 for the CUDA and OpenCL environments.

```
dim3 dimBlock(16, 16);                                              CUDA
dim3 dimGrid(IE / dimBlock.x, JE / dimBlock.y)
...
for (n=0; n<NSTEPS; n++)
  sim_calc<<<dimGrid, dimBlock>>>( ...
```

```
size_t global_work_size[2];                                        OpenCL
size_t local_work_size[2];
...
local_work_size[0] = 16;
local_work_size[1] = 16;
global_work_size[0] = IE;
global_work_size[1] = JE;
...
for (n=0; n<NSTEPS; n++)
  clEnqueueNDRangeKernel (....global_work_size, local_work_size, ...);
```

Fig. 11.2 Kernel launch in host code for CUDA and OpenCL frameworks

The kernel code to compute the E_z, H_x and H_y fields for both CUDA and OpenCL frameworks is shown in Figs. 11.3 and 11.4, respectively. Each thread is set to compute one element of the grid, the necessary data to compute the element is copied to local variables, then the result is computed and stored in the global memory. Notice that this approach requires copying the E and H matrices only once to the GPU's memory before kernel launch, and copying the matrix E back to the host at the end of the computation. Moreover, the kernel is launched NSTEPS times, which is the simulation duration.

For this particular algorithm the kernel execution is short, having a significant launch overhead. Therefore, regarding the code presented in this section an additional optimization can be made if the $IE \times JE$ problem fits in the GPU memory. In such a case, the for() loop of the NSTEPS iterations can be put inside the kernel, in order to reduce the kernel launch overhead.

```
                                                                    CUDA
  __global__ void sim_calc (float *Ez, float *Hx, float *Hy, int tstep)
  /*__global__ void sim_calc (float *Ez, float *Hx, float *Hy)*/
  {
    int n = tstep;
    /* int n */

    int blx = gridDim.x*blockIdx.y + blockIdx.x;
    int thx = blx*(blockDim.x*blockDim.y) + blockDim.x*threadIdx.y
              + threadIdx.x;

    float ez, hx, hx_1, hy, hy_1, ez_x, ez_y;    if (blockIdx.y==0 && threadIdx.y==0)
    /* for (n=0; n<NSTEPS; n++){ */                 hx_1 = Hx[gridDim.x*blockDim.y*
                                                         blockDim.x*(gridDim.y-1)+
    ez = Ez[thx]; hx = Hx[thx]; hy = Hy[thx];            blockDim.y*blockDim.x*
                                                         blockIdx.x+blockDim.x*
    // set the boundary conditions for Ez field          (blockDim.y-1)+threadIdx.x];
    ..........................
                                                 else
    ...........................                     hx_1 = Hx[thx-IE];
    __syncthreads();

    if (blockIdx.x == 0 && threadIdx.x == 10)    if (blockIdx.x==0 && threadIdx.x==0)
      ez = cosf(2.0f*d_pi*d_freq*n*d_dt);           hy_1 = 0.0;
    else                                         else
      ez = ez + d_cb*(hy-hy_1+hx_1-hx);             hy_1 = Hy[thx-1];

    Ez[thx] = ez;
    __syncthreads();                             if (blockIdx.x==gridDim.x-1 &&
                                                     threadIdx.x==blockDim.x-1)
    // set the boundary conditions                  ez_x = 0.0;
    // for Hx and Hy fields                       else
    ..........................                      ez_x = Ez[thx+1];

    ...........................                   if (blockIdx.y==gridDim.y-1 &&
    __syncthreads();                                 threadIdx.y==blockDim.y-1)
                                                     ez_y = Ez[blockDim.y*blockDim.x*
    hx = hx + d_db*(ez - ez_y);                           blockIdx.x + threadIdx];
    hy = hy + d_db*(ez_x - ez);                  else
    Hx[thx] = hx;                                   ez_y = Ez[thx+IE];
    Hy[thx] = hy;
    __syncthreads();
    /*}*/
  }
```

Fig. 11.3 CUDA kernel code

11.2.3 Experimental Results

To evaluate the impact of the proposed optimization step, the FDTD algorithm was run on two experimental setups: (a) Two Nvidia's GTX 580 GPUs with Fermi architecture [9], with 1.5 GB of memory, compute capability 2.0 and 512 cuda cores running at 1.54 GHz, and (b) one Nvidia's Tesla K20c GPU with Kepler architecture [4], with 5 GB of memory, compute capability 3.5 and 2,496 cuda cores running at 0.71 GHz. The results for the single-GPU execution are presented in Fig. 11.5 and show the obtained speedup of the different implementations using as

```
                                                          OpenCL
__kernel void sim_calc (__global float *Ez, __global float *Hx,
                        __global float *Hy, int tstep)
/*__kernel void sim_calc (__global float *Ez, __global float *Hx,
                        __global float *Hy)*/
{
  int n = tstep;
  /* int n; */

  int blx = get_num_groups(0)*get_group_id(1)+get_group_id(0);
  int thx = blx*(get_local_size(0)*get_local_size(1)) +
            (get_local_size(0)*get_local_id(1)) + get_local_id(0);

  float ez, hx, hy, hx_1, hy_1, ez_x. ez_y;

  /* for (n=0; n<NSTEPS; n++){*/
  ez = Ez[thx]; hx = Hx[thx]; hy = Hy[thx];

  // set the boundary conditions for Ez field
  ...........................

  ...........................
  barrier (CLK_GLOBAL_MEM_FENCE);

  if (get_group_id(0)==0 &&
     get_local_id(0)==0)
    ez = cos(2.0*pi*freq*n*dt);
  else
    ez = ez + cb*(hy-hy_1+hx_1-hx);

  Ez[thx] = ez;
  barrier (CLK_GLOBAL_MEM_FENCE);

  // set the boundary conditions for
  // Hx and Hy fields
  ...........................

  ...........................
  barrier (CLK_GLOBAL_MEM_FENCE);

  hx = hx + db*(ez -ez_y);
  hy = hy + db*(ez_x - ez);

  Hx[thx] = hx;
  Hy[thx] = hy;
  barrier (CLK_GLOBAL_MEM_FENCE);
  /*}*/
}
```

```
if (get_group_id(1)==0 && get_local_id(1)==0)
  hx_1 = Hx[get_num_groups(0)*
         get_local_size(0)*get_local_size(1)*
         (get_num_groups(1)-1) +
         get_local_size(0)*get_local_size(1)*
         get_group_id(0)+get_local_size(0)*
         (get_local_size(1)-1)+get_local_id(0)];
else
  hx_1 = Hx[thx-IE];

if (get_group_id(0)==0 && get_local_id(0)==0)
  hy_1 = 0.0;
else
  hy_1 = Hy[thx-1];
```

```
if (get_group_id(0)==get_num_groups(0)-1 &&
   get_local_id(0)==get_local_size(0)-1)
  ez_x = 0.0;
else
  ez_x = Ez[thx+1];

if (get_group_id(1)==get_num_groups(1)-1 &&
   get_local_id(1)==get_local_size(1)-1)
  ez_y = Ez[get_local_size(0)*
         get_local_size(1)*get_group_id(0)+
         get_local_id(0)];
else
  ez_y = Ez[thx+IE];
```

Fig. 11.4 OpenCL kernel code

baseline the execution time of the approach with two different kernels. Notice that the x axis represents the IE value, for the presented results $JE = IE$. The baseline execution time (in seconds) is presented in the figure at the top of each group of columns. It can be observed that for both frameworks using a single kernel reduces the execution time, and that by executing the for () inside the kernel we were able to obtain a speedup of up to 2.6. However, for the GTX 580 the achieved speedup is up to 2.3. This difference can be due to features introduces by compute capability 3.5, for instance global memory atomic operations have higher throughput on Kepler than on Fermi [3].

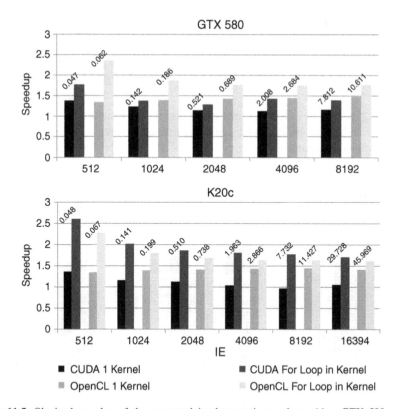

Fig. 11.5 Obtained speedup of the presented implementations, taken with a GTX 580 and a K20c GPUs, for CUDA and OpenCL frameworks. The baseline value, considering the original straightforward implementation is shown on the top of each group of columns in seconds

11.3 Scalable Implementation for Large Problems

While the FDTD implementation described in Sect. 11.2 allows achieving a substantial performance, its applicability is limited by the amount of memory of the GPU device. For large problem sizes where the GPU memory is too small to store all the required data, a different approach has to be used. This section addresses the development of a scalable approach, which splits the problem into slices along the y direction, as shown in Fig. 11.6. Such an approach, allows the data transfer between the CPU and the GPU to use contiguous memory blocks, and also enables coalesced memory accesses in the GPU.

For example, assuming a grid of size $IE \times JE$ and a GPU whose memory can only hold the data to compute a problem of size $d\_IE \times d\_JE$ (highlighted in Fig. 11.6a) where $d\_IE < IE$ and $d\_JE < JE$, the grid will be divided in chunks of $d\_IE \times d\_JE$ elements, in the following way: if T is the number of total bytes supported by the GPU and each element of the grid occupies B bytes, the total amount of memory required to compute one line is given by $IE \times B$. Thus, the maximum number of lines that can be computed by the GPU is given by

$$d\_JE = \left\lfloor \frac{T}{IE \times B} \right\rfloor,$$

where $\lfloor\ \rfloor$ represents the rounding operation to the smallest integer. This is highlighted in Fig. 11.6b. In this particular algorithm, there are however data dependencies when calculating each single element of the grid. Thus the dependency data for each chunk also needs to be passed to the GPU. With this purpose ghost cells are used as additional data to be transferred to the GPU. The ghost cells are highlighted in Fig. 11.6c.

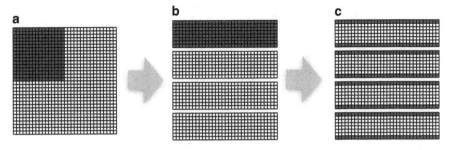

Fig. 11.6 Scalable implementation approach. (**a**) Representation of the whole problem and GPU's maximum supported size highlighted in *grey*. (**b**) Maximum number of supported elements are rearranged (grey elements style) in order address a coalesced chunk into the GPU and each chunk is computed at a time until the whole problem has been calculated. (**c**) The algorithm has data dependencies, for each chunk it is necessary to copy additional data highlighted in *grey* (the rows at the top and bottom of the chunk)

Each chunk can be calculated sequentially, with the respective memory transfer data. However, for improving efficiency, the computation can be overlapped with the memory transfer, reducing the execution time. For this implementation kernel and data transfer overlaps and dimensions of power of two will be used for simplicity. It is worth to notice that overlapping induces the size of the chunks to be smaller.

The size of data assigned to the GPU for each chunk is

$$d\_IE = IE,$$

$$d\_JE = \left\lfloor \frac{\frac{T}{Bbytes} - 2}{2} \right\rfloor.$$

where the subtraction by two is used to accommodate the ghost cells and the division by two is required for double buffering.

11.3.1 Implementation in CUDA

To implement the described parallelization approach, we take advantage of streams and non-blocking memory transfers. A CUDA stream represents a queue of GPU operations which can be kernel launches or data copies. The order by which operations are added to the stream specifies the order in which they will be executed. A GPU that supports overlapping possesses the capacity to simultaneously execute a CUDA C kernel while performing a copy between device and host memory.

```
cudaDeviceProp prop;                                    CUDA
int device;

cudaGetDevice (&device);
if (cudaGetDeviceProperties (&prop, device) == cudaSuccess){
   if (!=prop.deviceOverlap)
      printf("Device will not handle overlaps\n");
}
```

Fig. 11.7 Excerpt of code to query if the GPU supports the overlap feature

```
cudaStream_t stream[gpu_y_blocks];                      CUDA

for (i=0; i<gpu_y_blocks; i++)
   cudaStreamCreate (&stream[i]);
```

Fig. 11.8 Excerpt of code to create CUDA streams

To use the CUDA stream one should first query the device and check whether it supports the overlapping feature, using the `cudaGetDeviceProperties()` function, as shown in Fig. 11.7 After checking that the target device supports overlapping the streams to be used are created, which in this case it will be the number of chunks of the whole grid (see Fig. 11.8). We then proceed to data allocation on the host and the device, the sequence of computations illustrated in Fig. 11.9. In this case, generating the input and output buffers allocated on the host uses pinned memory (notice the call to `cudaMallocHost()`). The reason for using the page-locked host memory is that it makes copies faster and it is required for using the function.

The call to `cudaMemcpyAsync()` simply places a *request* to perform a memory copy into the `stream`. When the call returns, there is no guarantee that the copy has even started. The only guarantee is that the copy will be performed before the next operation placed into the same stream. It is also required that any host memory pointers passed to `cudaMemcpyAsync()` have been allocated by `cudaMallocHost()`. Moreover, notice that the angle-bracketed kernel launch also takes an optional stream argument `stream[j]`, indicating that the `stream[j]` is associated to this call.

```
float *Ez, *Hx, *Hy;                                               CUDA
float *d_Ez, *d_Hx, *d_Hy;
float *d_Ez_ghostCells, *d_Hx_ghostCells;
...
cudaMallocHost (&Ez, IE*JE*sizeof(float));
cudaMallocHost (&Hx, IE*JE*sizeof(float));
cudaMallocHost (&Hy, IE*JE*sizeof(float));

cudaMalloc (&d_Ez, d_IE*d_JE*sizeof(float));
cudaMalloc (&d_Hx, d_IE*d_JE*sizeof(float));
cudaMalloc (&d_Hy, d_IE*d_JE*sizeof(float));
cudaMalloc (&d_Ez_ghostCells, d_IE*sizeof(float));
cudaMalloc (&d_Hx_ghostCells, d_IE*sizeof(float));
...
for (n=0; n<NSTEPS; n++){
  for (j=0; j<gpu_y_blocks; j++){

    // Copy data to GPU
    cudaMemcpyAsync (d_Ez, &Ez[j*d_IE*d_JE], ... H2D, stream[j]);
    ...
    // Kernel launch
    sim_calc<<<dimGrid,dimBlock,0,stream[j]>>>(kernel arguments);
  }

  for (j=0; j<gpu_y_blocks; j++){
    // Copy data from the GPU
    cudaMemcpyAsync (&Ez[j*d_IE*d_JE], d_Ez, .... D2H, stream[j]);
    ...
  }

  for (j=0; j<gpu_y_blocks; j++)
    // Wait for the data copy to finish at the end of each iteration
    cudaStreamSynchronize (stream[j]);
}
```

Fig. 11.9 Excerpt of code with CUDA sequence of computations

Also notice that kernel launch is asynchronous, just like the preceding CPU to GPU memory copy (copies) and the trailing GPU to CPU memory copy (copies). Under this scheme, it is technically possible to end an iteration of the loop without actually having started any of the memory copies or kernel execution. As previously mentioned, it is guaranteed that the first copy placed into the stream will execute before the other following copies. Moreover, the last copy will be complete before the kernel starts, and the kernel will complete before the following copy starts.

When the for() loop has terminated, there could be still quite a bit of work queued up for the GPU to finish. To guarantee that the GPU is done with its computations and memory copies, it is important to synchronize it with the host, instructing it to wait for the GPU to finish before proceeding. This is accomplished by calling the cudaStreamSynchronize() function and by specifying the stream that we want to wait for.

Finally, notice that in the presented code we have used one stream for each chunk, where the inner for() loop computes each chunk of the problem for NSTEPS time-steps.

11.3.2   Implementation in OpenCL

This subsection presents the implementation of the scalable parallelization approach, considering the OpenCL framework. As described in the beginning of Sect. 11.3, it is necessary to take advantage of the non-blocking memory copy functions and the available events[1] as synchronization points, in order to develop an efficient implementation.

In OpenCL, when the context is successfully created, command queues can be created to allow commands to be sent to the compute devices associated with the context. While commands are placed into the command queues in order, OpenCL allows these queues to be executed out-of-order. In such a case, the execution dependencies can be guaranteed by using events.

Assuming that the device has been chosen, the context and programs has been created with success, and the kernels has been selected, the memory objects on the host and the device can be created as shown in Fig. 11.10.

```
float *Ez, *Hx, *Hy;                                    OpenCL
cl_mem d_Ez, d_Hx, d_Hy;
cl_context context;
...
Ez = (float *)calloc(IE*JE, sizeof(float));
Hx = (float *)calloc(IE*JE, sizeof(float));
Hy = (float *)calloc(IE*JE, sizeof(float));

d_Ez = clCreateBuffer (context, CL_MEM_READ_WRITE, ...);
d_Hx = clCreateBuffer (context, CL_MEM_READ_WRITE, ...);
d_Hy = clCreateBuffer (context, CL_MEM_READ_WRITE, ...);
```

Fig. 11.10 Excerpt of code with OpenCL buffers allocation

As shown in the figure, in this implementation, the host buffer is allocated as a normal buffer, while the device buffer is created using the clCreateBuffer(), and allowing read and write operations for the device.

As mentioned previously, the use of events allows to implement the overlapping of the memory transfer with the kernel execution. All clEnqueue* functions take three arguments: the number of events to wait on, a list of events to wait on, and an event that the command creates that can be used by another queue. Therefore, the correct order of the operations is established by these three arguments.

The steps to create the necessary events is shown in Fig. 11.11. Nine events are created: five for memory transfer into the GPU, one for the kernel execution, and three for collecting the data from the device. Typically one should create one event for each operation that will be executed. The sequence of computations will look like as shown in Fig. 11.12. In the presented code, the non-blocking memory transfer is made by setting the third parameter of the functions clEnqueueWriteBuffer() and clEnqueueReadBuffer() to CL_FALSE.

---

[1] Notice that in CUDA the definition of events refers to timing features.

```
cl_event write_events[gpu_y_blocks][5];                      OpenCL
cl_event kernel_events[gpu_y_blocks];
cl_event read_events[gpu_y_blocks][3];

for (j=0; j<gpu_y_blocks; j++){
  write_events[j][0] = clCreateUserEvent(context, &err);
  ...
  kernel_events[j] = clCreateUserEvent(context, &err);
  read_events[j][0] = clCreateUserEvent(context, &err);
  ...
}
```

Fig. 11.11 Excerpt of code with the OpenCL events creation

```
cl_command_queue cmd_queue;                                   OpenCL
cl_kernel kernel;

for (n=0; n<NSTEPS; n++){
  for (j=0; j<gpu_y_blocks; j++){

    // Copy data to the GPU
    clEnqueueWriteBuffer (cmd_queue, d_Ez, CL_FALSE, ....
                          &Ez[j*(d_IE*d_JE)], ....
                          &write_events[j][0]);
    ...

    // Set Kernel Arguments
    clSetKernelArg(kernel, 0, sizeof(cl_mem), &d_Ez);
    clSetKernelArg(kernel, 1, sizeof(cl_mem), &d_Hx);
    clSetKernelArg(kernel, 2, sizeof(cl_mem), &d_Hy);
    clSetKernelArg(kernel, 3, sizeof(cl_mem), &d_Ez_ghostCells);
    clSetKernelArg(kernel, 4, sizeof(cl_mem), &d_Hx_ghostCells);
    clSetKernelArg(kernel, 5, sizeof(cl_int), &n);

    // Kernel launch
    clEnqueueNDRangeKernel (cmd_queue, kernel, 2, NULL,
                            global_work_size, local_work_size,
                            5, write_events[j], &kernel_events[j]);
  }

  // Copy data from the GPU
  for (j=0; j<gpu_y_blocks; j++){
    clEnqueueReadBuffer (cmd_queue, d_Ez, CL_FALSE, ....
                         &Ez[j*(d_IE*d_JE)],
                         1, &kernel_events[j], &read_events[j][0]);
    ...
  }

  // Wait for the data copy from the GPU to finish
  for (j=0; j<gpu_y_blocks; j++)
    clWaitForEvents (3, read_events[j]);
}
```

Fig. 11.12 Excerpt of code with OpenCL sequence of computations

The `clEnqueueNDRangeKernel()` waits for the five `write_events` and the `clEnqueueReadBuffer()` waits for the `kernel_event`. Finally, in order to guarantee data coherence at the end of each iteration, a synchronization point is made with `clFinish()`, which it blocks host execution until all previously queued commands in `cmd_queue` have completed.

11.4 Exploring Multi-GPU Systems

Many modern computing environments now possesses multiple GPUs. Developing algorithms that allow exploring the full computing potential of such systems is fundamental. To use multiple GPUs, the first step is to retrieve the number of available devices. As shown before in CUDA, this can be done with the `cudaGetDeviceCount()`. For OpenCL case this is performed by using the `clGetDeviceIDs()` function. After gathering the number of available GPUs one must choose a particular GPU to perform each operation. In CUDA this is done as shown in Fig. 11.13.

In OpenCL, once the compute devices and their corresponding `device_id`(s) have been identified, the `device_id`(s) need to be associated with the context (see Fig. 11.14). Once the context is created, command queues can be created to

```
int gpu_num;                                        CUDA

cudaGetDeviceCount (&gpu_num)
cudaSetDevice (0);
...
cudaSetDevice (gpu_num-1);
```

Fig. 11.13 Excerpt of code with CUDA device set

```
cl_context context;                                 OpenCL
cl_device_id device_id[num_devices];
cl_context_properties properties[3];
...
// context properties list - must be terminated with 0
properties[0] = CL_CONTEXT_PLATFORM;
properties[1] = (cl_context_properties)platform_id;
properties[2] = 0;
...
context = clCreateContext (
            properties,   // list of context properties
            num_devices, // num of devices in the device_id list
            device_id,    //device id list
            NULL,       //pointer to the error callback function (if required)
            NULL,       // argument data to pass to the callback function
            &err        // return code
        );
```

Fig. 11.14 Excerpt of code with OpenCL context creation

allow commands to be sent to the compute device associated with the context. To use multiple GPUs, it is necessary to create one command queue for each GPU as shown in Fig. 11.15.

```
cl_command_queue cmd_queue[num_devices];                          OpenCL
...
for (j=0; j<num_devices; j++){
    cmd_queue[j] = clCreateCommandQueue (context, device_id[j],
                                         0, &err);
    if (err != CL_SUCCESS){
        printf("Error %i in clCreateCommandQueue\n", err);
        return -1;
    }
}
```

Fig. 11.15 Excerpt of code with OpenCL command queues creation for multiple GPUs

11.4.1 Scalable Implementation with Multiple GPUs in CUDA

The sequence of computations in the implementation presented in Sect. 11.3.1 but by using multiple GPUs looks like as shown in Fig. 11.16.

The main difference of this code when comparing to the one shown in Sect. 11.3.1, is the lines highlighted in the rectangles that have been added. These lines select a different GPU in each NSTEPS iteration. Moreover, additional buffers were allocated, two buffers for each used GPU, one for the E field and other of the H field.

11.4.2 Scalable Implementation with Multiple GPUs in OpenCL

In this subsection we present the code of the scalable implementation presented in Sect. 11.3.2 with the support for multiple devices in Fig. 11.17. The main differences when comparing with the code presented in Sect. 11.3.2 is the highlighted rectangles. Here, one command queue was created for each GPU. The commands are placed at each device command queue alternatively and also each device got its own field buffer.

```
for (n=0; n<NSTEPS; n++){                                    CUDA
  for (j=0; j<gpu_y_blocks; j++){
    dev_id = j%num;
    cudaSetDevice (dev_id);

    // Copy data to GPU
    cudaMemcpyAsync (d_Ez[dev_id], &Ez[j*d_IE*d_JE], .... H2D,
                     stream[j]);
    cudaMemcpyAsync (d_Hx[dev_id], &Hy[j*d_IE*d_JE], .... H2D,
                     stream[j]);
    ...
    // Kernel launch
    sim_calc<<<dimGrid,dimBlock,0,stream[j]>>>(kernel arguments);
  }

  for (j=0; j<gpu_y_blocks; j++){
    dev_id = j%num;
    cudaSetDevice (dev_id);

    // Copy data from the GPU
    cudaMemcpyAsync (&Ez[j*d_IE*d_JE], d_Ez[dev_id], .... D2H,
                     stream[j]);
    ...
  }

  for (j=0; j<gpu_y_blocks; j++){
    dev_id = j%num;
    cudaSetDevice (dev_id);

    //Wait for the data to finish at the end of each iteration
    cudaStreamSynchronize (stream[j]);
  }
}
```

Fig. 11.16 Excerpt of code with CUDA sequence of computations using multiple GPUs

11.4.3 Experimental Results

Using the experimental setup presented in Sect. 11.2.3, experimental results were obtained for the scalable implementation.

Figure 11.18 shows the results of the implemented algorithm with one and two GPUs. As it can be observed, by using multiple GPUs one can in fact reduce the execution time. It should however be noticed that in this particular case, the achieved speedup was not very significant. This is because ratio between computation and memory transfers is not very high. To overcome this issue, more advanced optimizations should be introduced in the program. Additionally, while in this case the multi-GPU system uses similar performance GPUs, their relative performance can be an issue. To efficiently utilize the computing power of devices with different processing performances, efficient scheduling algorithms should be used. Examples of such algorithms are presented in [1, 6, 12].

```
for (n=0; n<NSTEPS; n++){                                          OpenCL
  for (j=0; j<gpu_y_blocks; j++){
    dev_id = j%num_devices;

    // Copy data to the GPU
    clEnqueueWriteBuffer (cmd_queue[dev_id],d_Ez[dev_id],CL_FALSE,
                          .... &Ez[j*(d_IE*d_JE)], ....
                          &write_events[j][0]);
    ...

    // Set Kernel Arguments
    clSetKernelArg(kernel, 0,sizeof(cl_mem),&d_Ez[dev_id]);
    clSetKernelArg(kernel, 1,sizeof(cl_mem),&d_Hx[dev_id]);
    clSetKernelArg(kernel, 2,sizeof(cl_mem),&d_Hy[dev_id]);
    clSetKernelArg(kernel, 3,sizeof(cl_mem),&d_Ez_ghostCells[dev_id]);
    clSetKernelArg(kernel, 4,sizeof(cl_mem),&d_Hx_ghostCells[dev_id]);
    clSetKernelArg(kernel, 5,sizeof(cl_int), &n);

    // Kernel launch
    clEnqueueNDRangeKernel (cmd_queue[dev_id], kernel, 2, NULL,
                            global_work_size, local_work_size,
                            5, write_events[j], &kernel_events[j]);
  }

  // Copy data from the GPU
  for (j=0; j<gpu_y_blocks; j++){
    dev_id = j%num_devices;
    clEnqueueReadBuffer (cmd_queue[dev_id], d_Ez[dev_id],CL_FALSE,
                         .... &Ez[j*(d_IE*d_JE)],
                         1, &kernel_events[j], &read_events[j][0]);
    ...
  }

  // Wait for the data copy from the GPU to finish
  for (j=0; j<gpu_y_blocks; j++)
    clWaitForEvents (3, read_events[j]);
}
```

Fig. 11.17 Excerpt of code with OpenCL sequence of computations using multiple GPUs

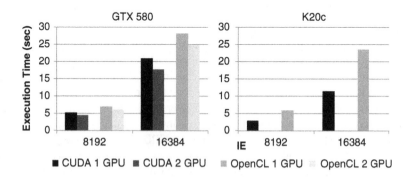

Fig. 11.18 Obtained results for a single and multiple devices

11.5 Conclusions

In this chapter we presented a parallel FDTD algorithm and proposed some tuning techniques for computing FDTD on GPUs. Optimizations were presented at a kernel level, where the purpose was to reduce the kernel launch overhead. Experimental results showed that the reduction of kernel launch overhead could achieved a speedup up to 2.6. Moreover, the computation and data transfer overlapping and the usage of multiple GPUs within the same system were also presented. The obtained results for multi-GPUs in this particular case did not show significant improvement, it requires further research to develop more advance optimization techniques. Nevertheless, all the obtained results showed that reduction in the execution times can be achieved by applying GPUs to compute the FDTD method.

Acknowledgements The work presented herein was partially supported by national funds through Fundação para a Ciência e a Tecnologia (FCT) under projects Threads (ref. PTDC/ EEA-ELC/117329/2010), P2HCS (ref. PTDC/EEI-ELC/3152/2012) and PEst-OE/ EEI/LA0021/2013, and also with the Ph.D. grant with reference number SFRH/BD/ 65636/2009.

References

1. Acosta, A., Corujo, R., Blanco, V., Almeida, F.: Dynamic load balancing on heterogeneous multicore/multiGPU systems. In: 2010 International Conference on High Performance Computing and Simulation (HPCS), pp. 467–476 (2010). doi:10.1109/HPCS.2010.5547097
2. Kuan, L., Tomas, P., Sousa, L.: A comparison of computing architectures and parallelization frameworks based on a two-dimensional FDTD. In: 2013 International Conference on High Performance Computing and Simulation (HPCS), pp. 339–346 (2013). doi:10.1109/HPCSim.2013.6641436
3. Nvidia: http://docs.nvidia.com/cuda/pdf/Kepler_Tuning_Guide.pdf. 7 Feb 2014
4. Nvidia: http://www.nvidia.com/content/PDF/kepler/NVIDIA-Kepler-GK110-Architecture-Whitepaper.pdf. 7 Feb 2014
5. Pratas, F., Trancoso, P., Stamatakis, A., Sousa, L.: Fine-grain parallelism using multi-core, cell/BE, and GPU systems: accelerating the phylogenetic likelihood function. In: International Conference on Parallel Processing, 2009 (ICPP'09), pp. 9–17. IEEE, Piscataway (2009)
6. Shirahata, K., Sato, H., Matsuoka, S.: Hybrid map task scheduling for GPU-based heterogeneous clusters. In: 2010 IEEE Second International Conference on Cloud Computing Technology and Science (CloudCom), pp. 733–740 (2010). doi:10.1109/CloudCom.2010.55
7. Taflove, A., Hagness, S.C.: Computational electromagnetics: The Finite-Difference Time-Domain Method, Third Edition. Artech House, (2005)
8. Taflove, A., Brodwin, M.: Numerical solution of steady-state electromagnetic scattering problems using the time-dependent Maxwell's equations. IEEE Trans. Microw. Theory Tech. **23**(8), 623–630 (1975). doi:10.1109/TMTT.1975.1128640
9. Wittenbrink, C.M., Kilgariff, E., Prabhu, A.: Fermi GF100 GPU architecture. IEEE Micro **31**(2), 50–59 (2011)
10. Yee, K., Chen, J.: The finite-difference time-domain (FDTD) and the finite-volume time-domain (FVTD) methods in solving Maxwell's equations. IEEE Trans. Antennas Propag. **45**(3), 354–363 (1997). doi:10.1109/8.558651

11. Zanjani, M., Akbari, A., Mirzaei, H., Shirdel, N., Gockenbach, E., Borsi, H.: Investigating partial discharge UHF electromagnetic waves propagation in transformers using FDTD technique and 3D simulation. In: 2012 International Conference on Condition Monitoring and Diagnosis (CMD), pp. 497–500 (2012). doi:10.1109/CMD.2012.6416187
12. Zhong, Z., Rychkov, V., Lastovetsky, A.: Data partitioning on heterogeneous multicore and multi-GPU systems using functional performance models of data-parallel applications. In: 2012 IEEE International Conference on Cluster Computing (CLUSTER), pp. 191–199. IEEE, Piscataway (2012)

Part III
Random Numbers and Monte Carlo Methods

Chapter 12
Pseudorandom Numbers Generation for Monte Carlo Simulations on GPUs: OpenCL Approach

Vadim Demchik

12.1 Introduction

The rapid development of computer technology during last decades has evoked intensive evolution of numerical methods. One of the most common classes of numerical algorithms is related to the Monte Carlo (MC) methods which relay on random numbers.

For several tasks, such as games or entertainment, special hardware, which produces random numbers, is often used. Such devices are based on the thermal noise in a resistor, the shot noise of a diode, illumination level, network traffic, user keystrokes or mouse movements. These devices are called true random number generators. The main flaws of such devices for MC simulations are low performance and impossibility to reproduce previous sequences of random numbers. That is why special algorithms, which can produce random numbers, are used in scientific applications instead of true random number generators. In this case we speak of pseudorandom number generators (PRNGs) in order to stress that random numbers are produced with the help of some determined algorithm. Below we will use this term.

Basic characteristics of any PRNG are its period, structure, size of generator state and algorithm complexity. One must clearly understand that there is no ideal PRNG, which is equally well suited to different tasks. Every PRNG has certain flaws and, while developing practical applications, it is very important to choose the generator which deficiencies can have a minimal impact on this particular task. The usage of a specific PRNG is always a balance between the statistical properties of the PRNG and its other characteristics. There are known cases when PRNG has a rather big period and easy algorithm, passes mathematical tests, but its usage in

V. Demchik (✉)
Dnipropetrovsk National University, Dnipropetrovsk, Ukraine
e-mail: vadimdi@yahoo.com

V. Kindratenko (ed.), *Numerical Computations with GPUs*,
DOI 10.1007/978-3-319-06548-9_12, © Springer International Publishing Switzerland 2014

computer simulations causes erroneous results. The notorious generator R250 [1] is an example. Its essential statistical defects were found 10 years afterwards in real MC simulations. In [2] Ferrenberg, Landau and Wong report essential discrepancies between known results of the exactly solvable Ising model and the results obtained in MC simulations with using of R250 PRNG. The source of these divergences (triplet correlations $\langle x_n x_{n-k} x_{n-250} \rangle$ around $k = 147$) was found several years later (see [3] and references therein). So, even if a generator passes artificial mathematical tests, it does not guarantee reliable results obtained with this generator. Besides, while choosing a generator for numerical experiment, one should not take into consideration only its productivity, the size of PRNG state or its simple implementation. The best way is to use one of the well-known generators which have been tested and applied by the community in the field [3].

The usage of graphics processing units (GPUs) for scientific simulations became an industrial standard in recent years. As known, GPU is massively parallel computational system with the SIMD architecture and very specific features (like high bandwidth memory, limited device memory, small register memory, etc.) The essential differences of GPU from classical computational facilities brought new requirements to the design and implementation of PRNGs for GPUs. During the fledgling years of the general purpose computing on graphics processing units (GPGPU) technology, random numbers were generated with CPU and then were copied into GPU device memory for further utilization. At present, this method is very low-efficient despite the enlargement of host-device bus bandwidth, although sometimes it can be used for the control of results.

In this chapter we describe general ideas, methods and tricks for PRNG implementation on GPUs with OpenCL framework by the example of the PRNGCL library. We restrict our discussion with PRNGs suitable for MC simulations only and do not consider PRNGs for cryptography (see [4] and references therein). Also we briefly review existing PRNG libraries for GPUs.

12.2   PRNGs on GPU

During the long history of computer simulations, a certain list of PRNGs, which are used in MC simulations, is formed. Such generators have well-known statistical properties, have been tested on many tasks and show a high performance.

One of the most frequently asked questions in the context of MC simulations is the choice of PRNG. The best practice is to use the generator exploited for producing the key results in the corresponding field of research. Any new generator employment requires additional testing the majority of known results because the new generator may possess some rarely appearing statistical defects. Utilization of the own generator without of serious theoretical background may cause substantial numerical discrepancies. L'Ecuyer, the author of the most comprehensive library of statistical tests TestU01, recommends to use MRG32k3a generator [5] as uniform PRNG in other cases. This generator is very robust and reliable, based on a solid theoretical analysis, and it also provides multiple streams and substreams.

In this section we briefly discuss a general structure of PRNG and methods of PRNG parallelization, denote the basic classes of PRNGs and survey the existing software for PRN generation on GPUs.

12.2.1   Structure of PRNGs

The present-day PRNGs are based on some algorithms. Let us consider such an algorithm. This algorithm produces the next generator state S_i from its previous state S_{i-1} and allows to obtain an item of PRN sequence X_i^{PRNG}, which corresponds to the state S_i of the generator

$$S_i = f(S_{i-1}), \quad X_i^{PRNG} = g(S_i). \tag{12.1}$$

Here and below we will use the upper index PRNG to identify the sequence produced by the generator PRNG. The maximum period of the generator, P^{PRNG}, is the length of the cyclic sequences produced by the PRNG and it is limited by the number of the states which can be represented by the PRNG. Thereby the algorithm is deterministic—its application to a given state of the generator leads to generation of the unique number and the unique next state. That is why the "randomness" can be provided only by the initial state of the generator. Good statistical properties of the generator are ensured both by the properties of the algorithm and a "good" choice of initial state. The choice of the initial state can also influence the PRNG period.

Generator state is a set of the *seed* (or *lag*) *table*, indices, flags and other PRNG runtime information. The size and structure of PRNG state are permanent for a specific PRNG. In some generators the last N numbers produced by the generator are used as the seed table. Generator period depends also on the seed table size. For optimal initialization of the seed table, as well as to reduce the number of initial parameters, a separate bootstrap procedure is often used. The bootstrap procedure is usually a simple PRNG with virtual small state, which fills the seed table according to the required criteria.

The general structure of PRNG is shown in Fig. 12.1. The work of generator starts with initializing its seed table—either by direct filling with initial seed values or by using bootstrap procedure. Any PRNG runtime stage conditionally can be divided into three phases:

- *initialization:* loading and preparing the previous PRNG state, carrying out all the preparatory operations for the PRNG
- *random number generation:* all the operations related to one or several PRNs production, updating the PRNG state
- *finalization:* storing the updated PRNG state for the next pass.

It is possible to store any PRNG state to continue PRN generation from the saved point.

There are two methods of PRN sequence generation: single PRN and batch PRN generation per one PRNG call. In the first method, PRN is produced and utilized directly by MC procedure (actually by means of inline procedures). This method can be applied to minimize the capacity of the global memory used. The usage of the single PRN generation method requires a direct correspondence of the number of MC kernel and PRNG kernel threads. The second method is to prepare PRNs by a procedure in the dedicated memory bucket with the following utilization of the PRNs. In this case the numbers of MC kernel and PRNG kernel threads could not coincide. It is obvious, that the second method allows to reach much higher performance due to single pass of initialization and finalization phases for the whole cycle of PRNs generation. This method is effective in the case of a relatively small amount of PRNs for MC procedure. In other case, most of the global memory is engaged only with the PRNs buffer, which shrinks the size of tasks we study. In order to make the PRNs buffer size smaller, the fragmentation of original task into several subtasks (for example, by dividing the whole lattice into several parts) and separate PRNs generation for the subtasks are used. Moreover, because of the hardware restrictions for the registered memory, the single PRN generation method can be unacceptable for the tasks with resource-intensive MC kernels. The batch PRN generation method allows to increase the productivity 1.3–5 times (see [6]).

To unify interfaces of PRNGs as well as to increase the productivity, we realize the batch PRN generation scheme in the library PRNGCL.

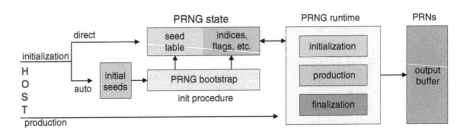

Fig. 12.1 The general structure of a PRNG

12.2.2 Basic Classes of PRNGs

Let us briefly introduce the following basic PRNG classes and present some members of these PRNG classes.

- linear congruential generators (LCG): one of the oldest and the most popular classes of PRNGs. It is based on the linear congruential integer recursion,

$$X_n^{LCG} = (aX_{n-1} + c) \bmod m, \tag{12.2}$$

where increment c and modulus m are desired to be positive coprime integers $(c < m)$ to provide a maximum period, multiplier a is an integer in the range $[2; (m - 1)]$. If increment $c = 0$ the LCG is often called the multiplicative linear congruential generator (MLCG), or Lehmer PRNG [7]. RANDU and Park-Miller (PM) [8] generators are the examples of LCG PRNGs:

$$X_n^{RANDU} = 65539 X_{n-1} \bmod 2^{31}, \tag{12.3}$$

$$X_n^{PM} = 16807 X_{n-1} \bmod (2^{31} - 1). \tag{12.4}$$

The maximum period of LCG is $P^{LCG} \leq (m - 1)$.

- feedback shift register generators (FSRG): another popular class of PRNGs. In 1965 Tausworthe introduced linear feedback shift register (LFSR) algorithm [9], based on the bit sequence

$$X_n^{LFSR} = \left(\sum_{i=1}^{r} a_i X_{n-i} \right) \bmod 2, \tag{12.5}$$

where $a_i, X_i = \{0; 1\}$, $r \leq n$. To obtain k-bit PRN Y_n from Eq. (12.5), one can group up k sequential bits,

$$Y_n^{LFSR} = \sum_{j=1}^{k} 2^{k-j} X_{kn+j-1}. \tag{12.6}$$

Such method is called the digital multistep method of Tausworthe. Another method proposed by Lewis and Payne [10] is the generalized feedback shift register (GFSR) algorithm. In GFSR scheme the bits in the positions j of the PRN are filled with the copy of initial one-bit recursion with some offsets $d_j \geq 0$,

$$Y_n^{GFSR} = \sum_{j=1}^{k} 2^{k-j} X_{n+d_j}. \tag{12.7}$$

R250 is an example of GFSR generators,

$$X_n^{R250} = X_{n-250} \wedge X_{n-103}. \tag{12.8}$$

Here $\wedge$ sign denotes an exclusive-or (XOR) logical operation, which is addition modulo 2 for one-bit variables. The maximum period of FSRG is $P^{LFSR} \leq 2^r - 1$.

- lagged Fibonacci generators (LFG): a popular class of PRNGs, which is based on the generalization of the well-known Fibonacci recurrence sequence

$$X_n^{LFG} = (X_{n-r} \odot X_{n-s}) \bmod m, \tag{12.9}$$

where r and s are called *"legs"*, $r \leq n$ and $1 < s < r$. RAN3 is one of the LFGs:

$$X_n^{RAN3} = (X_{n-55} - X_{n-24}) \bmod 10^9. \tag{12.10}$$

The LFG period for different operations $\odot$ is

$$P^{LFG} \leq \begin{cases} (2^r - 1)m/2 \text{ for } + \text{ or } - \\ (2^r - 1)m/8 \text{ for } \times \\ (2^r - 1) \qquad \text{ for } \wedge \end{cases} \tag{12.11}$$

- combined generators: a special and the widest class of PRNGs, which contains the features of the different PRNG classes. The multiple recursive generator (MRG) should be mentioned as the simplest extension of the LCG, which determined as the combination of the MLCGs

$$X_n^{MRG} = (a_1 X_{n-1} + a_2 X_{n-2} + \ldots + a_k X_{n-k} + c) \bmod m. \tag{12.12}$$

When $k > 1$, MRG is usually called MRG of the k order. The maximum period of the MRG is $P^{MRG} \leq m^k - 1$.

In contrast to the LCGs and FSRGs, which operates integer numbers only, the LFGs and some subclasses of combined generators allow to use a floating point numbers. It is particularly important when implementing on GPU-like computing devices.

12.2.3   Parallelization of PRNGs

The main difference of PRNG implementation for massively parallel systems is the necessity parallelization of computational streams. In addition, it is required to provide the statistical independency of PRNs within each subsequences as well as between the subsequences. Another requirement arises to enhance a high efficiency—all the threads must elapse of the same time, the threads divergence should be avoided or minimized. Besides, it is important that PRNG can be parallelized without loosing its efficiency.

The common methods of PRNs sequence parallelization are

1. random seeding: a widely using technique, which resides that all the threads use the same PRNG implementation, but each thread i operates its own PRNG state S_j^i. The initial PRNG state S_0^i is filled with different unique seeds.

$$S_0^i \to f(S_0^i) = S_1^i \to f(S_1^i) = S_2^i \to \ldots \to f(S_{j-1}^i) = S_j^i \to \ldots \tag{12.13}$$

$$X_1^i = g(S_1^i), \ X_2^i = g(S_2^i), \ \ldots, \ X_j^i = g(S_j^i), \ \ldots,$$

here and below the upper index denotes the PRNG thread index and the bottom index stays for the serial number of PRNG state. The main threat of the random seeding technique is the overlapping of PRN sequences produced by different threads. Nevertheless, for the majority of the actual generators, utilized in MC simulations, the probability of the overlaps is extremely low. The random seeding parallelization technique does not reduce PRNG performance, it is independent on the number of PRNG threads and is easy to be implemented for any PRNG

2. parametrization: a special technique for some PRNGs, which makes different threads utilize the same PRNG implementation with different parameters p_i for each thread. The obvious disadvantage of this method is a large number of independent unique parameters to be chosen, which makes this approach a rarely used on GPUs

$$S_0^i \to f(S_0^i, p_i) = S_1^i \to f(S_1^i, p_i) = S_2^i \to \ldots \to f(S_{j-1}^i, p_i) = S_j^i \to \ldots \quad (12.14)$$

$$X_1^i = g(S_1^i, p_i), \ X_2^i = g(S_2^i, p_i), \ \ldots, \ X_j^i = g(S_j^i, p_i), \ \ldots$$

3. block splitting: a method to split PRNs sequence into equal-size independent subsequences, each of which is produced by the corresponding thread. Initial PRNG states have to be computed. Main disadvantages of this method are the strong dependence on the number of PRNG threads and its usage within a limited class of PRNGs.

$$S_0^i \to f(S_0^i) = S_1^i \to f(S_1^i) = S_2^i \to \ldots \to f(S_{j-1}^i) = S_j^i \to \ldots \quad (12.15)$$

$$X_1 = g(S_1^1), \quad X_2 = g(S_2^1), \quad \ldots, X_i = g(S_K^1),$$
$$X_{K+1} = g(S_1^2), \ X_{K+2} = g(S_2^2), \ \ldots, X_{2K} = g(S_K^2),$$
$$X_{2K+1} = g(S_1^3), \ X_{2K+2} = g(S_2^3), \ \ldots, X_{3K} = g(S_K^3), \ldots$$

4. leapfrogging: a method similar to the block splitting technique, when each following member of PRNs sequence is produced by the next thread. The sequence is built by union of several items of subsequences in one. To perform the leapfrogging a PRNG must be able to skip a certain number of PRNs.

$$S_0^i \to f(S_0^i) = S_1^i \to f(S_1^i) = S_2^i \to \ldots \to f(S_{j-1}^i) = S_j^i \to \ldots \quad (12.16)$$

$$X_1 = g(S_1^1), \quad X_2 = g(S_1^2), \quad \ldots, X_i = g(S_1^N),$$
$$X_{N+1} = g(S_2^1), \ X_{N+2} = g(S_2^2), \ \ldots, X_{2N} = g(S_2^N),$$
$$X_{2N+1} = g(S_3^1), \ X_{2N+2} = g(S_3^2), \ \ldots, X_{3N} = g(S_3^N), \ldots$$

Properly speaking, the first two methods are artificial and do not guarantee full independence of the PRNG subsequences. Moreover, the great majority of existing PRNG libraries are optimized for using on single GPU device and parallelization on multi-GPU or GPU-cluster level is performed with random seeding method. Nevertheless, the first method seems to be the most applicable, since averaging of a considerable amount of data points is used in real numerical experiments to obtain the final result. A parallelization technique by combining several PRNGs with different parallelization methods is also often applied. However, we recommend checking the key results using different PRNGs.

12.2.4 Existing Implementations of PRNGs on GPUs

During the existence of GPGPU technology, a number of libraries has been developed to generate PRNs on GPUs. Most software are realized on the CUDA base as the most widespread platform for GPU programming. Many libraries have unique features and their application field. Some libraries are commercial software and their source code is closed. Nevertheless, it is possible to make the list of the PRNGs implemented in the libraries. The most comprehensive list of currently existing software packages for generating PRNs on GPUs is presented below.

CUDA Implementations

- CURAND: the most popular PRNG library for CUDA-ready GPUs, which was first released in 2010 [11]. It contains three PRNG implementations: XORWOW, MRG32k3a and MTGP32 Mersenne Twister. XORWOW is a combination of XORShift PRNG of period $2^{160} - 1$ and the Weyl generator of period 2^{32}. The period of XORWOW is $2^{192} - 2^{32}$. The main advantages are good integration with CUDA applications and high average performance. It should be noted that Saito and Matsumoto reported in [12,13] that XORWOW PRNG of CURAND is systematically rejected by three tests in the BigCrush battery of tests: Collision Over (test 7), Simplified Poker Test (test 27) and Linear Complexity Test (test 81). The rejection by test 7 is serious for MC simulations, since it is about the six most significant bits, whereas the rest two failed tests seem not to be very significant for usual MC. Another weak point of MTGP32 implementation in CURAND is the fixed limitation on independent PRN sequences. To increase the number of parallel streams, the parameter search should be performed. Unfortunately, it is very time consuming. For instance, finding a single parameter set for $k = 11,213$ can take up to an hour on the current hardware [14]. Thus, MTGP32 implemented in CURAND uses the predefined parameter set. The source codes for the library are unavailable
- Thrust: a parallel algorithms library [15]. Now it is a part of CUDA SDK, which contains the set of implementations of RAND0, RANLUX, RANLUX48, TAUS88 PRNGs

- NAG: a collection of numerical routines for GPUs [16]. It contains implementations of MRG32k3a and MT19937 PRNGs. The library is commercial software and its source codes are unavailable
- GASPRNG: GPU accelerated implementation of the Scalable Parallel Random Number Generators Library [17] (LFG, MLFG, PMLCG, LCG48, LCG64 and CMRG)
- TRNG: Tina's Random Number Generator Library [18] (LCG64, LCG64_shift, MRGn, MRGn_s, YARNn, YARNn_s, LFGn_xor, LFGn_plus, MT19937, MT19937_64)
- Random123: a library of cryptographic PRNGs [4] (AES, Threefish, Philox). A flexible implementation is applied in the library—due to the use of preprocessor directives and their dynamic redefinition the same code has both CUDA- and OpenCL-APIs
- PRAND: a library for pseudorandom number generation for modern CPUs and GPUs [19] (MRG32k3a, MT19937, LFSR113). The main feature of the library is the ability to parallelize PRNGs with up to 10^{19} independent substreams by the block splitting method
- MPRNG: a Massively Parallel Random Number Generation library [20] (MTGP, RANECU, TT800, PM, TAUS88, LFSR113, KISS07, DRAND48)
- GPU-rand: a template library for multi-GPU PRNGs [21] (LCG, LFG, Wichmann-Hill WHG)
- ShoveRand: a framework defining common rules to generate random numbers uniformly on GPUs [22] (MRG32k3a, MTGP32, TinyMT)
- MTGP: implementation of Mersenne Twister for Graphics Processors (MTGP) by authors of MT [14]. It contains both CUDA and OpenCL implementation of MT11213, MT23209 and MT44497 for 32-bit version and MT23209, MT44497, and MT110503 for 64-bit version. MTGP can generate 128 independent PRN sequences for each period. There is a variant of MTGP (MTGPDC, Dynamic Creator for MTGP) to generate more PRN subsequences.

OpenCL Implementations

- Random123: see previous paragraph "CUDA implementations"
- MTGP: see previous paragraph "CUDA implementations"
- OpenCLRNG: OpenCL implementation of Dynamic Creator Mersenne Twister (DCMT) [23]
- RANLUXCL: advanced implementation of RANLUX PRNG [24]
- MWC64X: OpenCL implementation of MWC64X PRNG [25]

As is obvious, the realization of MRG32k3a, MTGP and TAUS88 generators is dominated in most libraries. It should be noted that the parallelization procedure (block splitting and parametrization) exists for these generators, what makes these PRNGs attractive for implementation on massively parallel systems.

12.3 PRNGCL Construction

In this section we describe OpenCL implementation of a pseudorandom number generator by the example of PRNGCL library.

Nowadays the most popular programming languages for general-purpose computing on GPU are NVIDIA CUDA and OpenCL by the Khronos consortium. For compatibility with the wider range of multi-core devices we chose OpenCL as a software platform.

While developing the PRNGCL library, we are guided by the forthcoming key requirements:

1. implementation of classical PRNGs, which are used for actual MC simulations
2. portability and computational hardware-independence
3. independence of a given PRN sequence from computational devices
4. independence from the number of started PRNG threads
5. simple change of PRNG (by one parameter) for further PRNs utilization
6. optimization of PRNG implementations for GPU-like devices
7. a possibility of simple supplement of the library with other PRNG implementations

Below we will elaborate every point of these requirements and describe the main principles of their solutions.

12.3.1 OpenCL Tricks

For the library development we use some language- and GPU-architecture specific tricks to increase overall performance and portability.

12.3.1.1 Preprocessor Directives and Parameters Passing

To begin with, preprocessor directives are widely used in the library to define seldom changed parameters. OpenCL 1.0 standard and above allows to define preprocessor options directly through the compilation parameters of OpenCL programs by adding constructs like "-D name" or "-D name=definition" in compilation options (see clBuildProgram function in [26]). Here the constants defined in this way will take the values specified in the compilation options. For example, such parameters are the adjustable computational precision, luxury level for RANLUX PRNG, etc. In this case the runtime performance does not decrease. However, changing of such parameters requires rebuilding of the OpenCL program.

Other parameters that can be frequently changed (for example, a number of PRNs to be generated per thread) are passed directly into a kernel as arguments. There is some limitation to use the number of kernel arguments, which depends on hardware and SDK. Besides, using a big amount of arguments slightly reduces the kernel performance. If it is necessary to pass a bigger number of parameters into the kernel (for example, passing initial seed values for PRNG bootstrap), it is possible to use constant buffers. In terms of performance, a constant buffer is better to use than the regular memory object. However, for many applications the allowed maximum size of the constant buffer is also significantly less than the allowed size of the regular input-output memory object.

Conditional preprocessor directives #ifdef/#else are widely used in PRNGCL. This method allows to change the code flexibly depending on the processing precision or to perform certain operations (like luxurization in RANLUX PRNG, which depends on the luxury level).

Additionally, the operation system independence is provided by preprocessor redefinition of operation system-dependent functions (like input-output and timer operations) in source code of host program.

12.3.1.2   User-Defined Data Types

For flexible change of desired precision, we use own data types determined by parameters of OpenCL programs compilation: hgpu_float—the adjustable type for single or double precision float, hgpu_single for single precision and hgpu_double for double precision:

```
#define  hgpu_single  float
#ifdef DOUBLE_PRECISION_SUPPORTED
   #define  hgpu_double  double
   #ifdef PRECISION_DOUBLE
      #define  hgpu_float  double
   #else
      #define  hgpu_float  float
   #endif
#else
   #define  hgpu_float    float
   #define  hgpu_double   float
#endif
```

To ensure compatibility with computing devices, which do not support double precision, all of these data types are float data type on such devices. The same is true for vector data types. Using user-defined types does not decrease the performance of runtime phase of kernels. The host-side code should be adjusted according to the chosen precision—we employed it via void pointers to the memory objects of float or double data types.

12.3.1.3 Data Types and Structures

Our observations show that native data types can improve OpenCL kernel performance in compare with the data structures. This is due to the lack of efficiency of the present-day OpenCL compilers (by different vendors) in scalar code optimization. However, performance of kernels, which use data structures, can be increased with compilers development. That is why the most library source codes are optimized for `float4` data type.

12.3.1.4 Binaries Caching

OpenCL allows program executables to be built using either source codes (`clCreateProgramWithSource`), or using the precompiled device-sensitive binaries (`clCreateProgramWithBinary`)[26]. Preparation of binary program takes much less time, so it makes sense to keep the previously compiled program for following use. Some OpenCL SDKs contain internal caching system (NVIDIA CUDA SDK), but despite this tracking of changes in the dependent files (which are included with `#include` directive) it is not always correct. So, it is reasonable to use a separate caching procedure for OpenCL binaries.

12.3.1.5 Include Common Section

PRNGCL library is built according to the principle of all PRNG implementations independence—every PRNG implementation is a separate `.cl` file. However, as all PRNG implementations contain some common sections, they were carried out to a separate `.cl`-file. In particular, in this file there are description of precision definitions, user-defined data types and collection of functions for PRNs generation with double precision from two PRNs with single precision. Meanwhile a compiler option is added with path to the file with common program section, "`-I path`".

12.3.1.6 Memory-Access Optimization

To maximize memory throughput the coalesced memory access to the buffer with PRNG state is deployed—the neighboring threads access neighboring items in global memory. If particular implementation of PRNG requires several memory cell per one work-item, then such items spread in PRNG state with the interval, which is equal to the number of started work-items. It is also necessary to take into account the fact that for a wide range of OpenCL-compatible devices (especially, AMD GPUs) there is a limitation on the size of allocable memory for one memory object (half of total memory size).

The class of lagged generators requires storing PRNG state. Meanwhile PRNG state contains not only a seed table, but also a current index in it. For example, the seed table of RANMAR PRNG contains 97 float values, one float run parameter and two integer indices per each thread. It is obvious that in order to increase performance, it is reasonable to keep both the indices in PRNG state as float values. However, using float as indices is only possible while their converting into integer type. The type conversion is a "heavy" operation, so it is better to avoid it, if possible. That is why for storing such indices in PRNG state, it makes sense to use union data type to convert the type instantly.

12.3.1.7   Thread Divergence

Most OpenCL-compatible devices have a SIMD architecture—all processing elements execute a strictly identical instructions set. If a kernel contains a conditional operator with branches and if at least one of the work-group threads goes along the branch which is different from all other threads, then the remaining threads will wait for the full implementation of instructions in the branch by this tread. This effect is known as thread divergence and can significantly affect the performance. While developing PRNG implementations, it is commonly used a procedure of checking whether the variable resides in a given interval. It can be optimized by using vector data types and build-in relational function select(). If it can not be done, one should try to reduce the length of branches.

For further reading about general questions of OpenCL program optimization please refer to the books of Scarpino [27] and Tay [28].

12.3.2   Data Type Selection

Many PRNGs have realization for integer and floating-point arithmetics. However, the usage of integer arithmetic instead of floating-point arithmetic significantly reduces performance for many OpenCL-compatible devices [29]. Hence, it is better to use algorithms which are optimized for floating-point arithmetic.

It should be noted that PRNG implementation with integer arithmetics provides an important feature—portability, because it is not dependent on internal rounding. In some cases mad24() and mul24() build-in functions can be employed for operations with 24-bit integer values. Such functions use faster floating point hardware and can be executed on all compute units [26, 29]. However, the usage of 24-bit of integers significantly narrows the scope of their application.

Moreover, it seems to be reasonable to produce PRNs in float4 or double4 format. This is due to the following reasons:

- our observations show that the using of 4-vector data types allows to reach a great performance on most OpenCL-compatible devices. At the same time, even on devices with a scalar architecture due to rather effective OpenCL-compiler, within the drivers of correspondent devices, using 4-vector data types also allows to accelerate kernels slightly in compare with the case of scalar data types;
- as a rule, in real numerical experiments it needs more than one PRN for accomplishing each MC step

12.3.3 Double Precision

The architecture of modern GPUs is designed to obtain the best performance in single precision floating point operations. Top-end GPUs have performance on double precision arithmetics as a half of single precision one [30]. At the same time middle and low-end GPUs show much poorer performance in double precision floating point operations. For example, the ratio of single-to-double precision performance for NVIDIA Tesla K10 GPU reaches 24. Therefore the use of PRNGs, specially designed for the production of PRNs with double precision are not always effective.

Widely used recept, direct or indirect (for instance, by dividing integer 32-bit value by double precision value) converting single-to-double precision, does not allow to realize full double precision resolution. According to the IEEE 754 standard, the fractional part of floating point double precision number is stored in the lowest 52 bits (in the lowest 23 bits for single precision number). It is clear that operations with single precision value or even with 32-bit integers do not allow to realize all 2^{52} binary states. Thus it begs the obvious conclusion—to produce PRNs with double precision from several PRNs with single precision. This approach is PRNG-independent.

In PRNGCL library we offered and realized the general scheme when double precision PRNs are assembled from two PRNs with single precision. The idea of the scheme is rather simple—fulfil lower bits of output number with first PRN, and upper bits with the second one. At the same time, our procedure allows to keep the distribution uniformity of initial PRNs with single precision by discarding some values which are outside of the acceptable range of values [31].

12.3.4 PRNG State

One of the major constraints in the implementation of the GPU version of PRNGs is a low transfer rate between computing device and the host memory. This causes the necessity of permanent storage of seed tables directly in the compute device memory, and hence there are additional restrictions on the size and number of copies of the PRNG states. Another important constrain while GPU programming is a relatively low access rate of computing work-items to the global memory of computing device. Often it is advantageous to increase the number of arithmetic operations in the PRNG implementation to reduce the number of access to global memory in runtime—for example, instead of storing some indices in PRNG state they can be recalculated in runtime.

The standard technique to increase performance on data access operations is to utilize a local memory of compute unit for frequently used data. For example, PRNG state may be located in local memory for some PRNGs (like LFGs) in order to co-using by work-items in one compute unit. The usage of this method allows increase significantly performance of PRNG implementation and reduce the size of memory for PRNG state storing. Unfortunately, this approach has a significant drawback—PRNG implementation stays hardware-dependent because of different number of work-items per one compute unit on different devices. So, we do not use local memory in order to make hardware independent implementation.

12.3.5 PRNG Initialization and Portability

When developing a PRNG implementation on GPU for use in MC simulations, it is important to seek independence of PRNG output sequence from utilized hardware. On one hand, it is achieved by minimizing the possible rounding errors—by exclusion the use of atomic operations like MAD in the kernel. From the other hand, it is necessary to use external PRNG for generator initialization, but not build-in rand() function. In our implementation, we used XOR128 PRNG, as it provides portability, high productivity and has a small PRNG state. It allows to reproduce the same PRN sequence for any computing device and various operating systems. Besides, output PRN sequence generally depends on the amount of threads, which produce its generation. However, in our approach this parameter is also adjustable.

Some PRNGs like RANLUX, RANMAR and MT have a bootstrap procedure, which initializes PRNG seed table. This procedure is realized directly on the computing device and does not require additional memory objects transfer between host and computing device. Often, a bootstrap procedure has only one or two input parameters, which uniquely determine the PRNG initial state. Obviously, it is severely restricts the number of possible initial states of PRNG, but ensures a correct PRNG initialization. So, we introduced in PRNGCL library a unified initialization procedure for all PRNG threads based on only a single 32-bit unsigned integer number (randseries parameter). Moreover, the initialization scheme by only one parameter eliminates the problem of the PRNG initialization procedure dependence on the architecture of the computing device used (the number of available compute units and processing elements).

Another important question that arises while OpenCL programming—how many work-item should be started for particular kernel. If the inline version of PRNG implementation is used, then for each work-item of MC kernel the corresponding number of PRNG work-items must be started. There is a practice when several PRNG work-items may use the same PRNG seed table, but it is usually unsafe due to non-deterministic starting of work-items. While using a batch PRN generation method the number of PRNG work-items is in principle arbitrary. So, the number of PRNG work-items must be chosen in order to maximize the overall performance. As good variant we propose to use the number that is equal to CL_DEVICE_IMAGE3D_MAX_WIDTH (max width of 3D image in pixels) OpenCL parameter. In our opinion it provides the most optimal performance on a various hardware.

12.3.6 Testing of PRNs Sequence

Testing of PRNs is the standard way to qualify a PRNG. There are many statistical tests to check the randomness of PRNs sequence. In 1969 Donald Knuth proposed a set of statistical tests for PRNGs in famous multi-volume work "The Art of Computing Programming" [32]. The first suite of random number tests DIEHARD was written by Marsaglia in Fortran in 1990 [33]. Despite the fact that almost a quarter of a century has passed since the DIEHARD creation, it is still used for the necessary basic PRNGs checking. DIEHARD set of statistical tests is included in all new batteries of tests. Special requirements for PRNGs used for cryptographic purposes were a reason to create a separate tests set, offered by National Institute for Standards and Technology (NIST)–STS [34], mandatory for security applications. Another statistical tests set, DIEHARDER, initially developed by Brown [35], includes D. Knuth's, DIEHARD, NIST/STS, original and user-contributed tests. The most complete and hard to pass battery of tests at present day is TestU01, offered by L'Ecuyer [36]. Apart from separate statistical tests TestU01 includes several predesigned sets of tests (SmallCrush, Crush and BigCrush). Passing BigCrush test elapses about 7–10 h on modern CPUs.

There are two ways to test a PRNG with TestU01—to use a direct connection of implementation to the test of statistical properties or to use a file with previously prepared PRN sequence. Since the second method is universal for most batteries tests, we implemented it in the library PRNGCL.

A broad review of many CUDA PRNG implementations (LCG32, LCG64, MWC, LFG521, LFG1279, XORWOW, MTGP, XORShift/Weyl and Philo4x32) and results of its testing in real MC simulations (two-dimensional Ising ferromagnet model) could be found in [37].

Obviously, testing all the numbers producing by PRNG is not possible, so it is preferable to use several testing procedure starts to identify potential correlations between PRN subsequences. It should be understood that passed tests guarantees nothing, but checking is a necessary condition for any PRNG implementation.

12.3.7 PRNG Example: XOR7

As an example of PRNG implementation we provide here an OpenCL implementation of XOR7 PRNG.

Originally the class of XORshift generators was proposed by Marsaglia in [38]. XORshift PRNG class is one of the variants of feedback shift register generators. Let X_0^{XORshift} be a some initial k-bit row-state of XORshift and T is $k \times k$ nonsingular binary matrix which sets linear transformation. The n-th PRNG state may be derived through the following equation

$$X_n^{\text{XORshift}} = (x_1, x_2, \ldots, x_{k-1}, x_k)_n^{\text{XORshift}} = X_0 T^n. \tag{12.17}$$

To ensure the performance requirements Marsaglia proposed the special form of matrix T,

$$T = (I + L^a)(I + R^b)(I + L^c),\qquad(12.18)$$

where I is an identity $k \times k$ matrix, matrices L and R are $k \times k$ binary matrices

$$L = \begin{pmatrix} 0 & 0 & \dots & 0 \\ 1 & 0 & \dots & 0 \\ \vdots & \ddots & \ddots & \vdots \\ 0 & \dots & 1 & 0 \end{pmatrix}, \quad R = L^T = \begin{pmatrix} 0 & 1 & \dots & 0 \\ \vdots & \ddots & \ddots & \vdots \\ 0 & 0 & \dots & 1 \\ 0 & 0 & \dots & 0 \end{pmatrix},\qquad(12.19)$$

which effect shift of one to the left (L) and right (R), correspondingly,

$$(x_1, x_2, \dots, x_{k-1}, x_k)L = (x_2, x_3, \dots, x_k, 0),\qquad(12.20)$$
$$(x_1, x_2, \dots, x_{k-1}, x_k)R = (0, x_1, \dots, x_{k-2}, x_{k-1}).$$

So, if X_m is a k-bit state then L^a causes the new state $L^a X_m \equiv (X_m \ll a)$ as well as $(I + L^a)$ – the state $(I + L^a)X_m \equiv X_m \wedge (X_m \ll a)$. Here $\ll$ stays for a bitwise shift to the left operation.

Marsaglia lists in [38] all possible full-period triplets $[a, b, c]$ for 32-bit (648 combinations) and 64-bit (2,200 combinations) XORShift PRNG. In addition Marsaglia present a 128-bit PRNG XOR128 $[a, b, c] = [11, 19, 8]$, which is also implemented in PRNGCL library. The key feature of XOR128 PRNG is matching of PRNG state size with the size of GPU memory cell. The maximal period of XORshift generator is

$$P^{\text{XORshift}} \le 2^k - 1.\qquad(12.21)$$

In original paper [38] Marsaglia reported that XORShift PRNG passes the DIEHARD battery of tests of randomness. In [39] Brent showed that the sequences generated by this PRNG are identical to the sequences generated by certain linear feedback shift register (LFSR) generators using XOR operations. So, XORShift PRNG inherits all the good (and bad) theoretical properties of LFSR generators. Later Panneton and L'Ecuyer appoint that XORShift PRNG "spectacular failed" the SmallCrush and Crush battery of tests of TestU01 package [40]. They did not recommend to use this class of the generators, but propose own variant of the XORShift PRNGs, which passes all the tests in Crush,—seven-XORShift (XOR7) [40]. To improve the statistical robustness Panneton and L'Ecuyer proposed to increase the number of xorshifts. Besides, Brent also performed empirical studying of XORShift PRNGs with Magma computational algebra system to find good $[a, b, c]$ parameter sets [41]. He showed how XORShift PRNGs can be generalized by means of Weyl generator to give high-quality PRNGs with extremely long periods, greater than 10^{1232}.

XOR7 generator is determined by the following recurrent relation:

$$
\begin{aligned}
X_n^{\text{XOR7}} = {} & (I+L^{17})X_{n-1}^{\text{XOR7}} + (I+L^{10})X_{n-2}^{\text{XOR7}} + (I+R^9)(I+L^{17})X_{n-4}^{\text{XOR7}} \\
& + (I+R^3)X_{n-4}^{\text{XOR7}} + (I+R^{12})X_{n-5}^{\text{XOR7}} + (I+R^{25})X_{n-5}^{\text{XOR7}} \\
& + (I+R^3)(I+R^2)X_{n-6}^{\text{XOR7}} + (I+R^{27})X_{n-7}^{\text{XOR7}} \\
& + (I+R^{22})X_{n-7}^{\text{XOR7}} + (I+L^{24})(I+R^3)X_{n-8}^{\text{XOR7}},
\end{aligned}
\tag{12.22}
$$

here X_i^{XOR7} is a 32-bit word. So, the XOR7 PRNG state contains eight 32-bit words or two `uint4` items. Hence, it is very attractive to be implemented on GPU. The core of XOR7 PRNG, recursion (12.22), may be optimized and presented as following OpenCL function:

```
__attribute__((always_inline)) void
xor7step(uint4* seed1,uint4* seed2){
    uint t, y;
        t=(*seed2).w; t =t^(t<<13); y=t^(t<<9);
        t=(*seed2).x; y^=t^(t<<7);
        t=(*seed1).w; y^=t^(t>>3);
        t=(*seed1).y; y^=t^(t>>10);
        t=(*seed1).x; t =t^(t>>7);   y^=t^(t<<24);
    (*seed1).xyz=(*seed1).yzw;
    (*seed1).w  =(*seed2).x;
    (*seed2).xyz=(*seed2).yzw;
    (*seed2).w  =y;
}
```

Here `seed1` and `seed2` are 4-component unsigned integers, which contain the PRNG state. After applying the `xor7step` function to PRNG state the last `.w` component of `seed2` takes new X_n^{XOR7} value.

The OpenCL kernel `xor7` to produce XOR7 PRN sequence with single precision by using `xor7step` function is shown below:

```
__kernel void
xor7(__global uint4* seed_table,
     __global hgpu_float4* randoms,
     const uint N){
// Initialization
   uint giddst = GID;
   float4 r;
   float4 m = (float4) (4294967296.0f); // = 2^32
   uint4 seed1 = seed_table[GID];
   uint4 seed2 = seed_table[GID + GID_SIZE];
// Production
   for (uint i=0; i<N; i++) {
      xor7step(&seed1,&seed2); r.x=(float) seed2.w;
      xor7step(&seed1,&seed2); r.y=(float) seed2.w;
      xor7step(&seed1,&seed2); r.z=(float) seed2.w;
      xor7step(&seed1,&seed2); r.w=(float) seed2.w;
      randoms[giddst] = r / m;
      giddst += GID_SIZE;
   }
// Finalization
   seed_table[GID] = seed1;
   seed_table[GID + GID_SIZE] = seed2;
}
```

It contains three arguments: seed_table – uint4 buffer with XOR7 seed table, randoms – float4 buffer to store the resulting PRNs and N—the number of PRNs to be produced per one kernel run. GID—is a scalar value that defines a global ID of work-item (thread), GID_SIZE—is the total number of global work-items (threads). In order to convert the 32-bit unsigned integer output r into $[0; 1)$ interval the $m = 2^{32}$ multiplier is used. Each work-item stores own output PRN in randoms buffer pointed by giddst index. All PRNs produced by a single work-item in randoms buffer are separated by GID_SIZE samples generated by other work-items to optimize memory access.

According to the classification introduced in Sect. 12.2.1 the xor7 kernel can be divided into three parts: initialization, production and finalization. In the first part the initial adjustment of variables as well as loading of seed table into private memory are performed. In the second part the kernel generates PRNs and stores them in the output buffer randoms. In the latter kernel part the final PRNG state is saved into seed_table buffer.

The original version of XOR7 PRNG [40] does not have a bootstrap procedure for PRNG state special initialization, so it is initialized by the host through the direct filling of seed_table buffer with random numbers. After the preparing on the host side it is transferred into computing device. The random seeding method was applied to parallelize the XOR7 PRNG. In such scheme the number of work-items for xor7 kernel is not related to any particular value and may be chosen by any value.

The efficiency of an OpenCL kernel on different hardware is determined by arithmetics/memory-access operations ratio. On the other hand the overall performance of the kernel depends on the number of memory-accesses operations and their stretching with arithmetic instructions. Due to a few number of memory-access operations, this kernel is well-balanced with arithmetics operations. So, it demonstrates a good performance on different hardware.

There is also another fashion to store the XOR7 PRNG state—it may be stored in eight independent uint4 memory cells (for four threads at once). In that case it becomes possible to replace all operations in xor7step function directly with vector instructions. It is easy to check that the falling of the overall performance will be observed in that case.

12.3.8   Using PRNGCL Library

As an example of using PRNGCL library we choose a well-known problem of π number approximation using MC method. The concept of this method is quite simple—we take a square of side $a = 2R$, put inside it a circle of radius R and begin to place points inside the square. Geometrically, the probability P_{geom} that point falls in a circle is the ration of the circle and square areas: $P_{geom} = S_{circle}/S_{square} = \pi/4$. On the other hand, the probability P_{num} to hit a point in the circle can be calculated in numerical experiment—if we take N pairs (x_i, y_i) of PRNs uniformly distributed in the interval $[0; R)$ and count the number M of such pairs that hit the circle $(x_i^2 + y_i^2 \leq R)$: $P_{num} = M/N$. The probabilities P_{geom} and P_{num} should coincide in the large N limit: $\lim_{N \to \infty}(P_{geom} - P_{num}) = 0$ and hence $\pi \approx 4M/N$.

The kernel to calculate the probability P_{num} in $2N$ numerical attempts per one work-item is shown below,

```
__kernel void
PI(__global hgpu_float4* randoms,
   __global float* acceptance,
    const uint N){
  uint index = GID;
  float count=0.0;
  hgpu_float4 rnd,rnd2;
  uint4 seed1 = seed_table[GID];
```

(continued)

```
(continued)
   for (uint i=0; i<N; i++) {
      rnd = randoms[index];
      rnd2 = rnd * rnd;
      if ((rnd2.x+rnd2.y) <= 1.0) count += 1.0;
      if ((rnd2.z+rnd2.w) <= 1.0) count += 1.0;
      index += GID_SIZE;
   }
   acceptance[GID] += count;
}
```

Here `randoms` is `float4` memory buffer with PRNs, generated with any of
the implemented PRNGs. The `acceptance` buffer contains the number M of
successful attempts per one work-item to hit a point inside the circle. Reducing
the `acceptance` buffer over all work-items and averaging over $2N$ we get the π
approximation.

The pseudocode of host-side program is listed below

```
examplePI(int argc, char** argv){
   // Create and prepare OpenCL context
   parameters=get_parameters(argc,argv);
   context=context_select(parameters);    // new context

   // Create and prepare PRNG
   uint passes=1000;
   uint W=get_auto_instances(context);  // # of work-items
   uint N=1024;                         // # of samples per one work-item
      PRNG_set_default_instances(W);
      PRNG_set_default_samples(N);
   PRNG* prng=PRNG_new(PRNG_XOR7);         // create PRNG
   uint pID=PRNG_init(context,prng);       // init PRNG
   uint RID=prng->parameters->id_buffer_randoms;

   // Prepare, build and bind buffers to PI kernel
   char* src=file_read(SRC);               // read source code
   cl_program prg=program_new(src,context);
   cl_float* acc=calloc(W,sizeof(cl_float));
   uint AID=context_buffer_init(context,acc,TYPE_IO,
            W,sizeof(cl_float));  // create acceptance buffer
```

(continued)

(continued)

```
K=context_kernel_init(context,prg,"PI",1,{W});
context_kernel_bind_buffer(context,K,RID);
context_kernel_bind_buffer(context,K,AID);
context_kernel_bind_constant(context,K,&N);

// Produce PRNs and approximate PI passes times
for (uint i=0; i<passes; i++){
      PRNG_produce(context,pID); // produce PRNs into RID
      context_kernel_run(context,K);
// run PI kernel
      }

// Map result to host, reduce and print result
double result = 0.0;
cl_float* y=buffer_get_mapped(context,AID);
for (uint i=0; i<W; i++)
      result += y[i];
printf("PI=%f",2*result/(passes*W*N));
}
```

In this example we use an automatically adjusted number of work-items for PI kernel. On one hand, the usage of greater number of work-items increases a kernel performance, on the other hand—the usage of greater number of work-items leads to a greater compute device memory consumption. Thereby, as a number of work-items W we usually use a value which is determined by OpenCL parameter CL_DEVICE_IMAGE3D_MAX_WIDTH. Each work-item of PI kernel produces $4N = 4{,}096$ PRNs. It is obvious that both these parameters (W and N) can be adjusted manually.

Full sample source code with remarks is available in the PRNGCL library.

12.3.9 Performance

The average performance results for different OpenCL-ready computing devices for both single and double precision generation are presented in Figs. 12.2 and 12.3, correspondingly. We use six different graphics cards for benchmarks: AMD Radeon HD5870, HD6970, NVIDIA GeForce GTX560M, GTX560, GTX780 and Tesla K40m (with ECC turned on and off). Some of GPUs used here are the primary GPU devices installed in the system (i.e. they also provide visualization for the operating system) which lowers down the maximal performance of the system, but reflects more precisely the usual configuration of the GPU computational system.

A standard PRNG benchmarking procedure is the generation of some number of PRNs and measuring of elapsed time. All generated PRNs are stored in device global memory. Because of high performance of the PRNs generation the benchmarking procedure is started several times in order to increase the accuracy of measurements. In our case we generate PRNs for 3 s per each set of parameters and then measure the performance.

There are two parameters, which determine the total number of PRNs to be produced: number of PRNG work-items N and number of PRNs per one PRNG work-item M. As the number of PRNG work-items we use the number that is equal to the CL_DEVICE_IMAGE3D_MAX_WIDTH OpenCL parameter. The number of PRNG work-items is $N = 2048$ for most used GPUs. For example, the performance penalty may reach 100 % when using 4,096 work-items instead of 16,384 for Tesla K40m GPU. In our benchmarks all available device memory are used for PRNs generation. So, the maximum number of PRNs produced by one PRNG work-item is limited to size of available GPU memory. This parameter is also chosen as 2 in some power n, $M = 2^n$, to provide a best performance. To determine the optimum value of M we perform several benchmarks, which use a half of the previous M value. We found that the best performance occurs not for the highest possible value of $M^{highest}$, but for $M^{opt} = M^{highest}/2$ or for $M^{opt} = M^{highest}/4$. The performance falls with decreasing of M. The difference in performance when using $M = 1$ and M^{opt} may be 1500 %.

Among the generators there are two "toy" PRNGs: CONSTANT and PM (Park-Miller LCG). The first one is the simple kernel which just returns a given constant and may be used for debugging purposes of MC procedure. The second one has a very short period, which could be exhausted in millisecond on modern GPUs. The performance results are shown here for both of them for comparison only.

The PM and XOR128 generators have the same size of PRNG state (single 32-bit integer number) and perform the same number of memory-access operations. However, these generators show the different performances. It is caused by the following reason: the PM generator performs integer arithmetic operations on items of the seed table while the XOR128 uses the fast bitwise operations on them. The similar situation appears for XOR7 and RANECU, which contain two 32-bit integers in corresponding seed tables, but uses different arithmetic types.

Memory access is a major bottleneck of GPU-applications. It is confirmed once again by comparison of the performance results for different PRNGs on various GPUs. The PRNGs with the greater number of the memory operations demonstrate the worst performance results.

One of the key requirements in the developing of the PRNGCL library is full platform-independence. It reduces the ability to fine optimization directly for a specific hardware using special profilers. So, we left the opportunity to further source code optimization, for example, by splitting a kernel source code for particular computing devices with preprocessor directives.

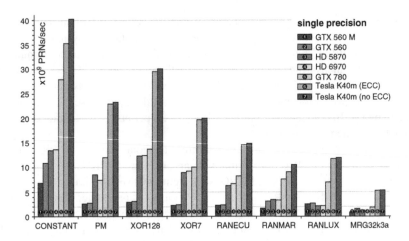

Fig. 12.2 Performance results for PRNGs in PRNGCL (single precision), ×10⁹ PRNs/s

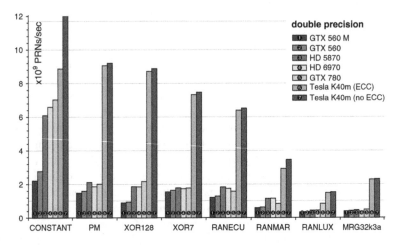

Fig. 12.3 Performance results for PRNGs in PRNGCL (double precision), ×10⁹ PRNs/s

12.4 Conclusions

Using PRNGs for MC simulations it is necessary to remember that it is based on some deterministic algorithms, and therefore one needs to be prepared that some statistical defects may occur in a given task. Back in 1951 John von Neumann said: *"Any one who considers arithmetical methods of producing random digits is, of course, in a state of sin"* [42]. It is important to use generators which are well tested in relation to a specific class of tasks [3]. Intensive development of computation performance on the background of limitation of PRNGs parallelization techniques also limits the scope of application of proven generators. It is important to demonstrate the independence of key results from PRNG used.

There is no "magic" PRNG, which can equally well work out all problems. If the generator shows good statistical properties in a specific problem, it means we are still within the scope of its applicability. Nevertheless, it does not guarantee that use of this generator for any other task will lead to the same positive result. Notorious generators R250 and RANDU can serve as examples.

GPGPU, as a new computing platform, brings its own requirements for the implementation of PRNG. These are additional requirements for parallelization mechanism of PRNGs by 1000 and more threads, limitation of the PRNG state size and limitation of the memory-access operations density. In addition, existing GPGPU-ready hardware is quite varied and those tricks, which lead to increased performance on particular device, can reduce the performance of the program on other devices. The maximum performance of software can be only achieved by using hardware-specific information, which narrows the optimal use of such program.

In this chapter a general structure of PRNGs and main parallelization methods are briefly described. A broad review of currently existing software packages for generating PRNs on GPUs is presented. Some tricks for PRNG implementation on GPUs with OpenCL framework by the example of the PRNGCL library are considered. The PRNGCL library contains implementation of a number of the most popular uniform generators.

For further reading, please refer the books by Gentle [43] and Niederreiter [44]. Additional information on PRN generation with GPUs are presented in [6, 19, 22, 37, 45, 46]. Complete bibliography concerning PRNGs could be found in [47].

We hope that more researchers will use, extend and contribute to the PRNGCL project. PRNGCL is available via GitHub: https://github.com/vadimdi/PRNGCL.

Acknowledgements Author thanks to my colleagues and friends Alexey Strelchenko for providing benchmarks for PRNGCL library on high-end NVIDIA GPUs, Natalia Kolomoyets, Alexey Gulov and Yevgen Syetov for essential help with the text preparation.

References

1. Kirkpatrick, S., Stoll, E.P.: A very fast shift-register sequence random number generator. J. Comput. Phys. **40**, 517–526 (1981)
2. Ferrenberg, A.M., Landau, D.P., Wong, Y.J.: Monte Carlo simulations: hidden errors from "good" random number generators. Phys. Rev. Lett. **69**, 3382–3384 (1992)
3. Janke, W.: Pseudo random numbers: generation and quality checks. In: Grotendorst, J., Marx, D., Muramatsu, A. (eds.) Quantum Simulations of Complex Many-Body Systems, vol. 10, p. 447. John von Neumann Institute for Computing, Julich (2002)
4. Salmon, J., Moraes, M., Dror, R., Shaw, D.: Parallel random numbers: as easy as 1, 2, 3. In: International Conference for High Performance Computing, Networking, Storage and Analysis (SC), pp. 1–12. http://www.deshawresearch.com/resources_random123.html (2011). Cited 6 Feb 2014

5. L'Ecuyer, P.: Uniform random number generators. In: Lovric, M. (ed.) International Encyclopedia of Statistical Science, pp. 1625–1630. Springer, Berlin/Heidelberg (2011)
6. Demchik, V.: Pseudo-random number generators for Monte Carlo simulations on ATI Graphics Processing Units. Comput. Phys. Commun. **182**, 692–705 (2011)
7. Lehmer, D.: Mathematical methods in large-scale computing units. Annu. Comput. Lab. Harvard Univ. **26**, 141–146 (1951)
8. Park, S., Miller, K.: Randoms number generators: good ones are hard to find. Commun. ACM **31**(10), 1192–1201 (1988)
9. Tausworthe, R.: Random numbers generated by linear Recurrence Modulo Two. Math. Comput. **19**, 201–209 (1965)
10. Lewis, T., Payne, W.: Generalized feedback shift register pseudorandom number algorithm. J. ACM **20**, 456–468 (1973)
11. NVIDIA, Corporation: CURAND Library: Programming Guide, p. 123. http://docs.nvidia.com/cuda/curand/index.html (2013). Cited 6 Feb 2014
12. Saito, M., Matsumoto, M.: Variants of Mersenne Twister Suitable for Graphic Processors. ACM Trans. Math. Softw. **39**(2), 12:1–12:20 (2012)
13. Saito, M., Matsumoto, M.: A deviation of CURAND: standard pseudorandom number generator in CUDA for GPGPU. In: Proceedings of 10th International Conference on Monte Carlo and Quasi-Monte Carlo Methods in Scientific Computing (2012)
14. Saito, M., Matsumoto, M.: Variants of Mersenne Twister Suitable for Graphic Processors. arXiv:1005.4973 [cs.MS], pp. 1–23. http://www.math.sci.hiroshima-u.ac.jp/~%20m-mat/MT/MTGP/index.html (2013). Cited 6 Feb 2014
15. Bell, N., Hoberock, J.: Thrust: a productivity-oriented library for CUDA. http://docs.nvidia.com/cuda/thrust/ (2013). Cited 6 Feb 2014
16. The NAG Library. The Numerical Algorithms Group (NAG), Oxford. http://www.nag.com/ (2013). Cited 6 Feb 2014
17. Gao, Sh., Peterson, G.: GASPRNG: GPU accelerated scalable parallel random number generator library. Comput. Phys. Commun. **184**(4), 1241–1249 (2013)
18. Bauke, H.: TRNG: Tina's Random Number Generator Library. http://numbercrunch.de/trng/ (2013). Cited 6 Feb 2014
19. Barash, L., Shchur, L.: PRAND: GPU accelerated parallel random number generation library: using most reliable algorithms and applying parallelism of modern GPUs and CPUs. arXiv:1307.5869 [physics.comp-ph], pp. 1–24. http://www.comphys.ru/barash/prand.tar.gz (2013). Cited 6 Feb 2014
20. Dammertz, H., Schied, C., Lensch, H.: Massively parallel random number generators. In: GPU Technology Conference. http://mprng.sourceforge.net/ (2010). Cited 6 Feb 2014
21. Szalkowski, D., Stpiczynski, P.: Template library for multi-GPU pseudorandom number recursion-based generators. In: IEEE Federated Conference on Computer Science and Information Systems (FedCSIS), pp. 515–519 (2013)
22. Passerat-Palmbach, J., Mazel, C., Bachelet, B., Hill, D.R.C.: Shoverand: a model-driven framework to easily generate random numbers on GP-GPU. In: Proceedings of International Conference on High Performance Computing and Simulation (HPCS), pp. 41–48. https://github.com/jHackTheRipper/ShoveRand (2011). Cited 6 Feb 2014
23. Arampatzis, G., Athanasopoulos, A.: Random Number Generator – Parallel Streams in OpenCL. http://mira.math.udel.edu/ParallelKMC/doku.php?id=projects:opencl_prng (2011). Cited 6 Feb 2014
24. Nikolaisen, I.: Bose-Einstein condensation in trapped bosons: a quantum Monte Carlo analysis using OpenCL and GPU programming. Masteroppgave, University of Oslo. https://bitbucket.org/ivarun/ranluxcl/overview (2011). Cited 6 Feb 2014
25. Thomas, D., Howes, L., Luk, W.: A comparison of CPUs, GPUs, FPGAs, and massively parallel processor arrays for random number generation. In: Proceedings of FPGA, pp. 63–72. http://cas.ee.ic.ac.uk/people/dt10/research/rngs-gpu.html (2009). Cited 6 Feb 2014
26. Munshi, A. (ed.): The OpenCL 1.0 Specification. Khronos OpenCL Working Group and others (2008)

27. Scarpino, M.: OpenCL in Action: How to Accelerate Graphics and Computation, 434 pp. Manning, Shelter Island (2012)
28. Tay, R.: OpenCL Parallel Programming Development Cookbook, 302 pp. Packt Publishing, Mumbai (2013)
29. Advanced Micro Devices, Inc.: AMD Accelerated Parallel Processing OpenCL Programming Guide (2013)
30. Comparison of Nvidia graphics processing units. http://en.wikipedia.org/wiki/Comparison_of_ Nvidia_graphics_processing_units. Cited 6 Feb 2014.
31. Demchik, V., Gulov, A.: Increasing precision of uniform pseudorandom number generators. arXiv:1401.8230 [cs.MS], pp. 1–4 (2014)
32. Knuth, D.: The Art of Computer Programming, vol. 2: Seminumerical Algorithms, 624 pp. Addison-Wesley, Reading (1969)
33. Marsaglia, G.: DIEHARD: a battery of tests of randomness. Technical report, Florida State University. http://www.stat.fsu.edu/pub/diehard/ (1996). Cited 6 Feb 2014
34. Rukhin, A., et al.: A Statistical Test Suite for the Validation of Random Number Generators and Pseudo Random Number Generators for Cryptographic Applications. NIST special publication 800-22, 131 pp. http://csrc.nist.gov/groups/ST/toolkit/rng/ (2010). Cited 6 Feb 2014
35. Brown, R., Eddelbuettel, D., Bauer, D.: DIEHARDER: A Random Number Test Suite. http://www.phy.duke.edu/~rgb/General/dieharder.php (2009). Cited 6 Feb 2014
36. L'Ecuyer, P., Simard, R.: TestU01: a C library for empirical testing of random number generators. ACM Trans. Math. Softw. 33, 1–40 (2007) [article 22]. http://www.iro.umontreal. ca/~simardr/testu01/tu01.html. Cited 6 Feb 2014
37. Manssen, M., Weigel, M., Hartmann, A.K.: Random number generators for massively parallel simulations on GPU. Eur. Phys. J. ST 210, 53–71 (2012)
38. Marsaglia, G.: Xorshift RNGs. J. Stat. Softw. 8(14), 1–6 (2003)
39. Brent, R.: Note on Marsaglia's Xorshift random number generators. J. Stat. Softw. 11(5), 1–5 (2004)
40. Panneton, F., L'Ecuyer, P.: On the Xorshift random number generators. ACM Trans. Model. Comput. Simul. 15(4), 346–361 (2005)
41. Brent, R.: Some long-period random number generators using shifts and xors. ANZIAM J. 48, C188–C202 (2006)
42. von Neumann, J.: Various techniques used in connection with random digits. Natl. Bur. Stand. Appl. Math. Ser. 12, 36–38 (1951)
43. Gentle, J.: Random Number Generation and Monte Carlo Methods, 399 pp. Springer, Berlin (2003)
44. Niederreiter, H.: Random Number Generation and Quasi-Monte Carlo Methods, 241 pp. SIAM, Philadelphia (1992)
45. Nandapalan, N., Brent, R., Murray, L., Rendell, A.: High-performance pseudo-random number generation on graphics processing units. Parallel Process. Appl. Math. 7203, 609–618 (2012)
46. Bradley, T., du Toit, J., Giles, M., Tong, R., Woodhams, P.: Parallelisation techniques for random number generators. In: Hwu, W.-M. (ed.) GPU Gems: Emerald Edition, pp. 231–246. Morgan Kaufmann, Amsterdam (2011)
47. Beebe, N.: A Bibliography of Pseudorandom Number Generation, Sampling, Selection, Distribution, and Testing, 631 pp. ftp://ftp.math.utah.edu/public_html/pub/tex/bib/prng.ps.gz (2013). Cited 6 Feb 2014

Chapter 13
Monte Carlo Automatic Integration with Dynamic Parallelism in CUDA

Elise de Doncker, John Kapenga, and Rida Assaf

13.1 Introduction

Problems in computational geometry, computational physics, computational finance and other fields give rise to computationally expensive integrals. This chapter will address applications of CUDA programming for Monte Carlo integration. A *Monte Carlo* method approximates the expected value of a stochastic process by sampling, i.e., by performing function evaluations over a large set of random points, and returns the sample average of the evaluations as the end result. The functions that are found in the application areas mentioned above are usually not smooth and may have singularities within the integration domain, which enforces the generation of large sets of random numbers.

The algorithms described here will be incorporated in the PARINT multivariate integration package, where we are concerned with the *automatic* computation of an integral approximation

$$Q[\mathscr{D}]f \approx \mathscr{I}[\mathscr{D}]f = \int_{\mathscr{D}} f(\mathbf{x})\, d\mathbf{x},$$

and an estimate or bound $\mathscr{E}[\mathscr{D}]f$ for the error $E[\mathscr{D}]f = |\mathscr{I}[\mathscr{D}]f - \mathscr{Q}[\mathscr{D}]f|$. Thus PARINT is set up as a *black-box*, to which the user specifies the integration problem including a requested accuracy and a limit on the number of integrand evaluations.

In its current form, PARINT supports the computation of multivariate integrals over hyper-rectangular and simplex regions. The user specifies the region $\mathscr{D}$ by the upper and lower integration limits in case of a hyper-rectangular region, or by the vertex coordinates of a simplex. The integrand $f : \mathscr{D} \to \mathbb{R}^k$ is supplied as a function written in C (or which consists of C-callable code). The desired accuracy

E. de Doncker (✉) • J. Kapenga • R. Assaf
Western Michigan University, 1903 W. Michigan Avenue, Kalamazoo, MI 49008, USA
e-mail: elise.dedoncker@wmich.edu; john.kapenga@wmich.edu; rida.assaf@wmich.edu

V. Kindratenko (ed.), *Numerical Computations with GPUs*,
DOI 10.1007/978-3-319-06548-9_13, © Springer International Publishing Switzerland 2014

is determined by an absolute and a relative error tolerance, ε_a and ε_r, respectively, and it is the objective to either satisfy the prescribed accuracy,

$$| Q[\mathscr{D}]f - \mathscr{I}[\mathscr{D}]f | \leq \mathscr{E}[\mathscr{D}]f \leq \max\{\varepsilon_a, \varepsilon_r | \mathscr{I}[\mathscr{D}]f |\}, \qquad (13.1)$$

within the allowed number of function evaluations, or flag an error condition if the limit has been reached. In the remainder of this chapter we will assume that the integrand f is a single-valued function ($k = 1$), and $\mathscr{D}$ is assumed to be the d-dimensional unit hypercube, $\mathscr{D} = [0, 1]^d$ (and will generally be omitted from the notation).

The available approaches in PARINT include adaptive, quasi-Monte Carlo and Monte Carlo integration. For N uniform random points $\mathbf{x}_i$, the Monte Carlo approximation,

$$Qf = \bar{f} = \frac{1}{N} \sum_{i=1}^{N} f_i, \qquad (13.2)$$

with $f_i = f(\mathbf{x}_i)$, is suitable for moderate to high dimensions d, or an irregular integrand behavior or integration domain. In the latter case, the domain can be enclosed in a cube, where the integrand is set to zero outside of $\mathscr{D}$.

The main elements of MC methods are the error bounding and the underlying *Pseudo-Random Number Generators (PRNGs)*, which are discussed in Sects. 13.2 and 13.3, respectively. In particular, Sect. 13.3 covers the implementation and use of parallel PRNGs in CUDA.

Section 13.5 describes *dynamic parallelism*, which became available with CUDA-5.0, for GPU cards supporting compute capability 3.0 and up. The recursive or "vertical" version in Sect. 13.5.1, and the iterative or "horizontal" version in Sect. 13.5.2 are utilized for an efficient GPU implementation of automatic integration in Sect. 13.5.3, which alleviates problems with memory restrictions and kernel launch overhead. Section 13.6 further addresses issues with accuracy and stability of the summations involved in the MC approximation of the result and the sample variance. In the Appendix we include results showing speedups and accuracy for a Feynman loop integral arising in high energy physics [7, 17]. Previously we covered problems from computational geometry in [5].

13.2 MC Convergence

In the following we give a derivation of the error estimate for the one-dimensional case [4]; a similar formulation can be given for $d > 1$.

Let $x_1, x_2, \ldots$ be random variables drawn from a probability distribution with density function $\mu(x)$, $\int_{-\infty}^{\infty} \mu(x)\, dx = 1$, and assume that the *expected value* of f, $\mathscr{I} = \int_{-\infty}^{\infty} f(x)\, \mu(x)\, dx$ exists. For example, $x_1, x_2, \ldots$ may be selected at random from a uniform random distribution in $[0, 1]$, $\mu(x) = 1$ for $0 \le x \le 1$, and 0 otherwise.

According to the Central Limit Theorem, the *variance*

$$\sigma^2 = \int_{-\infty}^{\infty} (f(x) - \mathscr{I})^2\, \mu(x)\, dx = \int_{-\infty}^{\infty} f^2(x)\, \mu(x)\, dx - \mathscr{I}^2 \tag{13.3}$$

allows for an integration error bound in terms of the probability

$$\mathscr{P}\left(Ef \le \frac{\lambda \sigma}{\sqrt{N}}\right) = PI + \mathscr{O}(\frac{1}{\sqrt{N}}) \tag{13.4}$$

where PI represents the *confidence level* $PI = \frac{1}{\sqrt{2\pi}} \int_{-\lambda}^{\lambda} e^{-x^2/2}\, dx$ as a function of λ. For example, $PI = 50\%$ for $\lambda = 0.6745$; $PI = 90\%$ for $\lambda = 1.645$; $PI = 95\%$ for $\lambda = 1.96$; $PI = 99\%$ for $\lambda = 2.576$. In other words we can state that, e.g., with probability or confidence level of 99% the error $E \le 2.567\sigma/\sqrt{N}$. The behavior of the error bound in (13.4) represents the $1/\sqrt{N}$ law of MC integration. Note that (13.4) assumes that σ exists, which relies on the integrand being square-integrable, even though that is not required for the convergence of the MC method [4].

To calculate the error estimate the sample variance estimate can be employed,

$$\bar{\sigma}^2 = \frac{1}{N-1} \sum_{i=1}^{N} (f_i - \bar{f})^2 = \frac{1}{N-1} \left(\sum_{i=1}^{N} f_i^2 - 2 \sum_{i=1}^{N} f_i \bar{f} + N \bar{f}^2\right)$$

$$= \frac{1}{N-1} \left(\sum_{i=1}^{N} f_i^2 - 2N \bar{f}^2 + N \bar{f}^2\right)$$

$$= \frac{1}{N-1} \left(\sum_{i=1}^{N} f_i^2 - N \bar{f}^2\right). \tag{13.5}$$

One can also make use of simple antithetic variates, This is known as a variance reduction method, where for each sample point $\mathbf{x}$ in (13.2), an evaluation is also performed at the point $\mathbf{x}^* = 1 - \mathbf{x}$ (symmetric with respect to the centroid of the cube) [9]. Thus (13.2) becomes

$$Qf = \frac{1}{N} \sum_{i=1}^{N} \frac{f_i + f_i^*}{2}, \tag{13.6}$$

and the variance estimate of (13.5) is updated as

$$\bar{\sigma}^2 = \frac{1}{N-1} \left(\sum_{i=1}^{N} \left(\frac{f_i + f_i^*}{2} \right)^2 - N \left(\frac{\sum_{i=1}^{N}(f_i + f_i^*)}{2N} \right)^2 \right).$$

In view of (13.4) one can then use

$$E \leq \frac{\lambda}{\sqrt{N(N-1)}} \left(\sum_{i=1}^{N} \left(\frac{f_i + f_i^*}{2} \right)^2 - N \left(\frac{\sum_{i=1}^{N}(f_i + f_i^*)}{2N} \right)^2 \right)^{\frac{1}{2}} \tag{13.7}$$

as an error estimate or bound (under certain conditions).

Whereas the above formulas for the variance estimate are correct, they can suffer from severe roundoff error due to cancellation. There are ways to address this; see Sect. 13.6.3. There are a number of additional variance reduction techniques that can be applied to Monte Carlo integration in addition to antithetic variates, such as control variates, importance sampling and stratified sampling. These can be included in a numerical integration package, such as PARINT, using the framework presented here, but are not the focus of the chapter and will not be discussed here.

13.3 Pseudo-Random Number Generators (PRNGs)

A *pseudo-random number generator (PRNG)* is a deterministic procedure to generate a sequence of numbers that behaves like a random sequence and can be tested for adherence to relevant statistical properties. Several PRNG testing packages are available [2, 18, 20]. Marsaglia's DIEHARD [2, 20] is nowadays being superseded by the more stringent tests in SmallCrush, Crush and BigCrush by L'Equyer [18] in the TESTU01 framework.

Whereas function sampling for Monte Carlo integration is considered *embarrassingly parallel*, some applications will involve large sets of evaluations, thus requiring efficient high quality random number streams on various multilevel architectures containing distributed nodes, multicore processors and many-core accelerators. In particular, good pseudo-random sequences are required for multiple threads or processes, and a much larger period of the sequence may be needed in view of much intensified sampling and larger problems. Choosing new PRNGs may also necessitate validation of streams of random numbers generated by different methods being used together. The brief overview of parallel PRNGs in this section is based on our AIAA SciTech 2014 paper [6].

Solutions for associating good quality streams of random numbers to many threads can be sought by [19]: (1) employing a PRNG with a long period and generating subsequences started with different seeds; (2) dividing a random number sequence with a long period into non-overlapping subsequences; and (3) using different PRNGs of the same class with different parameters.

Continuous progress has been made in this area, and there are several open source PRNG packages in wide use, such as the *Scalable Parallel Random Number Generators Library* (SPRNG) [32]. To assure statistically independent streams of random numbers, the SPRNG library incorporates particular parametrizations of: additive lagged-Fibonacci, prime modulus linear congruential, and maximal-period shift-register generators.

NVIDIA's CURAND package includes: an *XORShift* type of generator based on the XORWOW algorithm by Marsaglia [21], a *Combined Multiple Recursive* generator (MRG32k3a) originally proposed by L' Equyer[16] and a *Mersenne Twister* (MTGP11213) type generator which was ported to GPUs by Saito [28]. It also has support for Sobol quasi-random number generators (including scrambled and 64-bit PRNGs). A *quasi-random* sequence of multivariate points aims to fill the multivariate space evenly. The uniform, normal, log-normal and Poisson distributions are supported, as well as single and double precision. The CURAND library is freely available as part of the CUDA Toolkit [1], and distributed with it under the *End User License Agreement (EULA)*.

Regarding their use on HPC co-processors such as GPUs, a problem with SPRNG and many existing PRNGS, is that they were designed for architectures where fast memory is cheap so that large state arrays can be used. This is critical as on most HPC co-processors memory is often limited, and communication as a part of random number generation is undesirable. Therefore, for applications where each thread requires its own generator (as in highly parallel simulations of lattice models), parallel PRNGs with large state are at a disadvantage [19], such as XORWOW [21] and the standard Mersenne twister [22], which was originally used in the CUDA SDK but has been replaced by MTGP (see *Variants of Mersenne Twister suitable for Graphic Processors*) [28].

Manssen et al. [19] point to solutions for obtaining a small memory-load overhead by: (1) using good quality generators with very small state, which has been successful with the *counter-based, stateless generators (CBRNGs)* in the RANDOM123 library [29, 30]; or (2) sharing the state between the threads of a single (GPU) block. They further present a new massively parallel high-quality generator in this class, based on an XORShift/Weyl scheme for a very large word size, which performs favorably in their TestU01 suite comparison of (12) PRNGs for GPUs with respect to: the number of bits per thread, number of tests failed in SmallCrush, Crush and BigCrush, pass or failure in performing an application (Ising test), and performance (time) in terms of the number of random numbers generated per second. The XORShift/Weyl generator passes all tests in the rigorous TestU01 suite, and so do the Philox type generators from the RANDOM123 library.

13.3.1 Use of cuRAND

With the NVIDIA CURAND approach, the CPU initializes a PRNG and invokes a GPU function to generate the sequence directly in the GPU memory, where it is kept for subsequent use. This avoids the overhead of having the CPU generate the sequence and move it from main memory to the GPU memory. A potential drawback with this approach arises, however, in that GPU memory is limited (e.g., to a maximum of about 5 GB on the Tesla K20 cards [24]), so that the number of points needed to get the desired accuracy may exceed the available memory. In that case one may be tempted to solve the problem at hand by doing several kernel invocations, but that is undesirable because of the kernel launch overhead and the overhead of the wait for CURAND to complete the generation of the sequence.

A CUDA C program section allocating space for Ndim $= N * Dim$ floats in an array numbers on the device, and calling the CURAND functions is shown below. This section executes on the CPU before the GPU kernel is launched.

```
//  allocate Ndim floats on device in array numbers
cudaMalloc ((void **) &numbers, Ndim * sizeof(float));
//  create pseudo-random number generator
curandCreateGenerator (&gen, CURAND_RNG_PSEUDO_DEFAULT);
//  set seed
curandSetPseudoRandomGeneratorSeed (gen, 1234ULL);
//  generate Ndim random numbers on device
curandGenerateUniform (gen, numbers, Ndim);
```

Note that there are variations to the function *curandGenerateUniform* such as *curandGenerateUniformDouble*, which generates points in double precision.

13.3.2 Use of RANDOM123

As opposed to the static generation of the pseudo-random sequence at the beginning of the run by CURAND, the Philox type CBRNG of RANDOM123 allows generating the numbers on the GPU as needed, requires little memory and reduces the number of accesses to global memory. Furthermore it produces 2^{64} or more unique parallel streams of random numbers, each with a period of at least 2^{128}, and is found faster than CURAND on a single NVIDIA GPU [30].

For the CBRNG a statement of the form

```
result = CBRNGname(counter, key);
```

returns a (deterministic) value of *result* as a function of *key* and *counter*; i.e., using the same (*counter*, *key*) combination will always return the same *result*. The RANDOM123 library is implemented entirely in header files [29]. So all what is needed to start using the library is to *#include* it in the program source files, and guide the compiler to find its header files that are unpacked from the downloaded package.

Unlike CURAND, the RANDOM123 library requires no work by the CPU before the kernel launch. When the GPU kernel is launched, the counter and key of the Philox PRNG are initialized as follows:

```
philox4x32_key_t k = {{tid, 0xdecafbad}};
philox4x32_ctr_t c = {{0, 0xf00dcafe, 0xdeadbeef, 0xbeeff00d}};
```

where the use of the global thread index *tid*, which is unique to the thread, ensures the generation of unique random numbers by each thread. Next, the contents of the counter can be set by accessing the constant array v:

```
c.v[0] = ... ; /* some loop-dependent application variable */
```

and executing

```
philox4x32_ctr_t r = philox4x32(c, k);
```

This generates 4×32 bit random numbers stored in an array in r, which will be unique as long as c and k are not reused. One way to access the random numbers is:

```
float current_value = u01fixedpt_open_open_32_24(r.v[0]);
```

This returns the first random number in the array; to access the second one *r.v[0]* is replaced by *r.v[1]*, and so on for the third and the fourth numbers.

Note that *u01fixedpt_open_open* is a utility C function that converts 32- or 64-bit random integers to uniformly distributed random values. The 32_24 suffix refers to the conversion of 32-bit integers to floats determined with a 24-bit mantissa (24 is replaced by 53 for doubles); *_open_open* indicates that the output range is open at both ends.

13.4   MC CUDA Kernel

13.4.1   Methods

The Monte Carlo approximation is obtained in (13.2) as the average of the function values on a set of N uniformly distributed random points. In general, an integration rule approximation with function evaluations $f_i = f(\mathbf{x}_i)$ and weights w_i, $1 \leq i \leq N$ can be considered as the dot product $\mathbf{w} \cdot \mathbf{f} = \sum_{i=1}^{N} w_i f_i$. The MC "rule" has constant weights $w_i = \frac{1}{N}$, $1 \leq i \leq N$, but its accumulation can be accomplished in a CUDA kernel with a similar structure as a dot product [31].

For simplicity we first assume that the points have been generated by the CuRAND PRNG library and are available on the GPU device. The GPU kernel *MonteCarlo* shown below evaluates the sum of function values and the sum of squares needed for the computation of the integral and error estimates using antithetic variates according to (13.6) and (13.7), respectively. The kernel can be launched as:

```
//Launch the GPU Kernel
MonteCarlo<<<blocksPerGrid, threadsPerBlock>>>
            (numbers, dev_partial_q, dev_partial_e, N);
```

The function evaluation results are accumulated as follows:

1. Each thread on the GPU performs a small number of function evaluations.
2. All threads in a block reduce their results to the first thread in the block.
3. The first thread in the block stores the reduced result in a global array on the device.

4. The global array is copied back to the CPU for a final reduction, to sum up the partial results computed by the blocks.
5. The sum of the per-block partial results divided by N is the final result.

Thus the parameter *dev_partial_q* that is passed to the kernel *MonteCarlo* is stored in the GPU's global memory and will hold the per-block partial results. The computations for *dev_partial_e* are performed in a similar way. The integrand function f is implemented as a CUDA *device function* that takes a *point* as a parameter (represented as a *struct* containing the random coordinates in an array of dimension *dim*), and returns the function evaluation at that *point*.

The purpose of the *cache* arrays in the kernel is to utilize the GPU shared memory whose access is much faster than that of the GPU global memory where the array *numbers* is stored. The *cache* arrays are shared by all threads within the same block only.

In the kernel note that *threadIdx.x* represents the thread's index with respect to all indices within the current block, *blockIdx.x* gives the block's index with respect to all the blocks in the grid, and *blockDim.x* is the dimension of each block, which is the total number of threads in each block. Therefore, the global index of the thread with respect to *all* forked threads in *all* blocks is computed as *threadIdx.x + blockIdx.x * blockDim.x*.

Furthermore, as the number of evaluations may be much larger than the total number of threads, the while loop *while(tid < N)* allows assigning evaluations to the threads in a round-robin fashion; i.e., each thread performs evaluations starting with its index and going up in increments of *blockDim.x * gridDim.x*. Within the while loop we use *Kahan summation* as a roundoff error guard for the accumulations, which are performed in single (float) precision. For a detailed explanation of stability and condition numbers involving the summations see Sect. 13.6.

Implementing the kernel entirely in double precision may not be desirable because:

- It requires Compute Capability 2.0 or higher, which few NVIDIA desktop video cards currently support.
- It may require twice the compute time and memory.
- It may not be needed to obtain the requested accuracy.

The analysis outlined in Sect. 13.6 can be used to gauge the effect of various precision choices.

```
struct point {
        float coordinates[dim];
};

__global__ void MonteCarlo(float *numbers, float *partial_q,
                           float *partial_e, int N) {

    // Make use of the GPU shared memory
    __shared__ float cache[threadsPerBlock];
    __shared__ float error_cache[threadsPerBlock];
```

```
// tid holds thread's global index with respect to
// all GPU threads
int tid = threadIdx.x + blockIdx.x * blockDim.x;

point current_point;
point second_point; // Point for antithetic variates
float eval, evaluations = 0, qk = 0, y, t;
float sum_of_squares = 0, qksq = 0, ysq, tsq;

while (tid < N) {
    int i = 0;
    int count = 0;
    int tdim = tid * dim;
    float temp = 0;
    for(i = tdim; i < tdim + dim; i++) {
        // Set point coordinates with random numbers
        temp = numbers[i];
        current_point.coordinates[count] = temp;
        second_point.coordinates[count] = 1 - temp;
        count++;
    }
    // Call the CUDA device function
    eval = (f(current_point) + f(second_point));
    // Kahan summation for: evaluations += eval;
    y = eval - qk;
    t = evaluations + y;
    qk = (t - evaluations) - y;
    evaluations = t;

    // Kahan summation for: sum_of_squares += eval*eval;
    ysq = eval*eval - qksq;
    tsq = sum_of_squares + ysq;
    qksq = (tsq - sum_of_squares) - ysq;
    sum_of_squares = tsq;

    // Let each thread do a number of evaluations
    tid += blockDim.x * gridDim.x;
}

// Store the value obtained by each thread
cache[threadIdx.x] = evaluations;
error_cache[cacheIndex] = sum_of_squares;

// Synchronize threads in this block
__syncthreads();

// For reductions, threadsPerBlock must be a power of 2
// because of the following while loop
int i = blockDim.x/2;
while (i != 0) {
    if (cacheIndex < i) {
        cache[cacheIndex] += cache[cacheIndex + i];
        error_cache[cacheIndex]
                    += error_cache[cacheIndex + i];
```

```
    }
    __syncthreads();
    i /= 2;
}

// If this is the first thread of this block
if (threadIdx.x == 0) {
    // Store the results of this block in the global arrays
    partial_q[blockIdx.x] = cache[0];
    partial_e[blockIdx.x] = error_cache[0];
}
}
```

After the kernel finishes executing, the control returns to the CPU, which copies the contents of *partial_results* from the GPU to a local array declared on the CPU, by the following statements:

```
// Copy the partial results from the GPU back to the CPU
cudaMemcpy(partial_q, dev_partial_q,
           blocksPerGrid*sizeof(float), cudaMemcpyDeviceToHost);
cudaMemcpy(partial_e, dev_partial_e,
           blocksPerGrid*sizeof(float), cudaMemcpyDeviceToHost);
```

Here *partial_q* and *partial_e* are the local arrays that hold the partial results, and the number of bytes transferred is *blocksPerGrid * sizeof(float)* since each block generates a partial result. As shown below, the partial results are summed by the CPU, which returns the integral approximation *final_result*, and an error estimate *error_est* according to (13.7).

```
// Sum up the partial results
double q = 0, e = 0, final_result, error_est;

for (int i = 0; i < blocksPerGrid; i++) {
    q += partial_q[i];
    e += partial_e[i];
}

final_result = q/(2*N);
error_est = sqrt(fabs((e/4 - q*q/(4*N))/(N-1)/N));
```

13.4.2 Numerical Validation

One issue associated with the need for large numbers of evaluations is the occurrence of roundoff error. Another problem, specifically when using CURAND for generating the random numbers, is the restricted global memory on the GPU. A solution is not obtained by multiple kernel launches from the CPU in view of the kernel launch overhead. However, the memory restriction problem can be alleviated by using a RANDOM123 PRNG to generate the random numbers as needed on the GPU.

Figure 13.1 plots (on $\log_{10}$ scale) the absolute error (blue/diamonds curve) and estimated error (red/rectangles curve) of the MC approximation, as a function of ($\log_{10}$ of) the number of points N, for the integration of the function $f(\mathbf{x}) = 3x_0^2$ over the 12D unit cube. Thus the length of the pseudo-random sequence used is $12N$ for each integration. The estimated error is based on (13.7) with $\lambda = 1$. The function $\log_{10}(1/\sqrt{N}) = -\frac{1}{2}\log_{10} N$ is also plotted (green/triangles curve). The error adheres closely to the $1/\sqrt{N}$ behavior through $N = 10^{10}$; then is stagnant or increases slightly through $N = 10^{12}$.

We used a version of the kernel in Sect. 13.4.1, employing the Kahan summation technique for the accumulations in single precision. For comparison we also ran the program with Kahan summation replaced by using doubles for the summation variables (only). The accuracies as well as the times were very similar. When plotted, the results were in fact indistinguishable from those of Fig. 13.1.

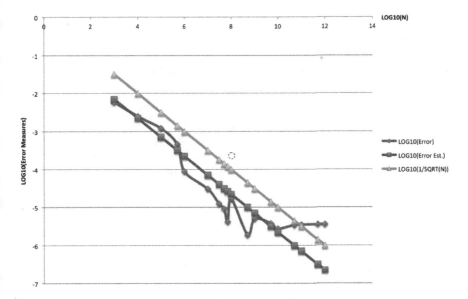

Fig. 13.1 Error behavior for MC approximation using Kahan summation. Shown are: $\log_{10}$(Error), $\log_{10}$(Error Estimate) and $\log_{10}(1/\sqrt{N})$ as a function of $\log_{10} N$

13.5 Dynamic Parallelism

An important feature with NVIDIA's release of CUDA-5.0, that works on GPU cards supporting compute capability 3.0 and up is *Dynamic Parallelism*. We will further refer to it as *DP*. From the NVIDIA Techbrief on *DP*, "additional parallelism can be exposed to the GPU's hardware schedulers and load balancers dynamically, adapting in response to data-driven decisions or workloads. Algorithms and programming patterns that had previously required modifications to eliminate recursion, irregular loop structure, or other constructs that do not fit a flat, single-level of parallelism can be more transparently expressed." [25] Consequently, *DP* may eliminate the need for additional kernel launches, and thus eliminates the extra kernel launch overheads that would be incurred. Basically a parent kernel can now invoke a child kernel, without CPU intervention, keeping the control completely on the GPU and eliminating the overhead of switching control between the GPU and the CPU between kernel calls. With respect to the implementation of MC, *DP* enables *chunking*, i.e., instead of launching one kernel with a huge number of points, the CPU can launch one kernel with a smaller number of points, which in turn can launch child kernels each with their chunk of a smaller number of points.

Furthermore, *DP* allows the number of threads to exceed the usual limit, since threads running within the parent grid can invoke a child grid, so now instead of having each thread take on an amount of work in the parent kernel, work can be off-loaded to the child kernels. A child grid needs to complete before the parent grid is considered complete.

13.5.1 Vertical (Recursive) Dynamic Parallelism

To make use of this, *Dynamic Parallelism* techniques were applied to the kernel using RANDOM123. As opposed to letting the kernel be launched by the CPU do all the work, this kernel does part of the work, say a chunk of *limit* = 200 million points, and then launches another kernel to do another chunk of the work, until the total number N is reached. This type of control is displayed in the parent kernel (pseudo code) below.

```
__global__ void MonteCarlo(float *partial_results, int runs) {

    // Make use of the GPU shared memory
    __shared__ float cache[threadsPerBlock];

    // tid holds thread's global index w.r.t. all GPU threads
    int tid = threadIdx.x + blockIdx.x * blockDim.x;

    float current_point;
    float second_point;
    float evaluations;

    //limit can be any number
    while (tid < limit) {

        Generate the current point using Random123;
        Get a second point as 1 - current_point;
        // Call the CUDA device function to evaluate it at
```

```
            //   the current points using Kahan summation and/or
            //   double accumulators
            evaluations = evaluations + f(current_point) +
                          f(second_point);
            // Let each thread do a number of evaluations
            tid += blockDim.x * gridDim.x;
        }
        // Store the value obtained by each thread
        cache[threadIdx.x] = evaluations;

        // Synchronize threads in this block
        __syncthreads();

        // For reductions, threadsPerBlock must be a power of 2
        // because of the following while loop
        int i = blockDim.x/2;
        while (i != 0) {
            if (cacheIndex < i) {
                cache[cacheIndex] += cache[cacheIndex + i];
            }
            __syncthreads();
            i /= 2;
        }
        // If this is the first thread of this block
        if (threadIdx.x == 0) {
            // Store the block's result in the global array
            partial_results[kernel_index * gridDim.x + blockIdx.x]
                        = cache[0];
        }
        // Recompute the original tid
        tid = threadIdx.x + blockIdx.x * blockDim.x;
        if(tid == 0) // the first thread of the current kernel
        {
            if(runs > 0) // and not all partitions have been run
            {
                MonteCarlo<<<gridDim.x,threadsPerBlock>>>
                    (partial_results, runs-1);
            }
        }
    }
}
```

Here *kernel_index* is the index of the current kernel, which is used in this case
to indicate the global index of the current block with respect to all blocks of all
launched kernels to store the block result in the correct position of the global array.
The variable *runs* stores the number of kernels that remain to be launched, based on
the number of chunks, computed by the CPU as *N/limit* and passed as a parameter
to the kernel, where it is decremented and passed on to the subsequent kernel
launched.

A drawback of vertical *DP* where the kernels are launched in depth one at a time, is the maximum recursion depth, which currently limits the number of recursive launches to 24 and may not be sufficient for some applications.

13.5.2 Horizontal (Iterative) Dynamic Parallelism

We implemented another approach where kernel calls are performed breadth-wise, thereby creating *DP* in a horizontal/iterative way. The first kernel invokes all other necessary kernels, and the runtime takes care of scheduling them over the resources. The differences in the control, from the kernel of Sect. 13.5.1, are shown in the section below.

```
// Recompute the original tid
tid = threadIdx.x + blockIdx.x * blockDim.x;
if(tid == 0) // The first thread of this kernel
{
    while(runs > 0) // not all partitions have run
    {
        MonteCarlo<<<gridDim.x,threadsPerBlock>>>
                (partial_results, 0);
    }

}
```

In this approach the *if(runs > 0)* is replaced by a *while* statement, since all the kernel launches take place in the beginning. The variable *runs* is passed as 0 to all subsequent kernels, to ensure that no other kernel launches more kernels than required.

Figure 13.2 gives the accuracy and estimated error obtained with horizontal *DP* using a chunk size of $200M = 2 \times 10^8$ for $2 \times 10^9 \leq N \leq 4 \times 10^{11}$, and a chunk size of $1B = 10^9$ for $6 \times 10^{11} \leq N \leq 10^{12} = 1T$. No dynamic parallelism was used for $N \leq 200M$. This version of the kernel used double precision accumulators. For the legend description see that of Fig. 13.1. Figure 13.2 shows good accuracy through N in excess of 10^{11} points.

13.5.3 Automatic Integration with Dynamic Parallelism

Automatic integration was introduced in Sect. 13.1. The integration algorithm works as a black-box, which (minimally) takes as input: the user-specified integrand function, domain, prescribed absolute and/or relative accuracy and a maximum number of function evaluations. With the notations of (13.1), it is the goal of the computations to improve the integral approximation $Q[\mathscr{D}]f$ and decrease the error estimate $\mathscr{E}[\mathscr{D}]f$ accordingly, until it no longer exceeds the tolerated error,

$$| Q[\mathscr{D}]f - \mathscr{I}[\mathscr{D}]f | \leq \mathscr{E}[\mathscr{D}]f \leq \texttt{error tolerance.} \qquad (13.8)$$

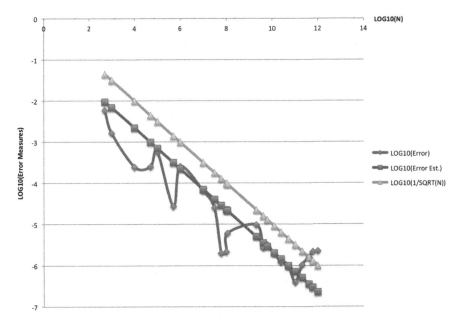

Fig. 13.2 Error behavior using *horizontal DP* and *chunking*. The plot shows (for integration of $f(\mathbf{x}) = 3x_0^2$ over the 12D unit cube): $\log_{10}(\text{Error})$, $\log_{10}(\text{Error Estimate})$ and $\log_{10}(1/\sqrt{N})$ as a function of $\log_{10} N$

$Q[\mathscr{D}]f$ and $\mathscr{E}[\mathscr{D}]f$ are returned when either (13.8) is achieved or the maximum number of function evaluations has been reached (cf., the *meta-algorithm* of Fig. 13.3).

> **while** (estimated error too large and
> evaluation limit not reached)
> Compute new result and error estimate

Fig. 13.3 Automatic integration meta-algorithm

We use the *Dynamic Parallelism* approach above to launch successive MC kernels on the GPU, and add a kernel (*metaMC*) implementing the controller of the meta-algorithm. The controller is launched from the CPU and runs in one block using one thread on the GPU. It sequences the MC kernel calls which perform chunks of evaluations, until either the prescribed accuracy has been attained, or the maximum number of chunks has been computed. Pseudo-code for the *metaMC* controller is given below.

```
__global__  void metaMC(float *partial_results,
              float *partial_errors, long long int N,
              float sequence_factor, float tolerance,
              float *gpu_results, int limit)
{
    //Initialize all variables
```

```
float partial_r = 0.0;
float partial_e = 0.0;
float final_result = 0.0;
float final_error = 0.0;
float error = tolerance+100.0; // just to enter loop
long long int totalN = 0;
float count = 1.0;

while(error > tolerance && count <= limit)
{
    MonteCarlo<<<blocksPerGrid, threadsPerBlock>>>
            (partial_results, partial_errors, N);
    // Wait for the kernel to return
    cudaDeviceSynchronize();

    //Keep track of the total number of points used
    // multiplied by 2 because of antithetic variates
    totalN += 2.0 * N;

    //Compute the value of N for the next kernel call
    N = N * sequence_factor;

    // Sum up the partial results returned by the
    // MonteCarlo kernel to the old values
    for (int i = 0; i < blocksPerGrid ;  i++) {
        partial_r += partial_results[i];
        partial_e += partial_errors[i];
    }
    // Compute the final result and error estimate
    final_result = partial_results/(totalN);
    final_error = sqrt(fabs((partial_errors/4.0 -
            partial_results*partial_results/(4.0*totalN))/
            (totalN-1.0)/totalN));

    // Set error to the correct value to be compared
    // with the tolerance next
    error = final_error;

    count++;
}
// Store the final results and the number of times
// MonteCarlo was run
gpu_results[0] = count;
gpu_results[1] = final_result;
gpu_results[2] = final_error;
}
```

13.6   Accuracy and Stability in Numerical Integration

Some comments on the numerical evaluation of integrals using the methods presented need to be pointed out. These comments focus on the Monte Carlo method for simplicity, but apply to all methods. For a more complete discussion see [14].

Table 13.1 IEEE 2008 floating point values

IEEE 754-2008	Common name	Data type	Base	Precision	e = Machine epsilon
binary32	Single precision	float	2	24[a]	$2^{-23} \approx 1.19\mathrm{e}{-}07$
binary64	Double precision	double	2	53[a]	$2^{-52} \approx 2.22\mathrm{e}{-}16$
binary80	Extended precision	_float80	2	64	$2^{-63} \approx 1.08\mathrm{e}{-}19$

[a]One bit is implicit

First, IEEE floating point on hosts and NVIDIA GPUs will be mentioned. Then the evaluation of the summation in (13.2) will be addressed, followed by problems in evaluating the formula for the sample variance in (13.5). A robust integration program, with the sizes of N now being used, must deal with these issues or the results can be completely misleading.

The analysis of numerical integration methods can be broken down into several parts: analytic errors in the method, errors in the computed points $\mathbf{x}_i$, errors in the computed function values $f_i = f(\mathbf{x}_i)$, and rounding errors in the rule's application. For Monte Carlo integration the analytics are simple, and the other errors can all be addressed by considering the value of machine epsilon (defined in Sect. 13.6.1 below), used in the analysis of summations.

13.6.1   IEEE Floating Point Arithmetic

CUDA applications often require floating point operations on both the GPU and the host CPU. The IEEE-fp standard [12, 13] is implemented on virtually all modern general computing architectures. For an introduction to some of the practical details of IEEE-fp computing see [10, 23].

NVIDIA CUDA GPUs with Compute Capabilities 2.0 and greater have single and double IEEE-fp 754 support, including computing *square root to nearest* and *Fused Multiply Add (FMA)* [33]. All four IEEE-fp rounding modes are supported with *"round to nearest, ties to even"* being the default.

Modern Intel and AMD processors also support IEEE-fp. Some also support the old x87 80-bit extended precision format, which is not IEEE-fp compliant and has not been recommended for applications since 2008. These processors do support various extended instruction sets, *SSE, SSE2, AVX, AVX2,...,* that provide additional precision and capabilities, such as FMA.

Table 13.1 provides the values of e, the *machine epsilon (relative machine precision)* [11], that will be used in the following sections. Although there are small differences in opinion about what these e values should be, the current values will suffice. Roughly, we usually expect the relative error in any arithmetic computation $(+, -, *$ or $/)$ to only introduce an error in the 7th decimal place for *single* operations and the 16th decimal place for *double* operations. These discussions will not make use of the exponent ranges in IEEE-fp.

Although the IEEE-fp standard and its implementations is a vast improvement over the situation that formerly existed, the goal of reproducible results (exactly the same binary results on any system on which a program is run) is very difficult to obtain [8], and not practical in most circumstances. There are a number of reasons for this in hardware, compilers, and even in using library functions, that generally cannot return the best value (e.g., the table maker's dilemma [23]).

The results from the same program on two different IEEE-fp compliant platforms, consisting of the hardware (Intel, AMD, CUDA) and compiler (including its flags), can totally disagree. This is a special concern for problems with large condition numbers (see the next section). A more practical goal is to try to produce reliable results with error estimates, and warnings when this may not be the case.

13.6.2   Computing Summations

The accurate evaluation of a summation, as in (13.2), is a classic problem. Higham [11] gives a good review of several methods for summation, and there is a lot of work still being done on the problem [8, 26, 27].

This is not a review of the current methods for summation. The presentation here focuses on a couple of methods that are well suited for simple CUDA implementations in numerical integration. The analysis in this section considers summations of 10^{14} function values using doubles. This can be modified for other cases, e.g., where floats are used throughout, or the accumulations (only) are done in doubles.

The CUDA architecture favors algorithms that:

- have very small global, block and thread memory footprints;
- do not have significant branching.

Many of the other known methods violate one or both of these. The general term *"online method"* implies the method can be applied in a single pass through the data (hence the data do not need to be stored), and does not require significant memory. Such methods are of interest in the CUDA environment.

First consider S_N, the exact value of the sum in (13.2).

$$S_N = \sum_{i=1}^{N} f_i,$$

and let $\hat{S}_N$ be the computed value of the sum; then E_N is the error, where

$$E_N = S_N - \hat{S}_N.$$

The interest here is in the relative error E_N / S_N, that is undefined when $S_N = 0$. When $S_N = 0$, a bound on the absolute error can still be sought [14]. A bound on the relative error indicates how many digits of precision can be lost due to roundoff. Under reasonable assumptions on the integrand, if the integral value is I, the expected value of S_N is NI. So if the integral is not 0, the value of $|S_N|$ is expected to be non-zero and increase with N. In what follows it is assumed that $S_N \neq 0$.

One way to accumulate the sum is naive summation as shown in Fig. 13.4.

```
recursive_summation(float *f, int N) {
        float sum = 0.0;      // the sum being accumulated
        int i;
        for(i = 0; i < N; i++) {
            sum += f[i];
        }
        return sum;
}
```

Fig. 13.4 Naive recursive summation

This is a member of a class of summation algorithms called recursive summation, where one element to be summed is selected at a time and accumulated. As another example of a recursive summation method, the data could be sorted and then summed. It is clear that naive summation can be done as an online method, but sorting first cannot.

Letting e be machine epsilon and assuming $N * e < 1$, (13.9) below is a bound on the relative error in $\hat{S}_N$ for naive summation [11],

$$\frac{|E_N|}{|S_N|} \leq ((N-1)e + \mathcal{O}(e^2)) \frac{\sum_{i=1}^{N} |f_i|}{|\sum_{i=1}^{N} f_i|}. \tag{13.9}$$

In (13.9) the factor

$$\kappa_s = \frac{\sum_{i=1}^{N} |f_i|}{|\sum_{i=1}^{N} f_i|} \tag{13.10}$$

is the *condition number* for the summation process. It is known that any summation method that uses a fixed precision and takes time independent of the data (it can be dependent on N) will have a factor κ_s in its condition number.

A condition number of a process is a measure of the sensitivity of the result to the input. Note that $\kappa_s = 1$ if all the f_i are positive, and the summation is well conditioned. κ_s can be very large if there is significant cancellation in the summation.

For positive f_i the naive summation using doubles for $N = 10^{14}$ results in a relative error bound of 0.01 or 2 digits of precision. If $\kappa_s = 10^5$ there might be serious problems. Note though that (13.9) gives the worst case situation in terms of the accumulation of the rounding errors.

```
Kahan_summation(float *f, int N) {
    float sum = 0.0;        // the sum being accumulated
    float c = 0.0;          // the compensation for lost precision
    float y, t;             // temporary variables
    int i;
    for(i = 0; i < N; i++) {
        y = f[i] - c;
        t = sum + y;
        c = (t - sum) - y;
        sum = t;
    }
    return sum;
}
```

Fig. 13.5 Kahan compensated summation

Next consider *Kahan compensated summation* (Fig. 13.5), proposed by Kahan and later analyzed by Kahan and Knuth [11, 15]. Again assuming $N * e < 1$, the relative error bound for Kahan summation is given by

$$\frac{|E_N|}{|S_N|} \leq (2e + \mathcal{O}(Ne^2)) \frac{\sum_{i=1}^{n} |f_i|}{|\sum_{i=1}^{n} f_i|}.$$

Here for positive f_i, Kahan summation using doubles for $N = 10^{14}$ results in a relative error bound of $2e$, nearly full double precision. If $\kappa_s = 10^5$ the relative error bound is still about 10^{-11}. As in naive summation, this is a worst case bound. It is clear that Kahan summation meets the two criteria given above for a good CUDA method. As a note, Kahan summation works best when $|sum| > |f_i|$ in Fig. 13.5.

If one considers using quad precision, $e \sim 10^{-34}$, then the naive summation error bound is close to the bound on Kahan summation using doubles. There is no quad hardware support currently in CUDA, and a complete software implementation would have several serious drawbacks. For these reasons, Kahan summation is suggested in threads when the f_i are doubles, and either Kahan summation or a double accumulator in naive summation is used when the f_i are single.

If applied directly to the f_i, instead of partial sums, the parallel reduction of the partial sum results of the threads in a block to a single value, presented in Sect. 13.4.1, is another known summation method, *pairwise summation* [11]. In pairwise summation the sum is the result of building a binary tree of intermediate results. Again assuming $N * e < 1$, the relative error bound for pairwise summation is given by

$$\frac{|E_N|}{|S_N|} \leq \left(\frac{e \log_2 N}{1 + e \log_2 N} + \mathcal{O}(Ne^2) \right) \frac{\sum_{i=1}^{N} |f_i|}{|\sum_{i=1}^{N} f_i|}. \tag{13.11}$$

When applied to the parallel reduction in Sect. 13.4.1, the f_i in (13.11) are the partial sums stored in *cache[threadsPerBlock]* and $N = threadsPerBlock$. Currently NVIDIA Compute Capability 3.5 has *maxThreadsPerBlock = 1,024*. So for the parallel reduction a bound for the roundoff error is $10e\kappa_s$, where the condition number is associated with the partial sums from each thread. That condition number can be much better than the original condition number for the f_i in each thread, where the majority of cancellation is expected to occur. Thus the parallel reduction may result in the loss of one decimal place in single or double precision.

If results for each block are transferred to the host for final assembly by naive summation, Kahan summation, or quad precision, the final result's roundoff error can also be bounded. For this final summation, the condition number should again be better than that of the f_i can be close to 1.

The roundoff error bounds presented are worst case bounds. If the roundoff errors are treated as unbiased random normally distributed variates, which is not unreasonable to assume with the default IEEE-fp rounding on CUDA Compute capability 2.0 or later, then the rounding errors can be treated as a random walk where the RMS error in S_N is then given by $E_{RMS} = \mathcal{O}(\sqrt{N}e)$. This implies that the result may be much more accurate than the bounds might indicate.

It is also possible to consider the same arguments made here when the computation on a thread is being done in single precision. In that case, the overall error in 10^{14} terms will be dominated by the errors in each single precision f_i.

A common three level summation (reduction) process has been described for integration on GPUs, in threads, in blocks and finally on the host. In using dynamic execution, or in combining results from many GPUs (say with MPI) there may be four or more levels, and the same analysis outlined here can continue. A last observation is that, using Kahan summation going from one level to the next, the compensation term of the partial sum could be carried up as well and used at the higher level. This can also be done in the pairwise summation.

In summary, using Kahan summation with a very little extra computation and memory over naive summation in threads, roundoff error can be controlled for sums of 10^{14} terms. The condition number for the result can also be provided and may be useful in suggesting that the integral should be transformed to avoid cancellation for better results. The bounding of roundoff in the higher levels of accumulation can be done as in the pairwise reduction for blocks and in reductions combining blocks.

13.6.3 Computing the Sample Variance

As with the computation of a summation, the computation of the sample variance in (13.5) has a long history [3, 11]. This section will simply point out some major issues and similarities to the summation problem. For more detail on CUDA implementations see [14]).

Letting the sample variance be $\sigma^2 = \frac{1}{N-1} SSE$, where SSE is given by

$$SSE = \sum_{i=1}^{N} (f_i - \bar{f})^2$$

$$= \sum_{i=1}^{N} f_i^2 - N\bar{f}^2, \tag{13.12}$$

it is clear that the second form in (13.12) can be implemented as an online method in one pass by accumulating both $\sum_{i=1}^{N} f_i^2$ and $\sum_{i=1}^{N} f_i$. However, this "textbook method" should in fact not be used in practice and often results in a very serious cancellation error in the final subtraction, even resulting in negative values for SSE!

The condition number for SSE is given by κ_{SSE} where

$$\kappa_{SSE} = \frac{||\mathbf{f}||_2}{\sqrt{SSE}}$$

and $\mathbf{f}$ is the vector of the f_i. It is easy to see that large values for the f_i with small variation between them will cause problems.

However, from the standpoint of using σ as an estimate of the error in Monte Carlo integration, the need for high precision in the SSE is not as important as it is in the summation. Only a few decimal places of accuracy will suffice.

Just as with the summation process, there are more stable online methods for computing running online values for SSE than the textbook method. Several online methods have been proposed using single point updates and block updates. Additionally, compensation can be done for extra precision if needed. Some of these methods fit into the CUDA architecture well, much like the corresponding summation methods, for single point updates on threads, pairwise parallel reduction on blocks, and the block reduction on a host [3, 11, 14].

13.6.4 Other Methods

There are other methods beyond those mentioned here for computing accurate summations and variances. Some of these can be reasonably implemented on CUDA architectures, such as double compensation, cascading, and software multiple precision to name a few. None of these appear to be as effective on CUDA architectures in conserving resources as Kahan compensation, which looks to be sufficient for numerical integration.

13.7 Conclusions and the Future

It is clear that CUDA processors can be very effective in numerical integration. Important high-dimensional integrals that were previously practically impossible to evaluate can now be dealt with routinely, and the development of reliable open source software to provide this support will be available shortly. This is possible because of three factors:

- the advances in CUDA hardware and software by NVIDIA,
- the advances in efficient pseudo-random number generation for CUDA, and
- analysis and application of error and roundoff controls.

Rapid advances are happening in all three of these areas. NVIDIA GPUs are planned and being released with more cores, memory and features. CUDA 6.0, now available in beta, has features for coordinating eight GPUs on a single host. Effective pseudo-random and hopefully new quasi-random (low-discrepancy sequence) number generation on CUDA continues to evolve. Finally, general frameworks for CUDA implementation on workstations, clusters and clouds have been demonstrated and are evolving.

Acknowledgements We acknowledge the support from the National Science Foundation under Award Number 1126438, and from NVIDIA for the award of our CUDA Teaching Center.

Appendix

As an application we treated a Feynman (two-)loop integral, which was previously considered in [7, 17]. This type of problem is important for the calculation of the cross section of particle interactions in high energy physics.

Table 13.2 lists the integral approximation, absolute error and error estimate, and the parallel time of a GPU computation on Kepler K20, using the MC kernel with RANDOM123 as the PRNG, and Kahan summation on the GPU. The results in the bottom part of the table (for $N \geq 4 \times 10^8$) are obtained using the horizontal dynamic parallelism strategy with a chunk size of 200 million. For the smaller values of N in the top part of the table, the chunk size is set equal to N, so no child kernels are launched. For these values of N, the execution time is compared to that of a corresponding sequential calculation where erand48() is called to generate the pseudo-random sequence on the CPU. Speedups of near full peak performance are observed. Note that the error decreases to 9.9e−08 at $N = 4 \times 10^{10}$. The integrand function is given below.

Table 13.2 Times and speedup results for a Feynman loop integral

N	Result	Abs. Err.	Err. Est.	Seq. Time (ms)	Par. Time (ms)	Speedup
10^3	7.725290e−02	8.1e−03	1.1e−02	3.4e−01	2.5e−01	1.37e+00
5×10^3	8.380065e−02	1.6e−03	4.8e−03	1.7e+00	2.5e−01	6.64e+00
10^4	8.317693e−02	2.2e−03	3.2e−03	3.4e+00	2.6e−01	1.31e+01
5×10^4	8.428590e−02	1.1e−03	1.6e−03	1.7e+01	2.8e−01	5.92e+01
10^5	8.421736e−02	1.1e−03	1.1e−03	3.3e+01	3.2e−01	1.04e+02
5×10^5	8.467325e−02	6.8e−04	5.3e−04	1.6e+02	6.1e−01	2.70e+02
10^6	8.497359e−02	3.8e−04	3.9e−04	3.3e+02	9.7e−01	3.38e+02
5×10^6	8.516875e−02	1.8e−04	1.8e−04	1.6e+03	3.6e+00	4.49e+02
10^7	8.535279e−02	1.4e−06	1.4e−04	3.3e+03	6.9e+00	4.71e+02
5×10^7	8.528537e−02	6.6e−05	5.9e−05	1.6e+04	3.3e+01	4.89e+02
10^8	8.534221e−02	9.2e−06	4.2e−05	3.3e+04	6.6e+01	4.95e+02
2×10^8	8.533084e−02	2.1e−05	3.0e−05	6.5e+04	1.3e+02	4.94e+02
4×10^8	8.532608e−02	2.5e−05	2.1e−05		2.6e+02	
10^9	8.531650e−02	3.5e−05	1.3e−05		6.6e+02	
2×10^9	8.533230e−02	1.9e−05	9.6e−06		1.3e+03	
4×10^9	8.534547e−02	5.9e−06	6.8e−06		2.6e+03	
10^{10}	8.535301e−02	1.6e−06	4.3e−06		6.6e+03	
4×10^{10}	8.535130e−02	9.9e−08	2.2e−06		2.6e+04	
10^{11}	8.534716e−02	4.2e−06	1.4e−06		6.6e+04	

```
__device__ float f(point p) {

    float x1,x2,x3,x4,x5,x6,x7;
    float dd,cc,dd3,t1,f0;

    x1 = p.coordinates[0];
    x2 = (1.0-x1)*p.coordinates[1];
    x3 = (1.0-x1-x2)*p.coordinates[2];
    x4 = (1.0-x1-x2-x3)*p.coordinates[3];
    x5 = (1.0-x1-x2-x3-x4)*p.coordinates[4];
    x6 = (1.0-x1-x2-x3-x4-x5)*p.coordinates[5];
    x7 = 1.0-x1-x2-x3-x4-x5-x6;

    t1 = x1+x2+x3;
    cc =(x1+x2+x3+x4+x5)*(x1+x2+x3+x6+x7) - t1*t1;
    dd = cc*(x1+x2+x3+x4+x5+x6+x7)
       -((x1*x2*x4 + x1*x2*x5 + x1*x2*x6 + x1*x2*x7
       + x1*x5*x6 + x2*x4*x7 - x3*x4*x6)
       + (x3*(-x4*x6 + x5*x7))
       + (x3*(x1*x4 + x1*x5 + x1*x6 + x1*x7 + x4*x6 + x4*x7))
```

```
          + (x3*(x2*x4 + x2*x5 + x2*x6 + x2*x7 + x4*x6 + x5*x6))
          + (x1*x4*x5 + x1*x5*x7 + x2*x4*x5 + x2*x4*x6
          + x3*x4*x5 + x3*x4*x6 + x4*x5*x6 + x4*x5*x7)
          + (x1*x4*x6 + x1*x6*x7 + x2*x5*x7 + x2*x6*x7
          + x3*x4*x6 + x3*x6*x7 + x4*x6*x7 + x5*x6*x7));
   dd3 = dd*dd*dd;
   if(dd3 == 0) return(0);

   f0 = 2.0*cc/dd3
        *(1.0-x1)*(1.0-x1-x2)*(1.0-x1-x2-x3)
        *(1.0-x1-x2-x3-x4)*(1.0-x1-x2-x3-x4-x5);
   return f0;
}
```

References

1. CUDA Library. http://www.nvidia.com/getcuda (last accessed May 2014)
2. Brown, R.: DIEHARDER. http://www.phy.duke.edu/~rgb/General/dieharder.php (last accessed May 2014)
3. Chan, T.F., Golub, G.H., LeVeque, R.J.: Updating formulae and a pairwise algorithm for computing sample variances. Technical Report STAN-CS-79-773, Stanford University ftp://reports.stanford.edu/pub/cstr/reports/cs/tr/79/773/CS-TR-79-773.pdf (1979)
4. Davis, P.J., Rabinowitz, P.: Methods of Numerical Integration. Academic, New York (1975)
5. de Doncker, E., Assaf, R.: GPU integral computations in stochastic geometry. In: VII Workshop Computational Geometry and Applications (CGA). Lecture Notes in Computer Science, vol. 7972, pp. 129–139 (2013)
6. de Doncker, E., Kapenga, J., Liou, W.W.: Open source software for Monte Carlo/DSMC applications. In: 55th AIAA/ASMe/ASCE/AHS/SC Structures, Structural Dynamics, and Materials Conference, The American Institute of Aeronautics and Astronautics (AIAA) (2014). doi:10.2514/6.2014-0348
7. de Doncker, E., Yuasa, F.: Distributed and multi-core computation of 2-loop integrals. In: 15th International Workshop on Adv. Computing and Analysis Techniques in Physics (ACAT 2013), Journal of Physics, Conference Series. To appear (2014).
8. Dremmel, J., Nguyen, H.D.: Fast reproducible floating-point summations. In: 2013 21st IEEE Symposium on Computer Arithmetic (ARITH), pp. 163–172 (2013)
9. Genz, A.: MVNPACK. http://www.math.wsu.edu/faculty/genz/software/fort77/mvnpack.f (2010)
10. Goldberg, D.: What every computer scientist should know about floating-point arithmetic. ACM Comput. Surv. **23**(1), 5–48 (1991)
11. Higham, N.J.: Accuracy and Stability of Numerical Algorithms, 2nd edn. SIAM, Philadelphia, Addison-Wesley (2002). ISBN 978-0-898715-21-7
12. IEEE Standard for Binary Floating-Point Arithmetic, ANSI/IEEE Standard 754-1985. Institute of Electrical and Electronics Engineers, New York (1985). Reprinted in SIGPLAN Notices **22**(2), 9–25 (1987)
13. IEEE Standard for Binary Floating-Point Arithmetic, ANSI/IEEE Standard 754-2008. Institute of Electrical and Electronics Engineers, New York (2008)
14. Kapenga, J., de Doncker, E.: Compensated summation on multiple NVIDIA GPUs. HPCS Technical Report HPCS-2014-1, Western Michigan University (2014)
15. Knuth, D.E.: The Art of Computer Programming, Volume 2, Seminumerical Algorithms, 3rd edn. Addison-Wesley (1998)

16. L' Equyer, P.: Combined multiple recursive random number generators. Oper. Res. **44**, 816–822 (1996)
17. Laporta, S.: High-precision calculation of multi-loop Feynman integrals by difference equations. Int. J. Mod. Phys. A **15**, 5087–5159 (2000). arXiv:hep-ph/0102033v1
18. L'Equyer, P., Simard, R.: A C library for empirical testing of random number generators. ACM Trans. Math. Softw. **33**, 22 (2007)
19. Manssen, M., Weigel, M., Hartmann, A.K.: Random number generators for massively parallel simulations on GPU (2012). arXiv:1204.6193v1 [physics.comp-ph] 27 April 2012
20. Marsaglia, G.: DIEHARD: a battery of tests of randomness. http://www.stat.fsu.edu/pub/diehard
21. Marsaglia, G.: Xorshift RNGs. J. Stat. Softw. **8**, 1–6 (2003)
22. Matsumoto, M., Nishimura, T.: Mersenne Twister: A 623-dimensionally equidistributed uniform pseudorandom number generator. ACM Trans. Model. Comput. Simul. **8**, 3 (2003)
23. Muller, J.-M., Brisebarre, N., de Dinechin, F., Jeannerod, C.-P., Lefevre, V., Melquiond, G., Revol, N., Stehle, D., Torres, S. Handbook of Floating-Point Arithmetic. Birkhäuser, Boston (2010). ACM G.1.0; G.1.2; G.4; B.2.0; B.2.4; F.2.1., ISBN 978-0-8176-4704-9
24. NVIDIA. Tesla Product Literature. http://www.nvidia.com/object/tesla_product_literature.html (last accessed May 2014)
25. NVIDIA. http://developer.download.nvidia.com/assets/cuda/files/CUDADownloads/TechBrief_Dynamic_Parallelism_in_CUDA.pdf (last accessed May 2014)
26. Rump, S.M., Ogita, T., Oishi, S.: Accurate floating-point summation part i: Faithful rounding. SIAM J. Sci. Comput. **31**(1), 189–224 (2008)
27. Rump, S.M., Ogita, T., Oishi, S.: Accurate floating-point summation part ii: Sign, k-fold faithful and rounding to nearest. SIAM J. Sci. Comput. **31**(2), 1269–1302 (2008)
28. Saito, M., Matsumoto, M.: Variants of Mersenne twister suitable for graphics processors. Trans. Math. Softw. **39**(12), 1–20 (2013)
29. Salmon, J.K., Moraes, M.A.: Random123: a library of counter-based random number generators. http://deshawresearch.com/resources_random123.html, and Random123-1.06 Documentation, http://www.thesalmons.org/john/random123/releases/1.06/docs (last accessed May 2014)
30. Salmon, J.K., Moraes, M.A., Dror, R.O., Shaw, D.E.: Parallel random numbers: as easy as 1, 2, 3. In: Proceedings of the International Conference for High Performance Computing, Networking, Storage and Analysis (SC11) (2011)
31. Sanders, J., Kandrot, E.: CUDA by Example - An Introduction to General-Purpose GPU Programming. Addison-Wesley, Reading (2011). ISBN: 978-0-13-138768-3
32. SPRNG: The scalable parallel random number generators library. http://www.sprng.org (last accessed May 2014)
33. Whitehead, N., Fit-Floreas, A.: Precision & performance: Floating point and IEEE 754 compliance for NVIDIA GPUs. http://developer.download.nvidia.com/assets/cuda/files/NVIDIA-CUDA-Floating-Point.pdf Nvidia developers (2011)

Chapter 14
GPU: Accelerated Computation Routines for Quantum Trajectories Method

Joanna Wiśniewska and Marek Sawerwain

14.1 Introduction

The Monte Carlo methods—starting the consideration from the first works of John von Neumann, Stanisław Ulam and Nicholas Metropolis [11, 12]—are widely used for calculations in the fields of: chemistry, astrophysics [15], quantum physics (e.g. [22, 25]) and also in the field of quantum computations.

There are many projects concerned about quantum computing simulation using Monte Carlo methods (a list of open source implementations is available at address http://www.quantiki.org/wiki/List_of_QC_simulators).

The mentioned list of projects contains presently a few popular packages using the Monte Carlo methods, in particular quantum trajectories [4] method (defined in further part of this chapter as QTM), for example: the package by Tan [18], QuTIP package [8] and also C++QED presented in [20]—the latest version of this package is described in [19]. It is possible to simulate quantum computing with an older solution [17] using quantum trajectories method as well.

Most listed packages do not use multi-core CPU and GPU systems. The exception is QuTIP package using available CPU cores during calculations for quantum trajectories (the GPU cores are not used). In [8] it is shown that for older solutions, e.g. [18], the parallel processing of quantum trajectories allows to obtain satisfactory linear acceleration. Using multi-core GPU systems for the quantum trajectories method causes the increase of efficiency and allows to obtain a better

J. Wiśniewska
Institute of Information Systems, Faculty of Cybernetics, Military University of Technology,
Kaliskiego 2, 00-908 Warsaw, Poland
e-mail: jwisniewska@wat.edu.pl

M. Sawerwain (✉)
Institute of Control and Computation Engineering, University of Zielona Góra,
Licealna 9, Zielona Góra 65-417, Poland
e-mail: M.Sawerwain@issi.uz.zgora.pl

V. Kindratenko (ed.), *Numerical Computations with GPUs*,
DOI 10.1007/978-3-319-06548-9__14, © Springer International Publishing Switzerland 2014

precision of computations in a shorter time—it is because the more trajectories are calculated the more precisely the statistical (average) trajectory is specified.

This chapter presents the implementation of quantum trajectories method with use of the CUDA technology. There is also shown an algorithm, which was implemented for GPU, and performance tests of obtained final solution. The quantum trajectories algorithm needs adequate methods for pseudorandom number generation (the CUDA Toolkit offers pseudorandom generators with good performance and fully optimized for a parallel environment) and for solving the initial value problem. There are two methods presented for the initial value problem (IVP): fourth-order Runge-Kutta method and fourth-order Backward Differentiation Formula. In further part of this chapter acronyms IVP or ODE (ordinary differential equation) will be used interchangeably.

14.2 The Quantum Trajectories Method

The quantum states' dynamics can be described as an evolution of a closed or an open quantum system. The closed system's evolution is used for instance in quantum circuits. The open system's evolution is considered when the external environment affects the computations. The mathematical details, describing closed and open systems' dynamics, are not needed to explain this chapter's aims—more detailed information can be found in [21] and [13].

However, it might be useful to mention that for closed systems the evolution is an unitary operation and it can be denoted as Schrödinger equation:

$$(A) \ i\hbar\frac{\partial}{\partial t}\Psi = \hat{H}\Psi, \quad (B) \ i\hbar\frac{d}{dt}|\psi\rangle = H|\psi\rangle \tag{14.1}$$

where (A) is the partial differential equation and (B) is used for the numerical simulations. In (B) H is a Hamiltonian expressing system's dynamics and $|\psi\rangle$ denotes the initial system's state.

If the influence of external environment should be added then the evolution of open quantum systems is given by the von Neumann equation:

$$\dot{\rho}_{tot}(t) = -\frac{i}{\hbar}[H_{tot}, \rho_{tot}(t)], \quad H_{tot} = H_{sys} + H_{env} + H_{int}, \tag{14.2}$$

where H_{sys} describes the dynamics of closed system, H_{env} is the environment's dynamics and H_{int} denotes the dynamics for environment-system interaction. The environment's influence can be removed from (14.2) by the operation of partial trace what is expressed by the Lindblad master equation:

$$\dot{\rho}(t) = -\frac{i}{\hbar}[H(t), \rho(t)] + \sum_n \frac{1}{2}\left[2C_n\rho(t)C_n^+ - \rho(t)C_n^+C_n - C_n^+C_n\rho(t)\right] \tag{14.3}$$

where C_n denotes so-called collapse operators representing the influence of external environment on a simulated system. Using a collapse operator on a quantum state causes state's change. However, simulating the behavior of quantum system in this case requires enormous memory capacity (growing exponentially according to the system's size).

At present, a method used to reduce the memory requirements is the Quantum Trajectories Method (QTM). The correct simulation of quantum system's behaviour needs calculating many single trajectories (as the natural consequence of Monte Carlo methods' random character). However, every trajectory may be simulated separately and this feature makes possible and efficient the parallel implementation for the quantum trajectories method.

Additionally, in a comparison to Lindblad master equation methods, where the density matrix formalism is used, the quantum trajectories method utilizes the wave function based on n-dimensional state's vector, so-called pure state. Generally, the number of vector's entries is exponential, but using sparse matrices allows efficient simulation of quantum system's behavior. In spite of the fact that the simulation based on the wave function concerns only one state of quantum system. Naturally, it seems to be a disadvantage of this solution because for Lindblad master equation methods the density matrix describes many different states of the same system and this causes higher demands on memory and computing resources. Using density matrices in most cases is not possible because of memory requirements—generally the size of density matrix is $2^n \times 2^n$. In a case of the quantum trajectories method, the collapse operator can be used to modify the state's vector and to monitor the influence of external environment on quantum state.

Summarizing, the system's evolution for QTM is described by the following Hamiltonian:

$$H_{\text{eff}} = H_{\text{sys}} - \frac{i\hbar}{2} \sum_i C_n^+ C_n, \tag{14.4}$$

The probability of so-called quantum jump, that means using a collapse operator C_n on current quantum state, is:

$$\delta p = \delta t \sum_n \langle \psi(t)|C_n^+ C_n|\psi(t)\rangle, \tag{14.5}$$

Whereas, the system's state after the collapse operation can be presented as:

$$|\psi(t+\delta t)\rangle = \frac{C_n |\psi(t)\rangle}{\sqrt{\langle \psi(t)|C_n^+ C_n|\psi(t)\rangle}} \tag{14.6}$$

If there are more than one collapse operator, the probability of using i-th operator is:

$$P_i(t) = \frac{\langle \psi(t)|C_i^+ C_i|\psi(t)\rangle}{\delta p} \tag{14.7}$$

In the process of simulation a random numbers generator is needed to ensure the probabilistic choice of adequate collapse operator. All mentioned calculations correspond to operations performed on matrices and vectors. In case of matrices they are usually band matrices, so they may be treated as sparse matrices to economize memory use and to speed up the calculations. The Compressed Sparse Row (CSR) format will be used because of many matrix-vector multiplication—it also gives an additional speed-up.

The above essential remarks constitute the idea of a following algorithm for a single quantum trajectory simulation which can be presented as four computational steps:

(I) the random value $r \in (0, 1)$ is computed, where r denotes the probability of quantum jump,

(II) the Schrödinger equation is integrated using Hamiltonian H_{eff} at time t in such way to make the state's vector norm equal or greater to r: $\langle \psi(t) | \psi(t) \rangle \geq r$,

(III) the quantum jump occurrence causes the system's state projection, in moment t, to one of the states given by Eq. (14.6). The operator C_n is selected to meet the following relation for adequate n:

$$\sum_{i=1}^{n} P_n(t) \geq r, \tag{14.8}$$

particular P_n values are specified by Eq. (14.7).

(IV) the projected state of a wave function is a new initial value for moment t; next, the new value of r is randomly selected and the procedure repeats the process of quantum trajectory simulation starting at the step (I)—more precisely: the simulation is performed for previously given value t.

The presented algorithm refers to the solutions described in other publications, e.g. in [8] and [1, 2].

14.3   Details of Implementation in CUDA C/C++

The Quantum Trajectories Method algorithm because of its basic property— calculating many unrelated trajectories—can be presented as a parallel algorithm consisting of two main steps:

- the first step—trajectories simulation—according to the method described in Sect. 14.2,
- the second step—averaging the obtained trajectories to compute the final trajectory.

The main task in a QTM algorithm is calculating quantum trajectories. The basic strategy for the implementation of quantum trajectories method is to calculate every trajectory without any correlations to one another. There are also some shared data e.g. Hamiltonian's definition, collapse operators' definition, time described as a variable in a form of list or table containing values of this variable. The mentioned types of data may be constant during the process of calculating the trajectories and during the whole simulation. Of course, the Hamiltonian and collapse operators may be considered as time-depended as well. In such case for each trajectory there will be a special variable representing mentioned structures. The presence of shared data means also the possibility of using so-called constant memory, which additionally simplify obtaining efficient implementation of QTM.

There should be a large area of local data in every thread for single trajectory pointed out for QTM implementation. The set of local variables contains also variables describing states of Pseudo-Random Number Generators (PRNGs). These states must be controlled during the simulation to provide different numbers' sequences in every thread (it is a very important feature for Monte Carlo methods, ensuring simulation's correctness). Unfortunately, a large number of local data, caused by many instances of procedure solving ODEs as well, means that it is worth to consider two following cases. The first case are simulations of small systems, when fast local memory resources available in units of the GPU are sufficient to run calculations efficiently. The second case are simulations of large systems, when it is necessary to use the main memory of the GPU. In both cases calculated trajectories are stored in the main memory. This approach allows to implement procedures using the capacities of GPU more efficiently.

Generally, the task scheme performed in a computation kernel during the quantum trajectories' calculating is shown in Fig. 14.1.

The second stage of QTM is averaging the set of obtained trajectories to one final trajectory. This step can be also efficiently implemented by the operation of reduction, which is presented in [5]. The whole process can be presented as a directed acyclic graph, shown in Fig. 14.2. Additionally, the solution described in [5] is considered to be very efficient.

14.3.1 Data Types and Auxiliary Functions

The Quantum Trajectories Method algorithm's implementation for GPU computation kernel needs a strictly specified data format. The data returned by the QTM algorithm is a sequence of expectation values—presented as real numbers obtained by using so-called expectation operator on system's state when the process of trajectories' calculating is already completed. It should be stressed that the entries of quantum state's vectors and the entries of matrices representing quantum operators are complex numbers. The CUDA package contains the adequate data type cuComplex (it provides the cooperation with types float and double), but the proposed implementation uses less advanced type for complex numbers because the additional code ensuring the numerical stability for the division or calculating absolute value of complex numbers is not necessary. This is justified by the

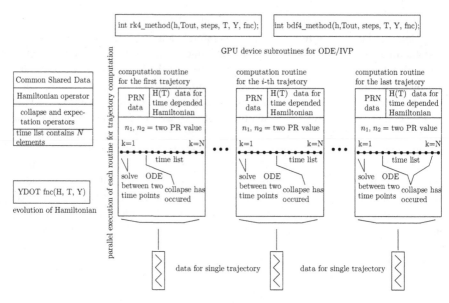

Fig. 14.1 The first stage's task scheme for quantum trajectories method in GPU computation system (PRN: pseudo-random numbers, H: Hamiltonian data, IVP: Initial Value Problem, RK4: the 4th order Runge-Kutta method, BDF4: the 4th order Backward Differentiation Formulae, T: time variable, Y: actual state of system)

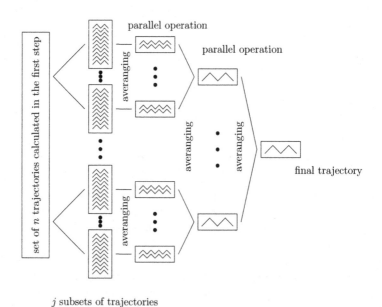

Fig. 14.2 The second stage's task scheme for quantum trajectories method in GPU computation system, averaging obtained trajectories after the first step

normalization of all calculated values, which allows to use simpler methods and results with system's higher efficiency, and also does not cause numerical instability. Although controlling the correctness of obtained results is recommended.

The definition of simpleComplex template used in proposed implementation is a typical definition based on a structure:

```
1  template <typename T> struct simpleComplex {
2       T re;
3       T im;
4  };
```

Naturally, for a user's convenience several additional operators (e.g. sum, multiplying, etc.) were overloaded in the simpleComplex template what allows clear presentation of arithmetic operations—both for host code and device code. The uVector template is defined similarly and it may be used for host code and GPU device code as well. To ensure the ease of use also the access operator for vector's elements was defined:

```
1  template <typename T, size_t v_size>
2  struct uVector {
3       unsigned int size;
4
5       T m[ v_size ];
6
7       __host__ __device__ T& operator[]
8               (const size_t idx) { return m[idx]; };
9       __host__ __device__ const T& operator[]
10              (const size_t idx) const { return m[idx]; };
11 };
```

To increase the code's performance the range is not verified.

Other definitions, e.g. the uMatrix template for full matrices and the uCRSMatrix template for sparse matrices, are based on uVector template. Using templates ensures easy changes between the types (from float or double to other type defined by the user) and simplify the process of memory management because dynamic memory allocation is no longer needed by GPU device. Especially smaller systems achieve higher performance when user-defined types are used instead of CUDA Toolkit types.

Creating own templates needs defining many auxiliary functions, e.g. for multiplying a CRS matrix by a vector. A typical implementation for this operation is given below:

```
1  typedef simpleComplex<double> SCD;
2
3  template<size_t v_size, size_t _S_ValueSize,
4           size_t _S_RowPtr, size_t _S_ColInd>
5  __host__ __device__ struct uVector< SCD , v_size>
6    mulCSRMatByuVec(
7       uCSRMatrix<SCD, _S_ValueSize, _S_RowPtr, _S_ColInd> m,
```

```
8           uVector< SCD, v_size> x)
9  {
10          size_t i, j;
11          uVector< SCD, v_size> y;
12          y.size = v_size;
13
14          for ( i=0; i < v_size; i++) {
15              y[i] = make_simpleComplex(0.0, 0.0);
16          }
17
18          for ( i=0; i < v_size; i++) {
19              for ( j=m.row_ptr[i] ; j < m.row_ptr[i+1] ; j++) {
20                  y[i]= y[i] + (m.values[j] * x[m.col_ind[j]]);
21              }
22          }
23
24          return y;
25  }
```

The above definition is constructed for the simpleComplex<double> type, which is necessary because with use of make_simpleComlex the initial values may be given as double numbers in this case.

A very important issue for implementing the QTM is the selection of Pseudo-Random Number Generator (PRNG). In this work the package cuRand [14], offering sufficient efficiency and proper statistical properties, from CUDA Toolkit was used. The available PRNGs in the cuRand package are XORWOW [10] and MRG32k3A [9, 16].

Using cuRand generator from the CUDA Toolkit is an easy task. There should be a global pointer, e.g. devStates, declared to keep the information about the states of generators used in every thread and there must be some global memory of GPU device allocated for proper use of PRNGs.

```
1  curandState *devStates;
2  cudaStatus = cudaMalloc(
3      (void **)&devStates, Ntrj * sizeof(curandState));
```

The next step is the initialization of each generator. The number of generators must correspond to the form of computing grid, which is considered as a line with Ntrj elements and Ntrj is an integer number equal to the quantity of calculated trajectories:

```
1  unsigned long long seedtime = time(0);
2  setup_kernel_for_curand <<<1, Ntrj >>>(devStates, seedtime);
```

The computation kernel's form is quite simple because it is based on the initialization of particular generators pointed by the indexes of threads—variable id.

```
1  __global__ void setup_kernel_for_curand (
2              curandState *state, unsigned long long seed) {
3      int id = threadIdx.x;
4      curand_init(seed, id, 0, &state[id]);
5  }
```

Considering the simulation of quantum system's dynamics with the use of QTM it may turn out that instead of generating pseudorandom numbers in every thread, the needed values can be prepared earlier: before the realization of QTM algorithm (e.g. with use of PRNG or physical sources of randomness [3, 7]). This approach does not affect significantly on data transfer but allows to free some local memory resources. As a result, the number of trajectories computed within one block can be increased.

14.3.2 Integration Methods

The second step leading to correct implementation of QTM is the selection of numerical methods for solving IVP. The implementations of the fourth-order Runge-Kutta method (RK4) and the fourth-order Backward Differentiation Formulae (termed as BDF4) were done. Both methods have to be adjusted for a parallel environment, so they were prepared as reentrant versions, this means that they will not be using any variables outside their own namespace. Due to the limitations put on this chapter, only the implementations for fixed step will be presented.

The implementation of the RK4 method, as a template, is shown at Listing 14.1. Using templates allows: easy types changing (i.e. between float or double) and specifying the size of structures which will be processed in the computational procedure. This approach is necessary because QTM computes the solution from a system of equations, where the number of equations equals to the size of analyzed quantum state.

Listing 14.1 The fourth-order Runge-Kutta method (without additional code for error detection like no convergence or exceeded iterations' number) as reentrant version in a form of device function for CUDA computation kernel

```
1   template <typename TYPE, size_t SIZE, size_t AlphaSize>
2   __device__ int rk4_method_for_mc (TYPE h, TYPE Tout, int steps,
3           simpleComplex<TYPE> &T,
4           uVector< simpleComplex<TYPE>, SIZE > &Y,
5           uVector< simpleComplex<TYPE>, SIZE > (*fnc)(
6               const simpleComplex<TYPE>&,
7               const uVector< simpleComplex<TYPE>, SIZE >& ) )
8   {
9     int j = 0;
10
11    uVector< simpleComplex<TYPE>, AlphaSize > k1, k2, k3, k4;
12
13    k1.size = AlphaSize; k2.size = AlphaSize;
14    k3.size = AlphaSize; k4.size = AlphaSize;
15
16    while( (j < steps) && (T.re < Tout) ) {
17          k1 = h * fnc(T, Y);
18          k2 = h * fnc(T + h/2.0, Y + k1/2.0);
19          k3 = h * fnc(T + h/2.0, Y + k2/2.0);
```

```
20          k4 = h * fnc (T + h, Y + k3 );
21
22          Y = Y + ((1.0/6.0) * (k1 + 2.0*k2 + 2.0*k3 + k4 ));
23
24          T = T + h;
25          j++;
26    }
27    return 0;
28  }
```

Naturally, the RK4 method can have a very compact implementation, which is shown in Listing 14.1. This is a consequence of using templates simpleComplex and uVector together with a rich set of overloaded operators, which makes the implementation of arithmetic operations on vectors and scalars very clear. It should be stressed that during the parallel trajectories' calculation many instances of RK4 method reside in a memory and for calculations' correctness the method needs additional vectors: k1, k2, k3, k4, which allocate memory available for every thread. For a small systems vectors k1, k2, k3, k4 may be placed in a register memory by the compiler's optimization process, because mentioned vectors are static variables.

The sets of parameters for methods RK4 and BDF4 are the same and require to supply: the width of integration's interval (h), the final value of time (Tout), a maximal number of iterations, an actual value of time variable, the present system's state Y. The very last parameter is a function specifying the state's evolution with use of previously given Hamiltonian.

Listing 14.2 shows the most important parts of the BDF4 method's implementation. To simplify the implementation, both methods (RK4 and BDF4) do not use the variable step integration.

Listing 14.2 The most significant parts of the fourth-order BDF method as reentrant version for CUDA computation kernel

```
1   template <typename TYPE, size_t SIZE, size_t AlphaSize>
2   __device__ int bdf4_autofac_method_for_mc (TYPE h, TYPE Tout,
3             int steps, simpleComplex<TYPE> &T,
4             uVector< simpleComplex<TYPE>, SIZE > &Y,
5             uVector< simpleComplex<TYPE>, SIZE > (*fnc )(
6               const simpleComplex<TYPE>&,
7               const uVector< simpleComplex<TYPE>, SIZE >& ) )
8   {
9     const int n = 8;
10    int istep, errLvl = 0, i, j;
11    int m_size = AlphaSize;
12    TYPE t;
13
14
15    uVector< simpleComplex<TYPE>, AlphaSize > YY[ n ];
16    uVector< simpleComplex<TYPE>, AlphaSize > D1[ n ];
17    // removed part
18    uVector< simpleComplex<TYPE>, AlphaSize > D4[ n ];
19
```

```
20
21    uVector< simpleComplex<TYPE>, AlphaSize > p, p1, c;
22    p.size = AlphaSize;
23    // removed part
24
25    uMatrix<TYPE, AlphaSize> J, IJ, id_mat;
26    J.rows = AlphaSize;
27    J.cols = AlphaSize;
28    // removed part
29
30    eye_of_matrix( id_mat );
31
32
33    zerovector( p ); zerovector( p1 ); zerovector( c );
34
35    for(i=0;i<n;i++) {
36      YY[i] = p; D1[i] = p; D2[i] = p; D3[i] = p; D4[i] = p;
37    }
38
39    for(i=0;i<AlphaSize;i++) {
40      YY[0][i] = Y[i];
41    }
42
43    errLvl = rk4_method_for_mc <TYPE,SIZE,AlphaSize>(h,
44        T.re + 4*h, 1, T, Y, fnc); YY[0] = Y;
45    // ... removed part ...
46    errLvl = rk4_method_for_mc <TYPE,SIZE,AlphaSize>(h,
47        T.re + 4*h, 1, T, Y, fnc); YY[4] = Y;
48
49
50    D1[0] = ( YY[1] − YY[0] );
51    // ... removed part ...
52    D4[0] = ( D3[1] − D3[0] );
53
54
55    i=4; istep=0;
56    while ( (istep < steps) && (T.re < Tout) ) {
57      p = YY[i] + D1[i−1] + D2[i−2] + D3[i−3] + D4[i−4];
58      t=1.0; j=1;
59      while ((t > 1e−16) && (j < 6)) {
60        c = (1.0/25.0) * ( 48.0*YY[i]−36.0*YY[i−1]+16.0 *
61                  YY[i−2]−3.0*YY[i−3] + 12.0
62                              * h * fnc( T + h, p) );
63        J = 12.0/25.0 * h * num_jacobian_vector <TYPE,
64                  SIZE,AlphaSize>( T + h, p, fnc) − id_mat;
65        inverse_of_matrix(J, IJ);
66
67        p1 = ( (IJ) * (p−c) ) + p;
68        t = norminf( p1 − p );
69        j = j + 1; p = p1;
70        T = T + h;
71      } // while ((t > 1e−16) && (j < 6))
72
```

```
73        YY[ i +1] = p ;
74
75        D1[ i ] = ( p − YY[ i ] );
76        // ... removed part ...
77        D4[ i −3] = (D3[ i −2] − D3[ i −3]);
78
79        i ++; istep ++;
80      } //( ( istep < steps ) && ( T. re < Tout ) )
81
82      for ( j =0; j < AlphaSize ; j ++) {
83                 Y[ j ] = YY[ i ][ j ];
84      }
85
86      return errLvl ;
87    }
```

The BDF4 method before starting the main calculations needs the estimation of value Y in the first four points—to obtain these values the BDF4 method uses RK4 method (Listing 14.1). The BDF method needs also the results of functions: num_jacobian_vector which calculates the numerical estimation of Jacobian's value; inverse_of_matrix to calculate inverse matrix.

It should be also noted that the solution of equations' system is saved in the array Y (lines 80–82). The partial solutions are placed in the array YY. When the main loop is over (lines 55–69), the variable i points the element of array YY which is the solution calculated by BDF4 method.

14.3.3 Quantum Trajectory Method: Implementation

There were auxiliary data structures, functions and two methods for solving differential systems of equations presented in previous sections. This allows to implement the Quantum Trajectories Method – an exemplary implementation, according to the description from Sect. 14.2 is given below.

The first part of the implementation – Listing 14.3 – is responsible for local variables' declaration, e.g. the variable h corresponds to the width of integration's step. The variables expressing time are: T, $T final$, etc. It is also necessary to declare state variables Y and vector P to collect the information about the probability of proper collapse operator's choice. It should be explained that LEAD_NUM_TYPE denotes basic data types (float or double) and LEAD_DIM corresponds to the size of variable Y and points out the main dimension of analyzed quantum state.

Listing 14.3 The first part of computational routine for QTM realisation is responsible for local variable declaration

```
1    __global__  void trj_sim ( curandState ∗state ,
2                      simpleComplex <LEAD_NUM_TYPE> ∗trj_data ) {
3        int i = 0, j =0, k_ons = 0,
4        odesolverstate = 0, ode_norm_steps = 5, cnt = 0;
5
```

```
6       int m = 5;
7       int counterIter ;
8
9       LEAD_NUM_TYPE a = (LEAD_NUM_TYPE)0.0;
10      LEAD_NUM_TYPE b = (LEAD_NUM_TYPE)10.0;
11      LEAD_NUM_TYPE h, hh, mu = 0.0, nu = 0.0,
12      // removed part
13      sump = 0.0;
14
15      int k = 0, cols = N;
16      int thidx = threadIdx.x;
17      int trj = (blockIdx.x * blockDim.x) + threadIdx.x;
18
19      uVector< simpleComplex<LEAD_NUM_TYPE>, N > P;
20
21      simpleComplex<LEAD_NUM_TYPE> T ;
22      // removed part
23
24      uVector< simpleComplex<LEAD_NUM_TYPE>, 2 > Y;
25      // removed part
26
27      Y.size = 2;
28      // removed part
29
30      simpleComplex<LEAD_NUM_TYPE>  ev ;
31      P.size = N;
32
33      for( k = 0 ; k < N ; k++) {
34          P[k].re=0.0;
35          P[k].im=0.0;
36      }
37
38      if (trj == 0) {
39          prepare_data_for_operators ();
40      }
41      __syncthreads ();
42
43      // the second part
44  }
```

A very important issue for implementing QTM in CUDA technology is the read-out of trajectory's number. An identification number of the thread and the block will be used to calculate the trajectory's number. For simplicity, the computing grid (see also Fig. 14.3) will be considered as a line sized $Ntrj \times 1$, where $Ntrj$ denotes the number of trajectories. It means that a single trajectory is calculated per block. The read-out of thread's number and assigning identifier to a trajectory is realized in the following way:

```
1  int thidx = threadIdx.x;
2  int trj = (blockIdx.x * blockDim.x) + threadIdx.x;
```

The second important issue is checking if the thread's identifier in block is equal to zero. If it is so, the function called as prepare_data_for_operators

in total n trajectories

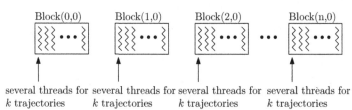

single thread for   single thread for   single thread for   single thread for
trajectory t=0   trajectory t=1   trajectory t=2   trajectory t=n

in total $n \cdot k$ trajectories

several threads for several threads for several threads for several threads for
k trajectories k trajectories k trajectories k trajectories

Fig. 14.3 The thread's allocation between blocks in presented method—shown implementation uses only one thread per block, however, if the quantity of local data is not too large, more threads may be activated in a one block

will initialize the instances of objects representing: Hamiltonian's form, the vector containing values of time, collapse operators and the operator used to calculate expectation value. Therefore, there is a synchronization command in the line 40, because all the threads have to wait until the process of initialization will be completed.

Listing 14.4 The second part of computational routine for QTM which computes a single trajectory

```
1   if( trj < Ntrj) {
2     curandState localState = state[trj];
3
4     mu = curand_uniform(&localState);
5     nu = curand_uniform(&localState);
6
7     T = make_simpleComplex((LEAD_NUM_TYPE)a, (LEAD_NUM_TYPE)0.0);
8
9     for( i = 0 ; i < LEAD_DIM ; i++) Y[i] = alpha[i];
10
11    ev = expect_cnv_csrdenmat(e_ops_0333, Y);
12    trj_data[trj * cols + 0] = ev;
13
14    counterIter = 0;
15    T=tlist[0];
16
17    for ( k = 1 ; k < N; k++)
18    {
19      a=T.re ;
```

```
20    b= tlist[k].re ;
21    h = (b − a) / m;
22
23    while( (T.re < tlist[k].re ) && counterIter < 255 )
24    {
25       T_prev = T;
26       Y_prev = Y;
27
28       norm2_prev = norm( Y );
29
30       odesolverstate = rk4_method_for_mc <LEAD_NUM_TYPE, 2, 2>(
31              h, tlist[k].re, 1, T, Y,
32              &user_evolution_fnc <LEAD_NUM_TYPE, 2> );
33
34       if( T.re > tlist[k].re )
35       {
36              T = T_prev;
37              Y = Y_prev;
38
39              odesolverstate = rk4_method_for_mc <LEAD_NUM_TYPE,
40                 2, 2>( h, tlist[k].re, m, T, Y,
41                         &user_evolution_fnc <LEAD_NUM_TYPE, 2> );
42       }
43
44       norm2_psi = norm( Y );
45
46       if(norm2_psi <= mu ) {
47              // removed code is a thrid part of this routine
48       } // if(norm2_psi <= mu )
49
50       counterIter++;
51    } // while(T.re < tlist[k].re)
52
53    out_psi = Y / normsqrt(Y);
54
55    ev = expect_cnv_csrdenmat(e_ops_0333, out_psi);
56    trj_data[trj * cols + k] = ev;
57    } // for ( k = 1 ; k < N; k++)
58
59    state[trj] = localState;
60    } // if(trj < Ntrj)
```

The second part of computational routine—presented at Listing 14.4—computes a single trajectory. In the lines 4–5 the PRNG is used to draw the values for the evaluation whether the collapse operator may be used. It should be mentioned, that the next drawing of these values takes place in the third part—Listing 14.5—which contains the code for checking if the collapse operator may be used.

The next actions are: initializing the state's value (line 9) and calculating the first expectation value (lines 11–12). The next points of the trajectory are calculated in the loop (lines 17–57). The points correspond to the moments in time—between them the system of differential equations is solved. This action was described in the step (II) of Sect. 14.2.

Listing 14.5 The third part of computational routine for QTM realization

```
 1   if ( norm2_psi <= mu )
 2   {
 3     T_final = T ;
 4     cnt = 0 ;
 5
 6     for ( k_ons =0; k_ons < ode_norm_steps ; k_ons ++)
 7     {
 8       T_guess = T_prev + log ( norm2_prev /mu) /
 9                     log ( norm2_prev / norm2_psi) * ( T_final −T_prev ) ;
10
11       Y = Y_prev ;
12       T = T_prev ;
13
14       a = T. re ;
15       b = T_guess . re ;
16       hh = (b − a ) / (LEAD_NUM_TYPE )m;
17
18       odesolverstate = rk4_method_for_mc <LEAD_NUM_TYPE, 2, 2>(
19                   hh, T_guess . re , m,    T, Y,
20                   &user_evolution_fnc <LEAD_NUM_TYPE, 2> );
21
22       norm2_guess = norm( Y ) ;
23       if ( abs (mu−norm2_guess ) < ode_norm_tol * mu )
24       {
25             break ;
26       }
27       else if ( norm2_guess < mu)
28       {
29             T_final = T_guess ;
30             norm2_psi = norm2_guess ;
31       }
32       else {
33                 T_prev = T_guess ;
34             Y_tmp = Y;
35             norm2_prev = norm2_guess ;
36       }
37       cnt = cnt + 1 ;
38     } // for ( k=1; k < ode_norm_steps ; k++)
39
40     if ( cnt > ode_norm_steps ) {
41         // The norm tolerance value is not reached .
42         // Increase accuracy of ODE solver or
43         // norm_steps is necessary .
44         odesolverstate = −1;
45         break ;
46     }
47
48     for ( j = 0 ; j < c_ops_size ; j++ )
49     {
50       Y_tmp = mulCSRMatByuVec ( c_ops [ j ], Y) ;
51       P[ j ]. re = norm ( Y_tmp );
52     } // for ( j = 0 ; j < c_ops_size; j++ )
```

```
53
54      P = P / sum(P).re;
55      sump = (LEAD_NUM_TYPE)0.0 ;
56
57      for ( j = 0 ; j < c_ops_size ; j++ )
58      {
59         if ( (sump <= nu) && (nu < sump+P[j].re) )
60         {
61              Y = mulCSRMatByuVec(c_ops[j], Y) ;
62         }
63              sump = sump + P[j].re ;
64      } // for ( j=0 ; j < c_ops_size; j++ )
65
66      mu = curand_uniform(&localState);
67      nu = curand_uniform(&localState);
68
69      normalize(Y);
70
71   } // if( norm2_psi <= mu )
```

The realization of step (III), i.e. using collapse operator can be seen in lines 48–64 at Listing 14.5. In analyzed procedure a variable odesolverstate is used for error handling, e.g. when the norm will be lower than established accuracy—lines 23 and 41 in Listing 14.5—what is connected to the precision of BDF4 and RK4 methods. The return code after execution of RK4 and BDF4 methods is also written in odesolverstate.

Running the above procedure, for previously introduced computing grid, shows the following code line:

```
1   trajectory_simulation <<<Ntrj,1>>>( devStates , trj_data_device );
```

At this point using a shared memory may be proposed, so in a one block of threads many trajectories can be calculated and the necessary data could be kept in the shared memory.

Using only one thread for every block allows to skip the problems with memory management and makes the implementation clear, but unfortunately reduces potential efficiency. It should be also considered, if the results for every trajectory should be written in the local memory, because after the process of calculations the results must be locally averaged or copied to the global memory.

For the discussed approach after the calculations the global variable trj_data_device contains the set of trajectories for further averaging, e.g. by GPU or by traditional processor. Generally, the operation of averaging Ntrj trajectories, where the size of every trajectory is given by N, may be presented as basic two "for" loops:

```
1   for ( k = 0 ; k < N; k++) {
2       tmp.re=0; tmp.im=0;
3       for ( trj = 0 ; trj < Ntrj ; trj++) {
4           tmp = tmp + trj_data_double[trj * cols + k];
5       }
```

```
6        tmp = tmp / (LEAD_NUM_TYPE) Ntrj ;
7        avg_trj[k] = tmp;
8  }
```

14.4 A Short Discussion About Performance

In this section the efficiency analysis is limited to the comparison between the solution presented in the chapter and the QuTIP package [8] supporting QTM. Only one example concerning the simulation of so-called unitary Hamiltonian will be considered:

$$H = \frac{2\pi}{10}\sigma_x, \quad \text{and} \quad \sigma_x = \begin{bmatrix} 0 & 1 \\ 1 & 0 \end{bmatrix}. \tag{14.9}$$

where σ_x represents Pauli operator X – so-called negation operator. The initial state is:

$$|\psi_0\rangle = |0\rangle = \begin{bmatrix} 1 \\ 0 \end{bmatrix}. \tag{14.10}$$

The collapse operator C_0 used during the simulation and the expectation value operator can be expressed as:

$$C_0 = \frac{5}{100}\sigma_x, \; E_0 = \sigma_z, \quad and \quad \sigma_z = \begin{bmatrix} 1 & 0 \\ 0 & -1 \end{bmatrix}. \tag{14.11}$$

where σ_z stands for Pauli sign operator.

Although the above structures are quite small, the simulation process for 50 trajectories, using the QuTIP package on PC unit equipped with Intel Code 2 Duo 8400 3.0 GHz, takes 3–4 s—using only one core. Working with two cores does not change the duration time significantly because the QuTIP package generates additional operations associated with two threads' support. Naturally, increasing the number of trajectories allows to notice the calculations' speed-up when more cores is used.

The result for graphics card Geforce 460 1 GB RAM when RK4 method is used equals 0.08 s—this means 50 times greater acceleration comparing to the single core calculation, mentioned above. It should be stressed that time 0.08 s was achieved using only one thread per block. For more complex BDF4 method the duration time is 0.2 s, so the acceleration is not so impressing—only ten times greater. Naturally, the calculations with use of CPU and GPU were carried out on double type numbers.

14.5 Conclusions and Future Goals

In comparison to other existing solutions, the proposed solution allows to achieve the acceleration for computations based on the quantum trajectories method. Especially for RK4 method, the precision of presented implementation is comparable to the precision of computations performed by serial or parallel solutions with use of traditional universal processors.

The next step in the presented software's evolution is implementing BBDF (Block Backward Differentiation Formulae) methods [6, 23, 24]. The BBDF methods should reduce the computations' time because they need less iterations to solve ordinary differential equations, maintaining the same stability as the backward differentiation formula methods. Using new technologies, like Dynamic Parallelism (implemented in the latest NVIDIA devices and accessible with CUDA C/C++ Toolkit ver. 5), might also help to obtain better efficiency for implemented methods.

Another important issue is developing a new solution for pseudorandom number generation. Instead of using pseudorandom number generators during the calculations, the pseudorandom numbers could be derived from previously prepared data set. This approach allows to free the local resources and also enables using physical devices and quantum sources of randomness e.g. [3] and [7].

Acknowledgements We would like to thank for useful discussions with the *Q-INFO* group at the Institute of Control and Computation Engineering (ISSI) of the University of Zielona Góra, Poland. We would like also to thank to anonymous referees for useful comments on the preliminary version of this chapter. The numerical results were done using the hardware and software available at the "GPU μ-Lab" located at the Institute of Control and Computation Engineering of the University of Zielona Góra, Poland.

References

1. Dalibard, J., Castin, Y., Molmer, K.: Wave-function approach to dissipative processes in quantum optics. Phys. Rev. Lett. **68**, 580 (1992)
2. Dum, R., Zoller, R., Ritsch, H.: Monte Carlo simulation of the atomic master equation for spontaneous emission. Phys. Rev. A **45**, 4879 (1992)
3. Frauchiger, D., Renner, R., Troyer, M.: True randomness from realistic quantum devices. arXiv:1311.4547 (2013)
4. Garraway, B.M., Knight, P.L.: Evolution of quantum superpositions in open environments: quantum trajectories, jumps, and localization in phase space. Phys. Rev. A **50**, 2548–2563 (1994)
5. Harris, M.: Optimizing Parallel Reduction in CUDA. http://developer.download.nvidia.com/compute/cuda/1_1/Website/projects/reduction/doc/reduction.pdf (2007)
6. Ibrahim, Z.B., Suleiman, M.B., Othman, K.I.: Fixed coefficients block backward differentiation formulas for the numerical solution of stiff ordinary differential equations. Eur. J. Sci. Res. **21**(3), 508–520 (2008)
7. ID Quantique SA: Quantis. http://www.idquantique.com/random-number-generators/products.html (2013)

8. Johansson, J.R., Nation, P.D., Nori, F.: QuTiP 2: a Python framework for the dynamics of open quantum systems. Comput. Phys. Commun. **184**(4), 1234–1240 (2013)
9. L'Ecuyer P., Simard R., Chen J.E., Kelton W.W.: An object-oriented random-number package with many long streams and substreams. Oper. Res. **50**(6), 1073–1075. http://pubsonline.informs.org/toc/opre/50/6 (2002)
10. Marsaglia, G.: Xorshift RNGs. J. Stat. Softw. **8**(14), 1–6 (2003)
11. Metropolis, N., Ulam, S.: The Monte Carlo method. J. Am. Stat. Assoc. **44**(247), 335–341 (1949)
12. Metropolis, N., Rosenbluth, A.W., Rosenbluth, M.N., Teller, A.H., Teller, E.: Equation of state calculation by fast computing machines. J. Chem. Phys. **21**(6), 1087–1092 (1953)
13. Nielsen, M.A., Chuang, I.L.: Quantum Computation and Quantum Information: 10th Anniversary Edition. Cambridge University Press, Cambridge (2010)
14. NVIDIA, CURAND Toolkit Documentation. http://docs.nvidia.com/cuda/curand/index.html (2013)
15. Pattabiraman, B., Umbreit, S., Wei-keng, L., Rasio, F., Kalogera, V., Memik, G., Choudhary, A.: GPU-accelerated Monte Carlo simulations of dense stellar systems. In: Innovative Parallel Computing, IEEE InPar 2012, San Jose, CA, pp. 1–10 (2012)
16. Saito, M.: A variant of Mersenne twister suitable for graphic processors. arXiv:1005.4973v2 (2010)
17. Schacka, R., Brun, T.A.: A C++ library using quantum trajectories to solve quantum master equations. Comput. Phys. Commun. **102**, 210–228 (1997)
18. Tan, S.M.: A computational toolbox for quantum and atomic optics. J. Opt. B Quantum Semiclassical Opt. **1**(4), 424 (1999)
19. Vukics, A.: C++QEDv2: the multi-array concept and compile-time algorithms in the definition of composite quantum systems. Comput. Phys. Commun. **183**, 1381–1396 (2012)
20. Vukics, A., Ritsch, H.: C++QED: an object-oriented framework for wave-function simulations of cavity QED systems. Eur. Phys. J. D **44**, 585–599 (2007)
21. Wyatt, R.E.: Quantum Dynamics with Trajectories. Springer, New York (2005)
22. Yang, B., Lu, K., Liu, J., Wang, X., Gong, C.: GPU accelerated Monte Carlo simulation of deep penetration neutron transport. In: Parallel Distributed and Grid Computing (PDGC), 2nd IEEE International Conference, pp. 899–904 (2012)
23. Yatim, S.A.M., Ibrahim, Z.B., Othman, K.I., Ismail, F.: Fifth order variable step block backward differentiation formulae for solving stiff ODEs. In: World Academy of Science, Engineering and Technology, vol. 38, pp. 280–282 (2010)
24. Yatim, S.A.M., Ibrahim, Z.B., Othman, K.I., Suleiman, M.B.: Numerical solution of extended block backward differentiation formulae for solving stiff ODEs. In: Proceedings of the World Congress on Engineering, WCE 2012, vol. I, London, 4–6 July 2012
25. Zhong, Z., Talamo, A., Gohar, Y.: Monte Carlo and deterministic computational methods for the calculation of the effective delayed neutron fraction. Comput. Phys. Commun. **184**(7), 1660–1665 (2013)

Chapter 15
Monte Carlo Simulation of Dynamic Systems on GPU's

Jonathan Rogers

15.1 Introduction

Monte Carlo is a highly flexible form of numerical quadrature used to solve a variety of mathematical problems. At a fundamental level, Monte Carlo methods use finite summations built from statistical samples to estimate definite integrals. In the field of dynamical systems, Monte Carlo is a technique used for prediction and uncertainty quantification. Given a system model, a set of sampling distributions is defined involving stochastic inputs, uncertainty in initial conditions, and/or uncertainty in system parameters. Simulations are then performed by sampling from these distributions and propagating a dynamic model. These so-called Monte Carlo simulations may be used to estimate the state probability density function (PDF), or moments of the PDF such as mean and variance, as a function of input distributions. As such, Monte Carlo simulation has become a standard tool in systems analysis in which output distributions may be estimated in a straightforward manner from a set of input distributions.

The popularity of Monte Carlo stems from both its flexibility in handling nonlinear, non-Gaussian systems and its relative ease of implementation. At the same time, Monte Carlo simulation suffers from poor computational scalability and is difficult to employ for high-dimensional systems. The Law of Large Numbers, upon which Monte Carlo is based, dictates that estimates of probability distributions obtained from Monte Carlo are only accurate for large numbers of samples [1]. For instance, it is well-known that if the expected value of a distribution is to be approximated via Monte Carlo, the accuracy of the estimated mean is proportional to $1/\sqrt{N}$ where N is the number of samples employed in the simulation. Thus if the number of Monte Carlo samples is quadrupled, uncertainty in the estimated mean is only halved.

J. Rogers (✉)
Woodruff School of Mechanical Engineering, Georgia Institute of Technology,
Atlanta, GA 30332, USA
e-mail: jonathan.rogers@me.gatech.edu

V. Kindratenko (ed.), *Numerical Computations with GPUs*,   319
DOI 10.1007/978-3-319-06548-9_15, © Springer International Publishing Switzerland 2014

While computational inefficiency is an important limitation, Monte Carlo simulation remains the standard approach to probabilistic modeling for a wide variety of dynamical systems. In the engineering community, control systems designers use Monte Carlo to evaluate robustness to environmental disturbances and/or modeling errors. Reliability engineers use Monte Carlo to evaluate mean time between failures given probabilistic models of component fatigue. In the physics community, Monte Carlo methods provide a technique for modeling radiation transport. In mathematical finance, Monte Carlo is commonly employed to evaluate expected value of investment options given sources of uncertainty and complex market forces.

Monte Carlo analysis became viable only through the advent of digital computers in the mid-twentieth century. Since sampling is the foundation of Monte Carlo, it relies heavily on the ability to generate pseudorandom numbers and compute long summations quickly, a capability which only digital computers can provide. Monte Carlo is unique as a systems analysis tool in that its accuracy improves as a function of computational throughput—more samples inherently leads to higher accuracy. As a result, Monte Carlo methods are well-disposed to take advantage of new high-throughput computing architectures such as Graphics Processing Units (GPU's).

The ideal Monte Carlo simulation represents a so-called "embarrassingly parallel" process in the sense that each sample is independent of each other sample [2]. For GPU's, in which maximum throughput is obtained when there are no serial dependencies between processing threads, Monte Carlo simulation is an extremely efficient data-parallel method that exposes the architecture's massive parallelism. Numerous researchers over the past decade have implemented GPU-based Monte Carlo simulations for a variety of scientific, engineering, and financial problems. A few examples include GPU-based Monte Carlo radiation transport [3, 4], stochastic control [5, 6], and option pricing [7].

This chapter will provide a recipe for implementation of Monte Carlo simulation on GPU's. The focus is on Monte Carlo simulation of dynamical systems, which are represented by a dynamic model, rather than Monte Carlo quadrature of explicit integrals. The chapter begins with an overview of the algorithmic implementation on the GPU, followed by a discussion of optimization and best practices. A case study is offered in which motion of a point-mass is simulated subject to viscous drag and gravity effects. Example GPU-based Monte Carlo simulation codes are provided for this example in both CUDA and OpenCL. These example codes are easily modified for other dynamical systems of interest.

15.2 GPU-Based Monte Carlo Algorithm Overview

The basic GPU-based Monte Carlo algorithm is depicted in Fig. 15.1. It is assumed that the Monte Carlo simulation is comprised of N simulations, where N is a user-defined parameter intended to provide the desired level of accuracy. First, initial conditions are established in device memory by the CPU (host) thread.

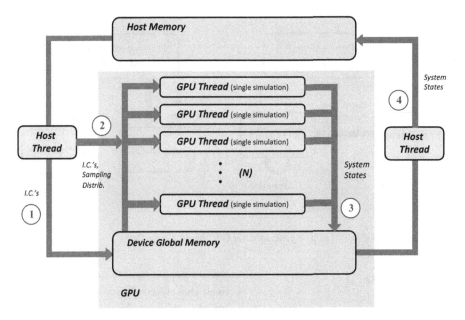

Fig. 15.1 Overview of Monte Carlo simulation on GPU. A single system trajectory is simulated by each GPU thread

These randomized samples may be generated directly on the GPU, or generated on the host and transferred to device memory (as discussed later). Also during initialization, space is allocated in GPU global memory for simulation results. Once samples are available in GPU memory, N GPU threads are launched by the host thread (labeled step 2 in Fig. 15.1). Each GPU thread is tasked to perform a single system simulation from the initial condition to the desired end condition. Depending on the chosen sampling method, each GPU thread may gather its assigned input parameters from device memory and compute a single simulation or, if random sampling was not performed before kernel launch, each GPU thread will generate its own random sample before simulating the system. Upon completion of a simulated trajectory, the GPU thread writes its output data to device global memory (labeled as step 3). Once all threads complete execution, a host memory copy is initiated to gather results from device memory to host memory, labeled as step 4. Pseudocode listings are provided in Listings 1 and 2 for host thread and GPU thread execution respectively.

Several implementation details are important to consider in order to maximize computational throughput and efficiency. A discussion of several of these considerations follows.

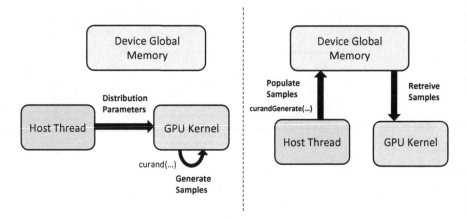

Fig. 15.2 Two methods of sample generation for GPU-based Monte Carlo

Listing 1 Host Thread
Pseudocode

Input Data: Sampling Distributions,
Initial Conditions, N
Result: Output Data from Monte
Carlo Simulation
start
1. Generate N sets of initial conditions according to sample distributions
2. Allocate device global memory for input data and simulation results
3. Launch N GPU threads
4. Gather output data from GPU global memory to host memory
return

15.2.1 Sampling Tasks

There are several ways to divide sampling tasks between the host and device. The most memory-efficient method of sampling is to sample directly within the GPU kernel itself using a device-side pseudorandom number generator such as that provided by the CURAND library [8]. In this case, the host sends GPU threads distribution parameters only (for instance, mean and standard deviation) either via

Listing 2   GPU Kernel
Pseudocode

> **Input Data:** Initial Conditions and System Parameters in Global Memory
> **Result:** Single System Trajectory Simulation
> **start**
> 1. **if** (samples present in global memory)
>   gather assigned sample
>   **else**
>   generate sample according to desired distribution
>   **end**
> 2. **while** (end condition not met)
>   integrate dynamic equations
>   **end**
> 3. Store final states in device global memory.
> **return**

transfer to global memory or pass-by-value on the kernel argument list. The GPU kernel then generates random samples itself prior to system simulation and stores them in local memory. The second method involves population of device memory with random samples by the host thread, using a GPU random number generation library. CURAND provides this capability through its host-side API. In this case, prior to kernel launch, the host thread generates random samples directly in device memory through appropriate library calls. Each of these two methods is depicted in Fig. 15.2. A final sampling method involves generating random samples first in host memory, and then copying these samples into device memory prior to kernel launch. This method may be required when input distributions are difficult to sample or in the case of Markov Chain Monte Carlo analysis. However, since the number of samples required is often quite large, copying random samples from host to device memory may result in significant latency and should be avoided where possible. In cases where host-side sampling is unavoidable, page locked memory should be used to minimize memory-copy latencies [9].

15.2.2   Storing Output Data

Each simulation must have space to write its own output data. Typically, output arrays are allocated in device memory by the host thread prior to kernel launch, and each GPU thread is assigned space in the output array to store states of interest. For many dynamical systems, the amount of data generated by Monte

Carlo simulation can become quite large, especially if intermediate values of the state must be stored. This problem is exacerbated in cases requiring extremely large sample sizes (millions or more). Due to the low bandwidth of the PCI-express bus used for memory transfers between host and device, transferring extremely large output data sets from the device to host may become problematic, especially in cases where Monte Carlo simulations must be run repeatedly or in a feedback manner. In general, it is best practice to record the minimum possible data from each simulated trajectory such that the goals of the Monte Carlo analysis can be met. For many problems it is sufficient only to output the final state, or a portion of the final state, without any of the intermediate data.

15.2.3   Maximizing Memory Throughput

Memory bottlenecks are a common reason for suboptimal performance of GPU-based Monte Carlo. Two elements of memory management are critical to performance optimization: efficient use of the GPU memory hierarchy (shared, local, and constant memory), and minimizing data transfers between host and device.

A common set of parameters are typically used repeatedly by all component simulations within a Monte Carlo run. These may take the form of system parameters, input histories, or constants employed in system modeling. For instance, in simulation of flight vehicles, the gravitational acceleration constant is used at each timestep by each GPU thread to compute gravitational force. Alternatively, in radiation transport codes the speed of light in a given medium may be required regularly by component GPU threads. In general, these common parameters should not be placed in global memory due to the potential for high latency. Constants that are known at runtime should be placed in constant memory when possible [9]. Alternatively, quantities that are not constant (or known at compile time) should be placed in shared memory. Shared memory latency is approximately 100 times less than global or uncached local memory which can lead to substantial runtime improvements for complex simulations. Care must be taken not to overload constant or shared memory resources for the specific GPU architecture being used.

Minimizing data transfers between host and device is another critical aspect of memory management. In the vast majority of Monte Carlo simulations, data should be transferred once prior to kernel launch (input data), and once following kernel execution (output data). The amount of data to be transferred depends on the sampling methodology and the type of output data required. Repeated data transfers during kernel execution should be either overlapped where possible or avoided altogether.

15.2.4 Maximizing Device Utilization

Maximum throughput is achieved when the GPU device, or devices, are occupied to the maximum extent. For Monte Carlo simulation, higher throughput means lower runtimes since the same number of samples are processed in a shorter amount of time. Two strategies are typically employed to maximize utilization of the device. First, the kernel execution configuration should be optimized for the device being used either through trial and error or based on metrics of kernel memory usage. Occupancy is defined as the number of warps that can run concurrently on a multiprocessor divided by the theoretical maximum number of warps that are allowed to run concurrently. Since memory resources on a multiprocessor are limited, occupancy is determined by a kernel's use of registers and shared memory. The optimal execution configuration therefore depends on kernel memory demands. More complex kernels requiring lots of registers and shared memory will run slower due to both reduced occupancy and register spilling (in which register resources on a multiprocessor are saturated and uncached local memory is used instead). For CUDA kernels, memory usage may be determined using the `nvcc` compiler option `-ptxas-options=-v`. The NVIDIA occupancy calculator uses these statistics to determine an optimal execution configuration for a given device. A useful rule of thumb is that the number of threads per block should be a multiple of the warp size to avoid wasting computational resources with underpopulated warps. It should be noted that optimizing the execution configuration may lead to significantly lower runtimes, and experimentation is encouraged.

Leveraging multiple GPU's for Monte Carlo simulation is rather straightforward due to the directly parallel nature of the algorithm. In this case, the N sample simulations are simply distributed across each GPU as appropriate. NVIDIA's Unified Virtual Addressing, which allows multiple GPU's to share the same virtual memory space, may be used to simplify memory transfers between the host and multiple devices.

15.2.5 Maximizing Instruction Throughput

Data-dependent divergent branching within a GPU kernel is well-known to significantly reduce throughput. When two GPU threads in a warp travel down different branches of execution, each thread must wait in a suspended state while the other completes the branch. This divides throughput by half during these instruction cycles. As a result, data-dependent divergent branching should be avoided whenever possible.

Unfortunately, realistic dynamic simulation often requires conditional branching and there are many instances where it is unavoidable. In such cases, it may be possible to arrange thread blocks such that all threads in a given block travel only down a single branch. Since all threads in a warp also reside in the same block, this guarantees that threads in the same warp will always travel down the same branch and the throughput penalty is thus eliminated. It should be noted that this arrangement of thread blocks may not always be achievable, especially for complex dynamic models in which a thread's path of execution may not be known *a priori*.

15.3   Cost-Benefit Considerations

The goal of implementing Monte Carlo on a GPU is nearly always to reduce runtime over CPU implementations which must execute sample simulations in a serial manner. By exposing the data-parallel aspect of Monte Carlo, GPU implementations are expected to outperform their CPU counterparts. However, this is not always the case. In fact, for any Monte Carlo simulations, sample sizes below a certain value will actually run faster on the CPU than the GPU. There are two primary causes for this. First, overhead memory transactions must be initiated for GPU-based simulations in terms of transferring data between the host and device. Current GPU's leverage the PCI-express bus for this purpose, which exhibits a maximum data transfer rate of only 3–6 GB/s. This is about ten times slower than the memory bandwidth of a high-end CPU, and about 30 times slower than the maximum memory bandwidth of the NVIDIA Tesla C2050. Thus, data transfer to and from the device imposes a bottleneck that may be severe enough in some cases to cost the GPU its data-parallel performance advantage.

The second reason why CPU runtimes may be lower than GPU runtimes is the capability of the individual processors themselves. On a per-core basis, CPU cores have higher clock rates and larger memory resources than GPU stream processors. As a result, a single trajectory simulation will take less time to run on a CPU core than a GPU stream processor (at least for high-end CPU's). This has important runtime implications for small sample sizes.

The combined effect of these two architectural characteristics is that a crossover point exists in terms of sample size which determines which processing architecture will yield better runtimes. For low sample sizes (small N) below the crossover point, the cost incurred from CPU to GPU data transfer outweighs the data-parallel benefit of the GPU, and CPU runtimes are lower. For large numbers of simulations (large N), data-parallel execution outweighs the cost of data transfer and GPU runtimes are lower. The position of this crossover point depends on the computational intensity and memory requirements of the individual kernel simulations. For highly computationally-intensive kernels, the crossover point will generally be observed at smaller values of N since the data-parallel aspect of GPU execution has a larger effect on runtime than data transfer latency. For kernels of low computational intensity, the crossover point will exist at higher values of N since data transfer has a larger effect on runtime than data-parallel execution. The sample size at which this crossover point is observed is a function of not only the simulation model, but also the degree to which the implementation is optimized and the specific GPU and CPU hardware being used.

An example runtime comparison is provided in Fig. 15.3, taken from [10]. Monte Carlo simulations of various sample sizes are run for a six-degree-of-freedom missile trajectory code. The CPU is an Intel Xeon dual-core 1.87 GHz processor, while the GPU is a NVIDIA Tesla C1060 with 240 cores clocked at 1.30 GHz. Note that for small sample sizes, below about 20, CPU runtimes are lower, while for sample sizes larger than 20 the GPU provides lower runtimes. Also notice that, unlike CPU runtimes, GPU runtimes increase at a slower rate with sample size due to the parallel nature of the architecture. At high sample sizes greater than 2,000, memory resources of the GPU are saturated and runtimes begin to increase at an equivalent rate to the CPU (but are still about 30 times lower than CPU runtimes for an equivalent sample size). The relatively small sample size at which the crossover point occurs in this example is due to the high computational intensity of the kernel. For less complex dynamic models, such as the one considered later in this chapter, the crossover point may occur at much higher values of N.

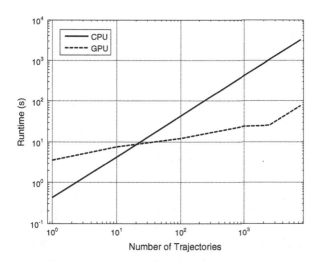

Fig. 15.3 Runtime vs sample size for missile trajectory Monte Carlo simulation (taken from [10])

15.4 Example: Point Mass Monte Carlo Simulation

An example implementation of a GPU-based Monte Carlo simulation is provided. This example simulates N point masses in atmospheric flight subject to gravity and simple viscous drag. Let the position of a point mass in inertial space be given by coordinates x, y, z. Furthermore, let $g = 9.8$ m/s^2 be the gravitational constant and $C_{DV} = 0.025$ N/m/s be the viscous drag coefficient. The dynamic model of this point mass is therefore given by the following six first-order ordinary differential equations:

$$
\begin{Bmatrix} \dot{x}_1 \\ \dot{x}_2 \\ \dot{x}_3 \\ \dot{x}_4 \\ \dot{x}_5 \\ \dot{x}_6 \end{Bmatrix} = \begin{bmatrix} 0 & 0 & 0 & 1 & 0 & 0 \\ 0 & 0 & 0 & 0 & 1 & 0 \\ 0 & 0 & 0 & 0 & 0 & 1 \\ 0 & 0 & 0 & -C_{DV}/m & 0 & 0 \\ 0 & 0 & 0 & 0 & -C_{DV}/m & 0 \\ 0 & 0 & 0 & 0 & 0 & -C_{DV}/m \end{bmatrix} \begin{Bmatrix} x_1 \\ x_2 \\ x_3 \\ x_4 \\ x_5 \\ x_6 \end{Bmatrix} + \begin{Bmatrix} 0 \\ 0 \\ 0 \\ 0 \\ 0 \\ g \end{Bmatrix}
$$

$$(15.1)$$

In Eq. (15.1), x_1, x_2, x_3 represent position coordinates x, y, z, while x_4, x_5, x_6 represent velocities $\dot{x}$, $\dot{y}$, $\dot{z}$. Also, m represents the mass of the object and is given by 0.145 kg. Equation (15.1) can be numerically integrated to solve for the trajectory of the point mass given a valid set of initial conditions. Note that a negative value of z represents positive height about ground.

The goal of this example Monte Carlo is to obtain an impact point dispersion pattern on the ground given initial velocity perturbations. Nominal initial conditions are given by $x_1 = x_2 = x_3 = 0$, $x_4 = 34$ m/s, $x_5 = 0$, and $x_6 = -30$ m/s. Initial conditions are subject to perturbations in the initial velocity states x_4, x_5, x_6. The distribution of these perturbation terms is assumed to be Gaussian with a standard deviation of 2 m/s. Both the model parameters and initial conditions are chosen as reasonable example values for a batted baseball. In this example simulation, the number of desired simulations N is set to 16,384.

15.4.1   Example CUDA Implementation

A CUDA implementation of this Monte Carlo simulation is provided in Listings 3–5. Listing 3 illustrates each device function with the exception of the kernel. Note that the mass, drag coefficient, gravitational acceleration, and simulation timestep are declared at file scope and placed in the GPU constant memory cache. These parameters are accessed each time the derivatives routine is called, and thus placing them in the constant memory cache reduces overall latency significantly. The derivatives routine, based on Eq. (15.1), computes the state derivative based on the current state (and potentially time for generic dynamic models). A numerical integrator such as RK-4 also must be defined as a device function to integrate the dynamic equations. Finally, Listing 3 shows a basic interpolation routine to compute the exact ground impact location of the point mass.

Listing 4 provides the GPU kernel function for the CUDA implementation. In this example, a nominal state is provided to each GPU kernel, and sampling is performed within the kernel itself using the CURAND library. First, a linear array index is computed from the thread indexing parameters, which is used for storing output data in global memory. The CURAND pseudorandom number generator is initialized and then exercised to generate initial velocity perturbations. Following state vector initialization, the point mass trajectory is propagated forward by repeatedly calling

```
#include <stdio.h>
#include <stdlib.h>
#include <curand_kernel.h>
#include <cuda.h>
#include <time.h>
#include <math.h>
#define NUMSTATES 6

// Constant variables in GPU constant memory
__device__ __constant__ double m = 0.145 ;            // kg
__device__ __constant__ double Cdv = 0.025 ;          // N/m/s
__device__ __constant__ double g = 9.8 ;              // m/s^2
__device__ __constant__ double timestep = 0.1 ;  // sec

// Derivatives routine for dynamic model
__device__ void derivative(double* state, double* increment){

    increment[0] = state[3] ;
    increment[1] = state[4] ;
    increment[2] = state[5] ;
    increment[3] = -Cdv*state[3]/m ;
    increment[4] = -Cdv*state[4]/m ;
    increment[5] = g-Cdv*state[5]/m ;
}

// Standard Runge-Kutta 4th Order Integrator
__device__ void RK4(double* state){

    [STANDARD RK-4 INTEGRATION ROUTINE]

}

// To interpolate trajectory points and find exact point of ground impact
__device__ void get_impact_point(double *state, double *oldstate){

    // Interpolate to find the exact impact point
    double weight = -state[2]/(state[2]-oldstate[2]) ;
    state[0] = weight*(state[0]-oldstate[0]) + state[0] ;
    state[1] = weight*(state[1]-oldstate[1]) + state[1] ;
    return ;
}
```

Listing 3 Device Functions for CUDA Example Implementation

the RK-4 integration routine, with the end condition specified as ground impact. Finally, the exact impact location is determined and the final (x,y) location is recorded in the output vector (located in global memory) using the flattened array index.

Listing 5 shows the host code for the example CUDA implementation. As shown in Fig. 15.1, nominal initial conditions are first defined and copied to GPU global memory. Storage locations are also allocated in GPU memory for simulation outputs, in this case ground impact locations. Next, kernels are launched and the velocity distribution (defined by a standard deviation) is passed by value to the kernel function. Following kernel execution, outputs are gathered back to host memory.

```
// GPU Kernel Function
__global__ void trajectory(double* init_state, double veloc_std, unsigned long seed,
double* dev_impacts_x, double* dev_impacts_y){

        double state[NUMSTATES] ;
        double oldstate[NUMSTATES] ;
        int i = 0 ;
        int idx = blockIdx.x*blockDim.x + threadIdx.x ;
        int idy = blockIdx.y*blockDim.y + threadIdx.y ;

        // Map two 2D indices to a single linear, 1D index
        int id_flat = idy * gridDim.x * blockDim.x + idx ;

        // Initialize CURAND random number generator
        curandState_t genState ;
        curand_init(seed, id_flat, 0, &genState) ;

        // Perturb nominal initial velocities with random numbers
        for(i = 0; i < NUMSTATES/2; i++){
                state[i] = init_state[i] ;          // Initial positions
        }
        for(i = NUMSTATES/2; i < NUMSTATES; i++){   // Initial velocities
                state[i] = init_state[i] + veloc_std*curand_normal(&genState);
        }

        // Propagate trajectory while point-mass is above ground
        while(state[2] <= 0.0){

                // Hang on to the previous state so we can interpolate at the end
                for(i = 0; i < NUMSTATES; i++){
                        oldstate[i] = state[i] ;
                }
                // Integrate trajectory one timestep forward
                RK4(state);

        }

        // Interpolate to find exact ground impact point
        get_impact_point(state, oldstate) ;

        // Write final state out to GPU global memory
        dev_impacts_x[id_flat] = state[0] ;
        dev_impacts_y[id_flat] = state[1] ;

        return;
}
```

Listing 4 GPU Kernel for CUDA Implementation

15.4.2 Example OpenCL Implementation

An example OpenCL implementation is provided in Listings 5–7. The primary
difference between the OpenCL and CUDA implementations is that random samples
are generated on the host (using a CPU-based random number generator), and then
transferred to GPU global memory prior to kernel launch. Kernel-based sampling
may of course be used in OpenCL, but this alternative approach is provided simply
as a demonstration.

```
int main(){

    int i ;
    double state_host[NUMSTATES] ;
    dim3 numBlocks(0,0,1) ;
    dim3 threadsPerBlock(0,0,1) ;

    // N
    int numRuns = 16384 ;

    // Velocity standard deviation for randomization
    double veloc_std = 2.0 ;

    // Nominal initial state
    state_host[0] = 0.0 ;
    state_host[1] = 0.0 ;
    state_host[2] = 0.0 ;
    state_host[3] = 34.0 ;
    state_host[4] = 0.0 ;
    state_host[5] = -30.0 ;

    // Copy nominal initial state to device memory
    double* state_dev ;
    cudaMalloc((void**)&state_dev, NUMSTATES*sizeof(double)) ;
    cudaMemcpy(state_dev, state_host, NUMSTATES*sizeof(double), cudaMemcpyHostToDevice) ;

    // Define execution configuration:  2D grid of thread blocks, with a total of numRuns threads
    threadsPerBlock.x = 8 ;
    threadsPerBlock.y = 8 ;
    numBlocks.x = 16 ;
    numBlocks.y = 16 ;

    // Allocate storage for impact location
    double* dev_impacts_x ;
    cudaMalloc((void**)&dev_impacts_x, numRuns*sizeof(double)) ;
    double* dev_impacts_y ;
    cudaMalloc((void**)&dev_impacts_y, numRuns*sizeof(double)) ;

    // Launch kernels
    trajectory<<<numBlocks, threadsPerBlock>>>(state_dev, veloc_std, time(NULL), dev_impacts_x,
    dev_impacts_y) ;

    // Gather impact points back to host
    double* host_impacts_x = (double*)malloc(numRuns*sizeof(double)) ;
    cudaMemcpy(host_impacts_x, dev_impacts_x, numRuns*sizeof(double), cudaMemcpyDeviceToHost) ;
    double* host_impacts_y = (double*)malloc(numRuns*sizeof(double)) ;
    cudaMemcpy(host_impacts_y, dev_impacts_y, numRuns*sizeof(double), cudaMemcpyDeviceToHost) ;

    // Write data to output file
    FILE* fp = fopen("output.txt","w") ;
    for(i = 0; i < numRuns; i++){
        fprintf(fp, "%.10f %.10f\n", host_impacts_x[i], host_impacts_y[i]) ;
    }
    fclose(fp) ;

    // Free allocated memory
    free(host_impacts_x) ;
    free(host_impacts_y) ;
    cudaFree(dev_impacts_x) ;
    cudaFree(dev_impacts_y) ;

    return 0 ;
}
```

Listing 5   Host Code for Example CUDA Implementation

Listing 6 provides the host source code for the OpenCL Monte Carlo implementation. Following specification of nominal initial conditions, several routine OpenCL initialization steps are performed (obtaining platform information, creating a context, etc.). These steps are omitted for brevity but are quite standard.

```
int main(void) {

    double state_host[NUMSTATES] ;
    double* host_impacts_x ;
    double* host_impacts_y ;
    double* xdot_rand_h ;
    double* ydot_rand_h ;
    double* zdot_rand_h ;
    int k,i ;
    cl_int ret ;
    int numRuns = 16384 ;        // N

    // Standard deviations for randomization
    double veloc_std = 2.0 ;

    // Initial state
    state_host[0] = 0.0 ;
    state_host[1] = 0.0 ;
    state_host[2] = 0.0 ;
    state_host[3] = 34.0 ;
    state_host[4] = 0.0 ;
    state_host[5] = -30.0 ;

    // Seed random number generator
    srand(time(NULL)) ;

    [GET PLATFORM, DEVICE INFORMATION; CREATE CONTEXT AND COMMAND QUEUE]
    ...
    [CREATE PROGRAM AND KERNEL OBJECT]


    // Create buffers
    cl_mem state_dev ;
    cl_mem dev_impacts_x ;
    cl_mem dev_impacts_y ;
    cl_mem xdot_rand ;
    cl_mem ydot_rand ;
    cl_mem zdot_rand ;

    // Generate numRuns sets of random initial states
    // (randomize velocity only)
    xdot_rand_h = (double*)malloc(numRuns*sizeof(double)) ;
    ydot_rand_h = (double*)malloc(numRuns*sizeof(double)) ;
    zdot_rand_h = (double*)malloc(numRuns*sizeof(double)) ;
    for(i = 0; i < numRuns; i++){
        xdot_rand_h[i] = gaussrand(veloc_std) ;
        ydot_rand_h[i] = gaussrand(veloc_std) ;
        zdot_rand_h[i] = gaussrand(veloc_std) ;
    }

    // Execution configuration
    size_t workSize[2] ;
    size_t threadsPerBlock[2] ;
    threadsPerBlock[0] = 8 ;
    threadsPerBlock[1] = 8 ;
    workSize[0] = 128 ;
    workSize[1] = 128 ;
```

Listing 6 Host Code for Example OpenCL Implementation

```
// Allocate buffers on GPU
   state_dev = clCreateBuffer(context, CL_MEM_READ_ONLY, NUMSTATES*sizeof(double), NULL, &ret);
   dev_impacts_x = clCreateBuffer(context, CL_MEM_WRITE_ONLY, numRuns*sizeof(double), NULL, &ret);
   dev_impacts_y = clCreateBuffer(context, CL_MEM_WRITE_ONLY, numRuns*sizeof(double), NULL, &ret);
   xdot_rand = clCreateBuffer(context, CL_MEM_READ_ONLY, numRuns*sizeof(double), NULL, &ret);
   ydot_rand = clCreateBuffer(context, CL_MEM_READ_ONLY, numRuns*sizeof(double), NULL, &ret);
   zdot_rand = clCreateBuffer(context, CL_MEM_READ_ONLY, numRuns*sizeof(double), NULL, &ret);

   // Copy sampled initial states to GPU global memory
   ret = clEnqueueWriteBuffer(c_queue1, state_dev, CL_FALSE, 0, NUMSTATES*sizeof(double),
   state_host, 0, NULL, NULL);

   ret = clEnqueueWriteBuffer(c_queue1, xdot_rand, CL_FALSE, 0, numRuns*sizeof(double), xdot_rand_h,
   0, NULL, NULL);

   ret = clEnqueueWriteBuffer(c_queue1, ydot_rand, CL_FALSE, 0, numRuns*sizeof(double), ydot_rand_h,
   0, NULL, NULL);

   ret = clEnqueueWriteBuffer(c_queue1, zdot_rand, CL_TRUE, 0, numRuns*sizeof(double), zdot_rand_h,
   0, NULL, NULL);

   // Set kernel arguments
   ret = clSetKernelArg(trajectory, 0, sizeof(cl_mem), (void *)&state_dev);
   ret = clSetKernelArg(trajectory, 1, sizeof(cl_mem), (void *)&dev_impacts_x);
   ret = clSetKernelArg(trajectory, 2, sizeof(cl_mem), (void *)&dev_impacts_y);
   ret = clSetKernelArg(trajectory, 3, sizeof(cl_mem), (void *)&xdot_rand);
   ret = clSetKernelArg(trajectory, 4, sizeof(cl_mem), (void *)&ydot_rand);
   ret = clSetKernelArg(trajectory, 5, sizeof(cl_mem), (void *)&zdot_rand);

   // Run kernel
   clEnqueueNDRangeKernel(c_queue1, trajectory, 2, NULL, workSize, threadsPerBlock, 0, NULL, NULL);

   // Read data
   host_impacts_x = (double*)malloc(numRuns*sizeof(double)) ;
   host_impacts_y = (double*)malloc(numRuns*sizeof(double)) ;

   clEnqueueReadBuffer(c_queue1, dev_impacts_x, CL_FALSE, 0, numRuns*sizeof(double), host_impacts_x,
   0, NULL, NULL);

   clEnqueueReadBuffer(c_queue1, dev_impacts_y, CL_TRUE, 0, numRuns*sizeof(double), host_impacts_y,
   0, NULL, NULL);
   clFinish(c_queue1);

   // Write to file
   FILE* fpout = fopen("output.txt","w") ;
   for(i = 0; i < numRuns; i++){
       fprintf(fpout, "%.10f %.10f\n", host_impacts_x[i], host_impacts_y[i]) ;
   }
   fclose(fpout) ;

   // Clean up
   [FREE MEMORY AND OPENCL OBJECTS]

   return 0;
}
```

Listing 6 (continued)

Following initialization, global memory buffers are created for the sampled initial velocity conditions and impact locations. Randomized initial velocity conditions are then generated on the host using a CPU-based pseudorandom number generator. These conditions are transferred to GPU global memory by enqueuing write statements. Kernel arguments are defined, and the kernel is enqueued. Finally, results are gathered from GPU memory by enqueuing read commands, and the output data is written to a file.

Listing 7 shows the device code portions for the OpenCL example. Note that this code must exist in a separate file from the host code according to the

```
#define NUMSTATES 6
#pragma OPENCL EXTENSION cl_khr_fp64 : enable

// Constant variables in GPU constant memory
__constant double m = 0.145 ;
__constant double Cdv = 0.025 ;
__constant double g = 9.8 ;
__constant double timestep = 0.1 ;

// Derivatives routine for dynamic model
void derivative(double* state, double* increment){...}

// Standard Runge-Kutta 4th Order Integrator
void RK4(double* state) {...}

// To interpolate trajectory points and find exact point of ground impact
void get_impact_point(double *state, double *oldstate) {...}

__kernel void trajectory(__global double* init_state, __global double* dev_impacts_x, __global double*
dev_impacts_y, __global double* xdot_rand, __global double* ydot_rand, __global double* zdot_rand){

            double state[NUMSTATES] ;
            double oldstate[NUMSTATES] ;
            int i = 0 ;
            int idx = get_global_id(0) ;
            int idy = get_global_id(1) ;

            // map the two 2D indices to a single linear, 1D index
            int id_flat = idy * get_num_groups(0) * get_local_size(0) + idx ;

            // Build our thread's randomized initial state from samples in global memory
            state[0] = init_state[0] ;
            state[1] = init_state[1] ;
            state[2] = init_state[2] ;
            state[3] = init_state[3] + xdot_rand[id_flat] ;
            state[4] = init_state[4] + ydot_rand[id_flat] ;
            state[5] = init_state[5] + zdot_rand[id_flat] ;

            // Propagate trajectory while point-mass is above ground
            while((state[2] <= 0.0)){

                        // Hang on to the previous state so we can interpolate at the end
                        for(i = 0; i < NUMSTATES; i++){
                                    oldstate[i] = state[i] ;
                        }

                        // Integrate trajectory one timestep forward
                        RK4(state);

            }

            // Interpolate to find exact ground impact point
            get_impact_point(state, oldstate) ;

            // Write final state out to GPU global memory
            dev_impacts_x[id_flat] = state[0] ;
            dev_impacts_y[id_flat] = state[1] ;
}
```

Listing 7 Device Code for Example OpenCL Implementation

OpenCL compilation model. The only function that differs substantially from the CUDA implementation is the kernel function itself. Note that the OpenCL code avoids random number generation within the kernel, instead retrieving precomputed random samples from GPU global memory prior to trajectory simulation.

Runtime performance is explored between the CUDA, OpenCL, and an equivalent host C++ implementation. Figure 15.4 shows an example Monte Carlo simulation output (impact points) for $N = 16,384$, while Fig. 15.5 shows runtime performance comparisons. Timing for these comparisons is based solely on GPU-related calculations, and thus for example does not include the OpenCL overhead of context creation, etc. There are several interesting features in the runtime comparison. First, CPU execution times grow at a constant linear rate with N as expected, while GPU execution times are generally lower and grow at a slower rate, at least for $N < 10^5$. Also note crossover points for both the CUDA and OpenCL implementations.

The CUDA code yields substantially better performance over the OpenCL implementation for $N < 5 \times 10^5$ since random samples are computed directly within the kernel itself and large memory transfers are not required before kernel launch. Thus the CUDA code represents a more optimized implementation. Beyond about 10^5 samples, GPU multiprocessor memory resources are exhausted and runtime

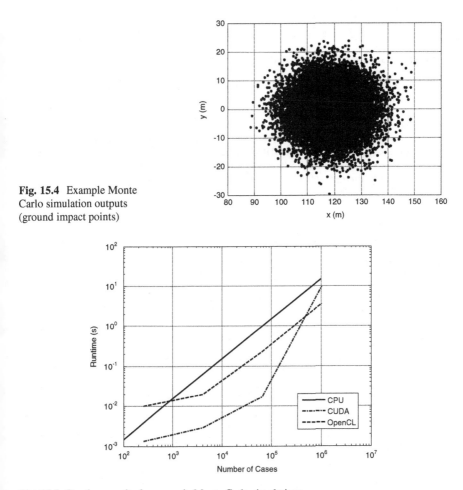

Fig. 15.4 Example Monte Carlo simulation outputs (ground impact points)

Fig. 15.5 Runtime results for example Monte Carlo simulations

increases rapidly with N. In general, for reasonable sample sizes the GPU exhibits 1–2 orders-of-magnitude runtime reduction over a serial CPU implementation which corroborates results obtained in the literature.

References

1. Dunn, W.L.: Exploring Monte Carlo Methods. Elsevier, Amsterdam (2012)
2. Moler, C.: Matrix computation on distributed memory multiprocessors. In: Hypercube Multiprocessors. Society for Industrial and Applied Mathematics, Philadelphia (1986). ISBN 0898712092
3. Liu, T., Ding, A., Xu, X.: GPU-based Monte Carlo methods for accelerating radiographic and CT imaging dose calculations: feasibility and scalability. Med. Phys. **39**(6), 3876 (2012)
4. Humphrey, A., Meng, Q., Berzins, M., Harman, T.: Radiation modeling using the Uintah heterogeneous CPU/GPU runtime system. In: Proceedings of the First Conference of the Extreme Science and Engineering Discovery Environment (XSEDE'12), vol. 4, pp. 4.1–4.8 (2012)
5. Rogers, J., Slegers, N.: Robust parafoil terminal guidance using massively parallel processing. J. Guid. Control. Dyn. **36**(5), 1336–1345 (2013)
6. Rogers, J.: GPU-Enabled Projectile Guidance for Impact Area Constraints. In: SPIE Defense, Security, and Sensing Symposium, Baltimore (2013)
7. Kolb, C., Pharr. M.: Option pricing on the GPU. GPU Gems 2, NVIDIA Corporation, Chapter 45 (2005)
8. CUDA Toolkit 4.2: CURAND Guide. NVIDIA Corporation, Santa Clara (2012)
9. CUDA C Programming Guide.: NVIDIA Corporation, Santa Clara (2012)
10. Ilg, M., Rogers, J., Costello, M.: Projectile Monte Carlo analysis using a graphics processing unit. In: AIAA Atmospheric Flight Mechanics Conference, Portland (2011)

Part IV
Fast Fourier Transform and Localized
n-Body Problems

Chapter 16
Fast Fourier Transform (FFT) on GPUs

Yash Ukidave, Gunar Schirner, and David Kaeli

16.1 Introduction to FFT

The Fast Fourier Transform (FFT) is one of the most common algorithms used in signal processing to transform a signal from the time domain to the frequency domain and vice-versa. It is named after the French mathematician and physicist named Joseph Fourier.

The FFT is a fast and efficient algorithm to compute the discrete fourier transform (DFT) and the inverse discrete fourier transform (IDFT). There are many distinct FFT algorithms involving a wide range of mathematics, from simple complex-number arithmetic, to group theory and number theory. Considerable research effort has been devoted towards optimizing FFT algorithms over the past four decades. The algorithm presented by Cooley and Tukey [2] reduced the algorithmic complexity when computing the DFT to $O(NlogN)$ from $O(N^2)$, which is viewed as a turning point for applications using the fourier transform.

16.2 How Does the FFT Work?

Every discrete function $f(x)$ of finite length can be described as a set of potentially infinite sinusoidal functions. This representation is known as the frequency domain representation of the function, defined as $F(u)$. The relationship between these two functions is represented by the Discrete Fourier Transform (DFT).

Y. Ukidave • G. Schirner • D. Kaeli (✉)
Northeastern University, Boston, MA, USA
e-mail: yukidave@ece.neu.edu; schirner@ece.neu.edu; kaeli@ece.neu.edu

V. Kindratenko (ed.), *Numerical Computations with GPUs*,
DOI 10.1007/978-3-319-06548-9__16, © Springer International Publishing Switzerland 2014

$$\mathscr{F}\{f(x)\} = F(u) = \sum_{x=0}^{N-1} f(x) W_N^{ux} \tag{16.1}$$

$$\mathscr{F}^{-1}\{F(u)\} = f(x) = \frac{1}{N} \sum_{u=0}^{N-1} F(u) W_N^{-ux} \tag{16.2}$$

where $W_N = e^{\frac{-j2\pi}{N}}$ is the *Twiddle Factor* and N is the number of input data points.

Equation (16.1) defines the forward DFT computation over a finite time signal $f(x)$ and the inverse DFT computation over a frequency function $F(u)$, as shown in Eq. (16.2).

The Fast Fourier Transform refers to a class of algorithms that uses divide-and-conquer techniques to efficiently compute the Discrete Fourier Transform of the input signal. For a one-dimensional (1D) array of input data size N, the FFT algorithm breaks the array into a number of equal-sized sub-arrays and performs computation on these sub-arrays. The process of dividing the input sequence in the FFT algorithm is called decimation. Two major classifications of the FFT algorithm cover all the variations of the computation based on the technique of decimation.

- **Decimation in Time (DIT):**
  The decimation in time (DIT) algorithm splits the N-point data sequence into two $\frac{N}{2}$-point data sequences $f_1(N)$ and $f_2(N)$, for even and odd numbered input samples, respectively.
- **Decimation in Frequency (DIF):**
  The decimation in frequency (DIF) algorithm splits the N-point data sequence into two sequences, of first $\frac{N}{2}$ data points and last $\frac{N}{2}$ data points respectively. The DIF algorithm does not consider the decimation as even and odd data points.

In this process of decimation, we exploit both the symmetry and the periodicity of the complex exponential $W_N = e^{\frac{-j2\pi}{N}}$, known as the *Twiddle Factor*. The algorithm using the divide-and-conquer technique was proposed by Cooley and Tukey. Algorithm 1 shows the pseudocode of the basic Cooley-Tukey FFT.

The Radix-2 Cooley-Tukey algorithm recursively divides the N-point DFT into two N/2-point DFTs, with a complex multiplication (the Twiddle Factor) in between. The division is done into two half length DFTs with even indexed and odd indexed samples, as described in Eq. (16.3). The recursive division continues until we achieve N 2-point signals.

$$\mathscr{F}\{f(x)\} = F(u) = \sum_{x(even)} F(x) W_N^{ux} + \sum_{x(odd)} F(x) W_N^{ux} \tag{16.3}$$

The butterfly computation transforms two complex input points to two complex output points to compute the FFT. The 2-point, Radix-2 FFT butterfly is derived by expanding the formula of the DFT for two point signal ($N = 2$), as shown in Eqs. (16.4)–(16.6). The 2-point, Radix-2 butterfly is also shown in Fig. 16.1.

Algorithm 1 Cooley-Tukey Radix-2 FFT

Require: Input data $A[0,\ldots,n-1]$
1: Recursive-FFT(A)
2: $n = length[A]$
3: **if** n = 1
4:     return A
5: **end**
6: $w_n = e^{2i\pi/n}$
7: $w = 1$
8: $A^{even} = (A[0], A[2], \ldots, A[n-2])$
9: $A^{odd} = (A[1], A[3], \ldots, A[n-1])$
10: $y^{even} = $ Recursive-FFT(A^{even})
11: $y^{odd} = $ Recursive-FFT(A^{odd})
12: **for** k = $0 \rightarrow \frac{n}{2}$ - 1
13:     $y_k = y_k^{even} + w y_k^{odd}$
14:     $y_{k+n/2} = y_k^{even} - w y_k^{odd}$
15:     $w = w w_n$
16: **end**
17: **return** y

$$\mathscr{F}\{f(x)\} = F(u) = \sum_{x=0}^{1} f(x) W_2^{ux} \qquad (16.4)$$

$$F(0) = f(0) + f(1) \qquad (16.5)$$

$$F(1) = f(0) - f(1) \qquad (16.6)$$

16.2.1 FFT as a Heterogeneous Application

Graphics Processing Units (GPUs) have been effectively used for accelerating a number of general-purpose computation. The high performance community has been able to effectively exploit the inherent parallelism on these devices, leveraging their impressive floating-point performance and high memory bandwidth of GPU. FFTs were one of the first algorithms ported to GPU [3]. The original GPU-based FFT code was implemented as a shader code executed in the graphics pipeline. Since

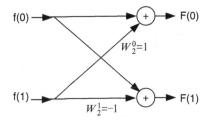

Fig. 16.1 Radix-2 FFT butterfly for a 2-point computation. The butterfly requires two complex additions and two complex multiplications

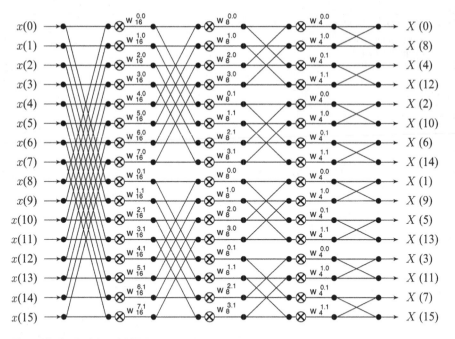

Fig. 16.2 Radix-2 based DIF FFT with 16 input data points

this time, GPU computing has moved forward aggressively, overcoming many of the limitations of programming in a shader-based language. The basic structure of the FFT makes it a good candidate for parallel execution on heterogeneous platforms. The use of heterogeneous programming models available in OpenCL and CUDA simplify the porting of mathematical computations such as FFT to a GPU [4, 6].

The FFT is a non-causal algorithm, since it is solely dependent of the data inputs. Thus, the FFT can be evaluated in parallel over all N input points by forming smaller subsequences of the problem. Each subsequence carries out the required number of FFT compute stages, which can be mapped to different computational cores (i.e., compute units) on the heterogeneous device. Figure 16.2 shows the computation of a Radix-2 DIF FFT, computed over 16 input data points. The input data points are transformed over four stages of compute using the FFT algorithm. This computation can be parallelized effectively on a GPU by creating many threads, each of which works on 2 input points at a given time for every stage. For the example presented in Fig. 16.2, the problem can be divided over stages using 8 (N/2) threads. For a large input data set, the FFT can be computed using a large number of threads. Thus, the parallel hardware present on a GPU can be effectively leveraged to efficiently compute the FFT.

Heterogeneous implementations of the FFT algorithm can be designed using OpenCL programming model. OpenCL provides functionality to represent the FFT computation for each work-item (i.e., thread). The memory model of OpenCL allows the programmer to map the entire input data for the FFT on the device for parallel execution on the available compute units (CUs) on the GPU.

16.3   Implementing FFT on GPU Using OpenCL

Next, we discuss a Radix-2 FFT implementation using the Cooley-Tukey algorithm [2]. We focus on the Radix-2 variant of the Cooley-Tukey algorithm due to its ease of implementation and numerical accuracy. The strategies presented here can be easily extended for higher-Radix cases.

Based on the structure of the Radix-2 Cooley-Tukey algorithm, we will design a OpenCL kernel for the FFT computation. We begin by describing our methodology to decompose the FFT into a number of simpler computations. Then we can use these simple operations as building blocks to produce the final result.

The code block shown in Sect. 16.2 presented an implementation of a Radix-2 FFT computation using a 2-point butterfly. The kernel takes as input complex inputs, which are represented using float2, a vector data type provided in OpenCL. The x and y components of the float2 vector represent the *real* and *imaginary* parts of the complex number, respectively. The kernel retrieves 2 input data points from global memory of the GPU and performs a 2-point FFT computation over them.

```
__kernel void FFT_2_pt( __global float2* d_in,
    __global float2* d_out)
{
    int gid = get_global_id(0);    // Global id of thread
    float2  in0, in1;
    in_0 = d_in[gid];              // Input data #0
    in_1 = d_in[gid+1];            // Input data #1

    /* 2-point FFT computation begins */
    float2  Var;
    Var = in_0;
    in_0 = Var + in_1;
    in_1 = Var - in_1;
    /* 2-point FFT computation ends */

    d_out[gid]   = in_0;           // Output data #0
    d_out[gid+1] = in_1;           // Output data #1
}
```

The 2-point FFT butterfly computation can be extended to a 4-point butterfly representation. The 4-point butterfly is shown in Fig. 16.3a. To produce results in the proper order, the input data points of the FFT computation are bit-reversed, as seen in Fig. 16.3a. To obtain coalesced memory accesses in the 4-point butterfly, the 2-point butterfly structures have to be rearranged in order to preserve spatial locality in the address stream, as shown in Fig. 16.3b. The 4-point FFT can compute two stages of the Radix-2 computation in one single pass. Hence, an 8-point FFT can be implemented using one pass of a 4-point FFT computation followed by one pass

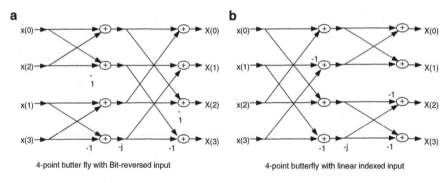

Fig. 16.3 A Radix-2 FFT using a 4-point butterfly with (**a**) bit-reversed input indexes and (**b**) linear input indexes

Fig. 16.4 Bit-reversed input indexes of a 8-point FFT

Thread 0				Thread 1			
0	2	4	6	1	3	5	7

Fig. 16.5 Input data mapping of a 8-point FFT computation to threads of a 4-point FFT kernel

of a 2-point FFT computation. If the computation is performed using only a 2-point FFT, it would require 3 passes of the 2-point FFT kernel. This would increase the kernel launch overhead on GPU and would also increase the number of accesses to global memory. Both of these factors can cause performance degradation. Hence, a large-scale FFT is computed using combinations of small FFTs.

To obtain coalesced memory accesses, the inputs to the FFT kernels should be indexed in a linear order. We have to develop a generic indexing scheme for all FFT kernels, since we want this code to work for different input sizes. We consider the example of the 8-point FFT computation using 4-point FFT and 2-point FFT kernels. The first two stages of the compute are implemented with 2 threads using a 4-point FFT. The last stage of the compute is implemented using 4 threads each computing a 2-point FFT. The inputs for a bit-reversed 8-point FFT are shown in Fig. 16.4.

Figure 16.5 shows how inputs are mapped to the two threads that compute the 4-point FFT to implement the 8-point FFT. We observe that the first data element required by each thread is located at an input index equal to the global id (gid) of the particular thread. We define a set of parameters in the kernel code to help us develop the desired indexing.

1. *gid*: Global Id of a thread
2. *gbl_size*: Work size (Total number of threads)
3. *N*: Number of input data points(complex)

4. **out_id** : The index for each output data point
5. **num_stage** : Parameter that defines output index
6. **kern_size**: Kernel Size

We use these parameters and our observations of the 4-point FFT to develop the input indexing of a 4-point FFT kernel. The code section for the input indexing is given below:

```
int gid = get_global_id(0);
int gbl_size = N/4;
float2 in_0, in_1, in_2, in_3;
uint kern_size = 4;

in_0 = d_in[(0*gbl_size)+gid];
in_1 = d_in[(1*gbl_size)+gid];
in_2 = d_in[(2*gbl_size)+gid];
in_3 = d_in[(3*gbl_size)+gid];
```

The output index order of the kernel needs to be sequential. To insure preserving spatial locality, we develop a generic indexing scheme for storing output data. The output index (out_id) is computed using the parameters defined above. The num_stage parameter must be set by the host code while executing the kernel. The value of num_stage is $2^{Stages\,completed\,by\,FFT}$ (i.e., the value of num_stage after the completion of the 4-point kernel is 4). After careful study of the 2-point and 4-point FFT computations, the following observations are made:

- In the first stage of the FFT computation, the output index out_id of each thread is computed as the product of the kernel-size(kern_size) and global id gid of that thread.
- For every stage of the FFT computation, each thread writes kern_size output points to the global memory with a stride equal to num_stage
- For the last FFT stage, the output index out_id of each work item is equal to the global id gid of the particular thread.

The value of the out_id can thus be calculated as follows for each FFT kernel:

```
out_id = (gid / num_stage) * num_stage * kern_size
         + (gid%num_stage);
```

A key portion of the FFT computation is the calculation of the Twiddle factor, which was defined in Sect. 16.2. The Twiddle factor depends on the size of the FFT computation and has to be computed at runtime during kernel execution.

$$W_N^{kn} = e^{\frac{-j2kn\pi}{N}} = cos\left(\frac{-2kn\pi}{N}\right) + jsin\left(\frac{-2kn\pi}{N}\right) \qquad (16.7)$$

The code used for computing the Twiddle factor using Eq. (16.7) is given below:

```
void twidle_factor(int k, float angle, float2 ip)
{
    float2 twiddle, var;
    twiddle.x = native_cos(k*angle);
    twiddle.y = native_sin(k*angle);

    var.x = twiddle.x*ip.x - twiddle.y*ip.y;
    var.y = twiddle.x*ip.y + twiddle.y*ip.x;

    ip.x = var.x;
    ip.y = var.y;
}
```

Twiddle factors are sinusoidal components, and are symmetrical by definition. Hence, developers can choose to hardcode the values of the Twiddle factors required by the FFT computation in the OpenCL kernel code.

The kernel code for the 2-point, 4-point and 8-point kernels is given below. The kernels are specialized implementations and use the input and output indexing scheme as described earlier.

Kernel Code for 2-Point FFT:

```
void FFT2_comp(float2 in_0, float2 in_1)
{
    float2 v0;
    v0 = in0;
    in0 = v0 + in1;
    in1 = v0 - in1;
}


__kernel void FFT_MS_2(__global float2* d_in,
                       __global float2* d_out,
                       __global uint* comp_stage,
                       uint num_data)
{
    uint N = num_data;
    uint gid = get_global_id(0);
    uint gbl_size = N/2;
    float2 in_0, in1;
    uint kern_size = 2;

    in_0 = d_in[(0*gbl_size)+gid];
    in_1 = d_in[(1*gbl_size)+gid];

    uint num_stage = comp_stage[0];
```

```
if (num_stage!=1)
{
  float angle = -2*PI*(gid)/(N);
  twidle_factor(1, angle, in1);
}

FFT2_comp(in_0, in_1);

uint I_dout = (gid/num_stage)*num_stage*kern_size
+(gid%num_stage);
d_out[(0*num_stage)+I_dout] = in_0;
d_out[(1*num_stage)+I_dout] = in_1;

}
```

Kernel Code for 4-Point FFT:

```
void FFT4_comp(float2 in_0, float2 in_1, float2 in_2,
float2 in_3) {
float2 v_0, v_1, v_2, v_3;
v_0 = in_0 + in_2;
v_1 = in_1 + in_3;
v_2 = in_0 - in_2;
v_3.x = in_1.y - in_3.y;
v_3.y = in_3.x - in_1.x;
in_0 = v_0 + v_1;
in_2 = v_0 - v_1;
in_1 = v_2 + v_3;
in_3 = v_2 - v_3;
}


__kernel void  FFT_MS_4( __global float2* Data_in,
      __global float2* Data_out,
      __global uint* comp_stage,
      uint num_data)
{
  uint N = num_data;
  int gid = get_global_id(0);
  int gbl_size = N/4;
  float2 in_0, in_1, in_2, in_3;
  uint kern_size = 4;
  in_0 = d_in[(0*gbl_size)+gid];
  in_1 = d_in[(1*gbl_size)+gid];
```

```
in_2 = d_in[(2*gbl_size)+gid];
in_3 = d_in[(3*gbl_size)+gid];
uint num_stage = comp_stage[0];
if (num_stage!=1)
{

float angle = -2*PI*(gid)/(N);
twidle_factor(1, angle, in_1);
twidle_factor(2, angle, in_2);
twidle_factor(3, angle, in_3);
    }

FFT4(in_0, in_1, in_2, in_3);
uint I_dout = (gid/num_stage)*num_stage*kern_size+
(gid%num_stage);
d_out[(0*num_stage)+I_dout] = in_0;
d_out[(1*num_stage)+I_dout] = in_1;
d_out[(2*num_stage)+I_dout] = in_2;
d_out[(3*num_stage)+I_dout] = in_3;

}
```

Kernel Code for 8-Point FFT:

```
void FFT8(in_0, in_1, in_2, in_3, in_4, in_5, in_6,
in_7) {
float2 v_0, v_1, v_2, v_3, v_4, v_5, v_6, v_7;
float2 s0, s1, s2, s3, s4, s5, s6, s7;
v_0 = in_0 + in_4;
v_1 = in_1 + in_5;
v_2 = in_2 + in_6;
v_3 = in_3 + in_7;
v_4 = in_0 - in_4;
v_5 = in_1 - in_5;
v_6.x = in_2.y - in_6.y;
v_6.y = in_6.x - in_2.x;
v_7.x = in_3.y - in_7.y;
v_7.y = in_7.x - in_3.x;
s0 = v_0 + v_2;
s1 = v_1 + v_3;
s2 = v_0 - v_3;
s3 = v_3 - v_1;
s4 = v_4 + v_6;
s5.x = v_5.y - v_7.y;
s5.y = v_7.x - v_5.x;
s6 = v_3 - v_6;
```

```
s7.x = v_7.y - v_5.y;
s7.y = v_5.x - v_7.x;
in_0 = s0 + s1;
in_1 = s0 - s1;
in_2 = s2 + s3;
in_3 = s2 - s3;
in_4 = s4 + s5;
in_5 = s4 - s5;
in_6 = s6 + s7;
in_7 = s6 - s7;
}


__kernel void  FFT_MS_8( __global float2* d_in,
      __global float2* d_out,
      __global uint* comp_stage,
      uint num_data)
{
  uint N = num_data;
  int gid = get_global_id(0);
  int gbl_size = N/8;
  float2 in_0, in_1, in_2, in_3, in_4, in_5, in_6, in_7;

  uint num_stage = comp_stage[0];
  uint kern_size = 8;
  in_0 = d_in[(0*gbl_size)+gid];
  in_1 = d_in[(1*gbl_size)+gid];
  in_2 = d_in[(2*gbl_size)+gid];
  in_3 = d_in[(3*gbl_size)+gid];
  in_4 = d_in[(4*gbl_size)+gid];
  in_5 = d_in[(5*gbl_size)+gid];
  in_6 = d_in[(6*gbl_size)+gid];
  in_7 = d_in[(7*gbl_size)+gid];

  if (num_stage!=1)
    {
    float angle = -2*PI*(gid%num_stage)/(num_stage*
    kern_size);
    twidle_factor(1, angle, in_1);
    twidle_factor(2, angle, in_2);
    twidle_factor(3, angle, in_3);
    twidle_factor(4, angle, in_4);
    twidle_factor(5, angle, in_5);
    twidle_factor(6, angle, in_6);
    twidle_factor(7, angle, in_7);
```

```
    }
FFT8_comp(in_0,  in_1,  in_2,  in_3,  in_4,  in_5,  in_6,
in_7);
uint I_dout = (gid/num_stage)*num_stage*kern_size+
(gid%num_stage);

d_out[(0*num_stage)+I_dout] = in_0;
d_out[(1*num_stage)+I_dout] = in_1;
d_out[(2*num_stage)+I_dout] = in_2;
d_out[(3*num_stage)+I_dout] = in_3;
d_out[(4*num_stage)+I_dout] = in_4;
d_out[(5*num_stage)+I_dout] = in_5;
d_out[(6*num_stage)+I_dout] = in_6;
d_out[(7*num_stage)+I_dout] = in_7;
}
```

When using hard-coded FFT kernels (e.g., 16-point and 32-point) for large data sets, each kernel will require a large number of registers per thread on the GPU. This can cause a reduction in number of active threads scheduled on the GPU due to unavailability of resources, which leads to degradation in execution performance of the application.

Developers should track the number of registers per compute unit consumed by the kernel, especially when using large-sized FFT kernels.

16.4   Implementation of 2D FFT

We have learned about the implementation of a 1D-FFT kernel on the GPU in Sect. 16.3. Next, we proceed to implement the 2D-FFT computation kernel.

2D FFT computations are generally used in media applications, including video and image processing. Such applications have input data arranged in a 2D matrix format, with data stored using either *row-major* or *column-major* format. A 2D FFT is defined as follows:

$$\mathscr{F}\{f(x,y)\} = F(u,v) = \frac{1}{MN} \sum_{y=0}^{M-1} \sum_{x=0}^{N-1} f(x,y) W_N^{ux} W_M^{vy} \qquad (16.8)$$

where M and N are the dimensions of the input matrix.

Equation (16.8) describes the 2D FFT computation. As observed, the 2D-FFT requires two summations over two dimensions. This computation can be implemented leveraging the 1D-FFT kernel as a building block. This method of computing the 2D-FFT is shown in Fig. 16.6.

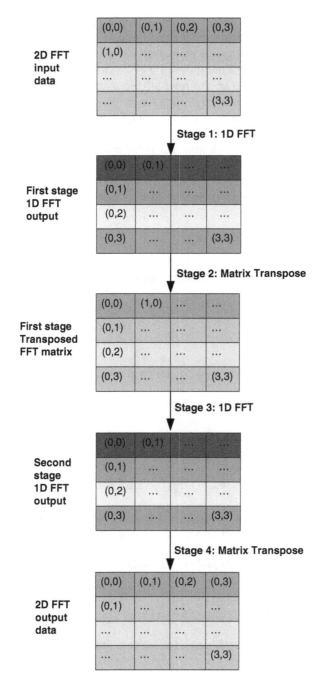

Fig. 16.6 Stages to compute a 2D FFT using a 1D FFT and a matrix transpose as building blocks

Figure 16.6 outlines the process of computing a 2D FFT for a 4 × 4 matrix. The first stage of the transform performs an in-place 1D FFT for each row of the input data matrix. The second stage performs a transpose over the output of the first stage. At this point we have a transposed 1D FFT matrix. The third stage performs another in-place 1D FFT over the rows of the matrix. The fourth stage again performs a transpose over the matrix using the output of the third stage, producing the resultant matrix. The output matrix has 2D FFT data output in its natural order (non-transposed).

The matrix transpose is an additional step that must be performed to compute a 2D FFT. This transpose can be performed on the GPU device using the same data buffers used for computing the FFT stages. This avoids the overhead of moving data back and forth between the CPU and the GPU. A basic matrix transpose kernel is provided below.

Kernel Code for Matrix Transpose:

```
#define BLOCK_SIZE 16
__kernel void  Matrix_Transpose(__global float2* d_in,
                                __global float2* d_out,
                                int width,
                                int height)
{

  // Read the matrix block into local memory

  __local float2 block[BLOCK_SIZE * (BLOCK_SIZE + 1)]

  unsigned int id_x = get_global_id(0);
  unsigned int id_y = get_global_id(1);

  if((id_x < width) && (id_y < height))
  {
    unsigned int in_index = id_y * width + id_x;
    int I_din = get_local_id(1)*(BLOCK_SIZE+1)
              + get_local_id(0);
    block[I_din]=  d_in[in_index];
  }

  barrier(CLK_LOCAL_MEM_FENCE);

  // Write transposed matrix block back to global
  memory

  id_x = get_group_id(1) * BLOCK_SIZE + get_local_id
  (0);
  id_y = get_group_id(0) * BLOCK_SIZE + get_local_id
  (1);
```

```
if((id_x < height) && (id_y < width))
{
    unsigned int index_out = id_y * height + id_x;
    int I_dout = get_local_id(0)*(BLOCK_SIZE+1)
                 + get_local_id(1);
    d_out[index_out] = block[I_dout];
}
}
```

The use of **double precision** data types is supported by setting a pragma directive in the OpenCL kernel code. Data types such as double, double2, double4, double8 and double16 can be used by specifying #pragma OPENCL EXTENSION cl_khr_fp64 : enable in the kernel code. Use of double precision data types can affect the execution performance of the FFT application when compared to using the single precision version.

16.5 Performance Evaluation of FFT

In this section, we provide a performance evaluation of the FFT computation described in the previous sections, as run on both AMD and Nvidia GPUs.

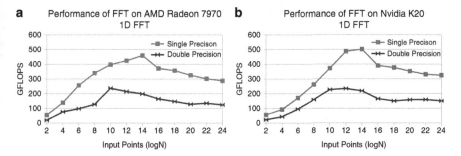

Fig. 16.7 Execution performance (in GFLOPS) for the 1D FFT, as run on a (**a**) AMD Radeon 7970 GPU and a (**b**) Nvidia K20 GPU

The evaluation is performed on a system consisting of an Intel Core i7 3770k processor as the CPU host. The AMD Radeon 7970 discrete GPU and Nvidia K20 discrete GPU are used to accelerate execution of FFT. The performance of the GPU is measured in Floating Point Operations per Second *FLOPS*. For an input of size N, the FFT performs $(5 \times N \log_2(N))$ floating point operations (FLOPs) [7].

Figure 16.7a shows the performance of the 1D FFT using single and double precision on AMD platform. Figure 16.7b shows the same for the Nvidia platform.

Figure 16.8 shows the performance of the 2D FFT as run on a Nvidia K20 and a AMD Radeon GPU. The 2D FFT uses 2 1D FFT computations and 2

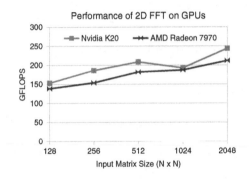

Fig. 16.8 Execution performance of 2D FFT (Single Precision) on Nvidia K20 and AMD Radeon 7970 GPU

transpose computations to carry out the transform. We can notice the added overhead of launching the transpose in the kernels for the 2D FFT, as compared to the performance of the 1D FFT.

GPU vendors provide a highly optimized FFT library for their devices. The library provides customized APIs for computing 1D and 2D FFT on the GPU. Developers can choose to use these optimized libraries to reduce code production time. **cuFFT** by Nvidia and **clAmdFFT** by AMD are optimized libraries developed in CUDA and OpenCL, respectively [1, 5].

In this chapter was have presented 1D and 2D implementations of an FFT. We have discussed some of the tradeoffs when mapping these implementations to GPUs. We have also provided samples runs of these codes developed in OpenCL when run on state-of-the-art GPUs.

Appendix

Host code for FFT application

```
/* FFT.c */

#include <stdio.h>
#include <stdlib.h>
#include <time.h>
#include <math.h>
#include <string.h>
#include <CL/cl.h>

/* Function to check and handle OpenCL errors */
int check_status(int status, char* message) {
  if (status != CL_SUCCESS) {
    printf(message);
    printf("\n ERR: %d\n", status);
```

```
      fflush(NULL);
      exit(-1);
   }
}

/* Define custom constants*/
#define MAX_SOURCE_SIZE (10000000)

int main(int argc, char** argv) {
  /* OpenCL Variables */
  cl_uint num_data = 0;
  cl_uint num_stage = 0;
  cl_uint stages_reqd = 0;
  cl_float2* input = NULL;
  cl_float2* output = NULL;
  cl_float* coeff = NULL;
  cl_float* temp_output = NULL;
  cl_platform_id platform_id = NULL;
  cl_device_id device_id = NULL;
  cl_context context;
  cl_command_queue command_queue;
  cl_uint ret_num_platforms;
  cl_uint ret_num_devices;
  cl_mem input_buffer;
  cl_mem output_buffer;
  cl_mem comp_stage_buffer;
  cl_kernel kernel_2;
  cl_kernel kernel_4;
  cl_kernel kernel_8;
  cl_event event;
  cl_int ret;

  /* Custom Variables */
  unsigned int i = 0;
  unsigned int count = 1;
  unsigned int local;
  unsigned int launch_8, launch_4, launch_2;
  unsigned int iters;
  char *platformVendor;
  size_t platInfoSize;

  /* Command Line based Initialization */
  if (argc > 1) {
    num_data = atoi(argv[1]);
    iters = atoi(argv[2]);
```

```c
}

/* Allocate input memory */
input = (cl_float2 *) malloc(
    num_data * sizeof(cl_float2));
output = (cl_float2 *) malloc(
    num_data * sizeof(cl_float2));

/* Generate Input Data */
for (i = 0; i < num_data; i++) {
  input[i].x = Random();
  input[i].y = Random();
}

/* Load the kernel source code into the array source_str */
FILE *fp;
char *source_str;
size_t source_size;

fp = fopen("FFT.cl", "r");
if (!fp) {
  fprintf(stderr, "Failed to load kernel.\n");
  exit(1);
}
source_str = (char*) malloc(MAX_SOURCE_SIZE);
source_size = fread(source_str, 1, MAX_SOURCE_SIZE, fp);
fclose(fp);

/* Get platform id */
ret = clGetPlatformIDs(1, &platform_id,
    &ret_num_platforms);
check_status(ret, "Error: Platform ID\n");

/* Get device id */

ret = clGetDeviceIDs(platform_id, CL_DEVICE_TYPE_GPU, 1,
    &device_id, &ret_num_devices);
check_status(ret, "Error: Create Program\n");
printf("\nNo of Devices %d", ret_num_platforms);

/* Get platform info */
clGetPlatformInfo(platform_id, CL_PLATFORM_VENDOR, 0,
    NULL, &platInfoSize);
platformVendor = (char*) malloc(platInfoSize);
clGetPlatformInfo(platform_id, CL_PLATFORM_VENDOR,
```

```
        platInfoSize, platformVendor, NULL);
printf("\nVendor: %s\n", platformVendor);
free(platformVendor);

/* Create an OpenCL context */
context = clCreateContext(NULL, 1, &device_id, NULL,
    NULL, &ret);

/* Create a command queue */
command_queue = clCreateCommandQueue(context, device_id,
    CL_QUEUE_PROFILING_ENABLE, &ret);

/* Define and create memory buffers on the device for each vector */
input_buffer = clCreateBuffer(context, CL_MEM_READ_ONLY,
    sizeof(cl_float2) * num_data, NULL, &ret);
output_buffer = clCreateBuffer(context,
    CL_MEM_READ_WRITE, sizeof(cl_float2) * num_data,
    NULL, &ret);
comp_stage_buffer = clCreateBuffer(context,
    CL_MEM_READ_ONLY, sizeof(cl_uint), NULL, &ret);

/* Create a program from the kernel source */
program = clCreateProgramWithSource(context, 1,
    (const char **) &source_str,
    (const size_t *) &source_size, &ret);
check_status(ret, "Error: Create Program\n");

/* Build program */
ret = clBuildProgram(program, 1, &device_id, NULL, NULL,
    NULL);

/* Dump the program build info to a string */
size_t paramValueSize = 1024 * 1024,
    param_value_size_ret;
char *paramValue;
paramValue = (char*) calloc(paramValueSize,
    sizeof(char));
ret = clGetProgramBuildInfo(program, device_id,
    CL_PROGRAM_BUILD_LOG, paramValueSize, paramValue,
    &param_value_size_ret);
printf(" \n\n %s \n\n", paramValue);
check_status(ret, "Error: Build Program\n");

/* Create the OpenCL kernel for 8-point, 4-point and 2-point FFT */
kernel_8 = clCreateKernel(program, "FFT_MS_8", &ret);
```

```
check_status(ret,
    "Error: Create kernel 8. (clCreateKernel)\n");

kernel_4 = clCreateKernel(program, "FFT_MS_4", &ret);
check_status(ret,
    "Error: Create kernel 4. (clCreateKernel)\n");

kernel_2 = clCreateKernel(program, "FFT_MS_2", &ret);
check_status(ret,
    "Error: Create kernel 2. (clCreateKernel)\n");

/* Set the arguments of the kernel_8, kernel_4, kernel_2 */
ret = clSetKernelArg(kernel_8, 0, sizeof(cl_mem),
    (void *) &input_buffer);
ret = clSetKernelArg(kernel_8, 1, sizeof(cl_mem),
    (void *) &output_buffer);
ret = clSetKernelArg(kernel_8, 2, sizeof(cl_mem),
    (void *) &comp_stage_buffer);
ret = clSetKernelArg(kernel_8, 3, sizeof(cl_uint),
    (void *) &num_data);

ret = clSetKernelArg(kernel_4, 0, sizeof(cl_mem),
    (void *) &input_buffer);
ret = clSetKernelArg(kernel_4, 1, sizeof(cl_mem),
    (void *) &output_buffer);
ret = clSetKernelArg(kernel_4, 2, sizeof(cl_mem),
    (void *) &comp_stage_buffer);
ret = clSetKernelArg(kernel_4, 3, sizeof(cl_uint),
    (void *) &num_data);

ret = clSetKernelArg(kernel_2, 0, sizeof(cl_mem),
    (void *) &input_buffer);
ret = clSetKernelArg(kernel_2, 1, sizeof(cl_mem),
    (void *) &output_buffer);
ret = clSetKernelArg(kernel_2, 2, sizeof(cl_mem),
    (void *) &comp_stage_buffer);
ret = clSetKernelArg(kernel_2, 3, sizeof(cl_uint),
    (void *) &num_data);

/* Initialize Memory Buffer */
ret = clEnqueueWriteBuffer(command_queue, input_buffer,
    1, 0, num_data * sizeof(cl_float2), input, 0, 0,
    &event);
check_status(ret, "Error: write input Buffer\n");
```

```
eventList->add(event);

/* Decide the local group formation for kernel_8 */
size_t globalThreads_8[0] = { num_data / 8 };

/* Decide the local group formation for kernel_4 */
size_t globalThreads_4[0] = { num_data / 4 };

/* Decide the local group formation for kernel_2 */
size_t globalThreads_2[0] = { num_data / 2 };

/* Setup Multi-stage FFT launch counters */
stages_reqd = log2(num_data);
launch_8 = stages_reqd;
launch_4 = 0;
launch_2 = 0;

/* Calculate 8-point FFT kernel launch count */
while (launch_8) {
  if ((launch_8) % 3) {
    launch_8--;
  } else
    break;
}
launch_8 = launch_8 / 3;

/* Calculate 4-point and 2-point FFT kernel launch count */
switch (stages_reqd - launch_8) {
case (2):
  launch_4 = 2;
  break;
case (1):
  launch_2 = 1;
  break;
default:
  break;
}

/* FFT compute Loop */
while (count < stages_reqd) {
  num_stage = (1 << count);
  if (launch_8) {
    ret = clEnqueueWriteBuffer(command_queue,
        comp_stage_buffer, 1, 0, sizeof(cl_uint),
        &num_stage, 0, 0, &event);
```

```
    check_status(ret, "Error: write Buffer\n");

    /* Execute the 8-point FFT kernel*/
    ret = clEnqueueNDRangeKernel(command_queue,
        kernel_8, 1, NULL, globalThreads_8,
        localThreads_8, 0, NULL, &event);
    check_status(ret, "Error: Run kernel 8\n");
    count += 3;
    launch_8--;
  } else if (launch_4) {
    ret = clEnqueueWriteBuffer(command_queue,
        comp_stage_buffer, 1, 0, sizeof(cl_uint),
        &num_stage, 0, 0, &event);
    check_status(ret, "Error: write Buffer\n");

    /* Execute the 4-point FFT kernel */
    ret = clEnqueueNDRangeKernel(command_queue,
        kernel_4, 1, NULL, globalThreads_4,
        localThreads_4, 0, NULL, &event);
    check_status(ret, "Error: Run kernel 4\n");
    count += 2;
    launch_4--;
  } else if (launch_2) {
    ret = clEnqueueWriteBuffer(command_queue,
        comp_stage_buffer, 1, 0, sizeof(cl_uint),
        &num_stage, 0, 0, &event);
    check_status(ret, "Error: write Buffer\n");
    /* Execute the 2-point FFT kernel */
    ret = clEnqueueNDRangeKernel(command_queue,
        kernel_2, 1, NULL, globalThreads_2,
        localThreads_2, 0, NULL, &event);
    check_status(ret, "Error: Run kernel 2\n");
    count += 1;
    launch_2--;
  }
}

/* Get the output buffer */
ret = clEnqueueReadBuffer(command_queue, output_buffer,
    CL_TRUE, 0, num_data * sizeof(cl_float2), output, 0,
    NULL, &event);

ret = clFlush(command_queue);
ret = clFinish(command_queue);
```

```
/* Release all resources */
ret = clReleaseKernel(kernel_8);
ret = clReleaseKernel(kernel_4);
ret = clReleaseKernel(kernel_2);
ret = clReleaseProgram(program);
ret = clReleaseMemObject(input_buffer);
ret = clReleaseMemObject(output_buffer);
ret = clReleaseMemObject(comp_stage_buffer);
ret = clReleaseCommandQueue(command_queue);
ret = clReleaseContext(context);
free(input);
free(output);
return 0;
}


}
```

References

1. AMD: clAmdfft, OpenCL FFT library from AMD (2013)
2. Cooley, J., Tukey, J.: An algorithm for the machine calculation of complex Fourier series. Math. Comput. **19**, 297–301 (1965)
3. Moreland, K., Angel, E.: The FFT on a GPU. In: Proceedings of the ACM SIGGRAPH/EURO-GRAPHICS Conference on Graphics Hardware, pp. 112–120 (2003)
4. Munshi, A.: The OpenCL 1.2 Specification. Khronos OpenCL Working Group, Beaverton (2012)
5. Nvidia: Cufft library (2010)
6. NVIDIA: CUDA Programming Guide, Version 5 (2012)
7. Van Loan, C.: Computational Frameworks for the Fast Fourier Transform. Society for Industrial and Applied Mathematics (1992). doi:10.1137/1.9781611970999. http://epubs.siam.org/doi/book/10.1137/1.9781611970999

Chapter 17
A Highly Efficient FFT Using Shared-Memory Multiplexing

Yi Yang and Huiyang Zhou

17.1 Introduction

Modern many-core graphics processor units (GPUs) rely on thread-level parallelism (TLP) to deliver high computational throughput. To mitigate the impact of long latency memory accesses, besides TLP, software managed on-chip local memory is included in state-of-art GPUs. Such local memory, referred to as shared memory in NVIDIA GPUs and local data share in AMD GPUs, has limited capacity. Therefore, efficient utilization of shared memory is critical for many GPGPU (general purpose computation on GPUs) applications. The Fast Fourier Transform (FFT), a classic algorithm widely used in many scientific domains, is such an example. In an optimized parallel implementation of 1-dimension (1D) FFT [1] on GPUs, as the outputs of one thread are the inputs to others, shared memory is used to store these temporary data to reduce off-chip memory accesses.

State-of-art GPUs manage shared memory in a relatively simple manner. When a group of threads (called a thread block or a workgroup) is to be dispatched, shared memory is allocated based on the aggregate shared memory usage of all the threads in the thread block (TB). When a TB finishes execution, the allocated shared memory is released. When there is not sufficient shared memory for a TB, the TB dispatcher is halted. The major limitation of the abovementioned shared memory management is that the allocated shared memory is reserved throughout the lifetime of a TB, even if it is only utilized during a small portion of the execution time. This limitation reduces the number of TBs that can concurrently run on a GPU, which may impact the performance significantly as there may not be sufficient threads to hide long latencies of operations such as memory accesses.

In this chapter, we first present a naïve FFT implementation on GPGPUs. We show such an implementation has a large amount of off-chip global memory

Y. Yang • H. Zhou (✉)
North Carolina State University, Raleigh, NC, USA
e-mail: yangyi@gmail.com; hzhou@ncsu.edu

V. Kindratenko (ed.), *Numerical Computations with GPUs*,
DOI 10.1007/978-3-319-06548-9_17, © Springer International Publishing Switzerland 2014

accesses and fails to utilize computing resource efficiently. We then use the code from AMD SDK [1] as an example to show how to overcome these limitations. The key idea is to use a single thread to carry out multi-point FFT computations and leverage shared memory to exchange data among threads. However, the shared memory usage may reduce the number of threads running concurrently on each SM of GPUs due to the limited capacity of shared memory. To address this challenge, we made an important observation that as shared memory is primarily used for data exchange among threads for FFT, it is actually utilized for a short amount of time compared to the overall execution time of a TB. Based on this observation, we propose novel ways to time-multiplex shared memory so as to enable a higher number of TBs to be executed concurrently. Our software approaches work on existing GPUs and they essentially combine two or more original TBs into a new TB and introduce if-statements to control time multiplexing of the allocated shared memory among the original TBs.

Our experimental results on NVIDIA GTX480 and Tesla K20c GPUs show that our approaches can improve the FFT performance significantly. On an NVIDIA GTX 480 GPU, our FFT kernel outperforms the vendor-tuned library NVIDIA CUFFT V4.0 by 21 % for a 1 k-point FFT with a batch size of 2,048. On an NVIDIA Tesla K20c GPU, our FFT kernel outperforms CUFFT V5.0 by 58 % for the same inputs.

The remainder of the chapter is organized as follows. In Sect. 17.2, we present a brief background on GPGPU architecture and a naïve implementation of 1D FFT on GPGPUs. We discuss the limitation of such a naïve implementation and use the 1D FFT code from AMD SDK to illustrate some key optimizations for FFT in Sect. 17.3. In Sect. 17.4, we show that the optimized FFT code uses large amount of shared memory to reduce off-chip memory access and we highlight the characteristics of its shared memory data usage. In Sect. 17.5 we present our two software approaches to time-multiplex shared memory to overcome the limitation of aggressive usage of shared memory. The experimental methodology is addressed in Sect. 17.6 and the results are presented in Sect. 17.7. Related work is discussed in Sect. 17.8. Finally, Sect. 17.9 concludes this chapter.

17.2 Background

State-of-art GPUs use many-core architecture to deliver high computational throughput. One GPU consists of multiple streaming multiprocessors (SMs) in NVIDIA GPU architecture or computer units (CUs) in AMD GPU architecture. Each SM/CU in turn includes multiple streaming processors (SPs) or thread processors (TPs). Threads running on GPUs follow the single-program multiple-data (SPMD) model and are organized in a hierarchy. A GPU kernel is launched to a GPU with a grid of thread blocks (TBs) using the NVIDIA CUDA terminology [2], which are called workgroups in OpenCL [3]. Threads in a TB form multiple warps, with each running in the Single Instruction Multiple Data (SIMD) mode. One or more TBs run concurrently on one SM, depending on the resource requirement of a TB.

On-chip shared memory is a critical resource for GPGPU applications. Shared memory provides a mechanism for threads in the same TB to communicate with each other. It also serves as a software managed cache so as to reduce the impact of long-latency memory accesses. Since each SM has a limited amount of shared memory, for many GPGPU applications, the shared memory usage of a TB determines how many TBs can run concurrently, i.e., the degree of thread level parallelism (TLP), on an SM. Besides shared memory, the register usage of each thread is another critical factor to determine the number of threads that can run concurrently. In state-of-art GPUs, shared memory is managed as follows. When a TB is to be dispatched to an SM, the TB dispatcher allocates shared memory based on the aggregate usage of all the threads in the TB. The allocated shared memory is released when the TB finishes execution. When there is not sufficient resource available in an SM, the resource is not allocated and no TB will be dispatched to the SM.

Between shared memory and RFs, current GPUs have higher capacity in RFs. For example, on NVIDIA GTX285 GPUs, each SM has 16 kB shared memory and a 64 kB RF. On NVIDIA GTX480 GPUs (i.e., the Fermi architecture), each SM has a 128 kB RF and a 64 kB hybrid storage that can be configured as a 16 kB L1 cache + 48 kB shared memory or a 48 kB L1cache + 16 kB shared memory. The latest NVIDIA GPU, GTX680 (i.e., the Kepler architecture), has the same size of shared memory per SM as GTX480 and a 256 kB RF. With a high number of SPs and a larger RF in each SM, the Kepler architecture is designed to host more concurrent thread blocks/threads in each SM than the Fermi architecture, thereby increasing the pressure on shared memory. On AMD HD5870 GPUs, each CU contains 32 kB shared memory (called local data share) and a 256 kB RF. As a result, for many GPGPU applications, shared memory presents a more critical resource to limit the number of TBs/threads to run concurrently on an SM.

With the promising performance of GPGPUs, many algorithms have been adapted to GPGPUs. One such algorithm is the Fast Fourier Transform (FFT). Figure 17.1 shows the pseudo code of a naïve implementation for 1D 1 k-point FFT on GPGPUs. In order to compute 1 k-point FFT, the kernel GPU_FFT2 needs to be invoked ten times, i.e., ten passes, and in each pass GPU_FFT2 performs a 2-point FFT. The output array of a previous invocation will be the input to next kernel innovation. The value of K in the function declaration is doubled in each invocation from 1 to 1,024. Since each thread performs a 2-point FFT, there are 512 threads for 1 k input elements in a TB. In Fig. 17.1, we can see that, GPU_FFT2 first loads two complex numbers to the array v from global memory, and then performs the twiddle-factor computation. The device function FFT2 is called to calculate the FFT results of two complex numbers. Finally, the results are stored to global memory to finish a pass.

```
#define N 1024
__device__ inline void FFT2 (float2& v0, float2& v1) {
        float2 v = v0;
        v0.x = v.x + v1.x;
        v0.y = v.y + v1.y;
        v1.x = v.x - v1.x;
        v1.y = v.y - v1.y;
}
__global__ void GPU_FFT2(float2* input, float2* output, int K) {
        float2 v[2];
        int gid = threadIdx.x;
        float angle = -2*PI*(gid%K)/(K*2);
        for (int r = 0; r < 2; r++) {
                v[r] = dataI[gid+r*N/2];
                float2 tw, vr;
                tw.x = __cosf(r*angle);
                tw.y = __sinf(r*angle);
                vr.x = tw.x*v[r].x - tw.y*v[r].y;
                vr.y = tw.x*v[r].y + tw.y*v[r].x;
                v[r].x = vr.x;
                v[r].y = vr.y;
        }
        FFT2 (v0, v1);
        int ind = (gid/K)*K*2+(gid%K);
        for (int r=0; r<R; r++) {
                output [ind +r*K] = v[r];
        }
}
int main( int argc, char** argv) {
    ......
    int flip = 1;
    for (int K = 1; K< 1024; K=K*2 ) {
        GPU_FFT2<<< 1, 512 >>>(flip?in:out, flip?out:in, K);
        flip = !flip;
    }
    ......
}
```

Fig. 17.1 The pseudo code of a 1 k-point FFT implementation The GPU_FFT2 kernel is invoked ten times from the main function. Each time GPU_FFT2 performs 2-point FFT with a total of 512 threads. The output array of previous invocation will be the input to the next invocation

17.3 Optimizations for FFT on GPGPUs

The code example in Fig. 17.1 is straightforward and easy to understand. However, it does not utilize GPU resource efficiently to achieve high performance due to the following limitations. First, each thread only performs a 2-point FFT at one time, and therefore a large number of kernel invocations is needed. Second, in each invocation, the kernel has to read the input array from and write the output array to global memory, which has much longer latency and lower bandwidth than on-chip memory.

In order to overcome these limitations, many optimizations have been proposed in previous studies [1, 4]. Here, we summarize the common optimizations. First, instead of performing a 2-point FFT, each thread can perform a 4-point, 8-point, 16 (or more)-point FFT. The benefit of using a thread to compute a multi-point FFT is to reduce the number of passes and the number of off-chip memory accesses. Since a multi-point FFT may be computed using registers only, the workload of each thread is determined by the register requirement. For example, NVIDIA GTX 480 and 680 GPUs limit the number of registers per thread to be less than 64, therefore each thread can perform at most a 16-point FFT due to the register usage for data and addresses. Second, shared memory is used to support inter-thread communication and enable single kernel invocation for a 1 k-point FFT. To do so, threads in a TB store the per-thread FFT computation results into shared memory, perform synchronization, read data from shared memory, perform another multi-point FFT, and so on. This procedure can be repeated until a thread requires the output from a thread that is located in a different TB.

Next, we demonstrate these optimizations using the code example adapted from the AMD SDK [1]. The kernel shown in Fig. 17.2 is used to perform 1 k-point FFT using a TB containing 64 threads. The kernel has two inputs arrays: one for the real part and the other for the imaginary part of the inputs. The results of the kernel computation will be overwritten to the input array, and therefore there is no additional output array.

Inside the kernel, we can see there are seven code sections as commented in the code. The first code section is used to read data from global memory, and the last one is used to store data to global memory. The remaining five sections correspond to five passes with each pass performing four 4-points FFTs. The first FFT pass uses the function FFT_P1, which is different from the function FFT_P used in the second, third, and fourth passes. The reason is that float4 is used to load data from global memory and the x, y, z, w fields of a float4 require different values of twiddle factors in the first pass. For other passes, the data is loaded from shared memory and the same twiddle factors are shared. The fifth pass is different as it does not need to compute twiddle factors. The parameter of FFT_P is used to compute the twiddle factors.

Within each pass, there is data interchange through shared memory, i.e., the loadFromSM and the saveToSM functions. The parameters of loadFromSM or saveToSM are used to determine where to load/store the data in shared memory.

Compared to the implementation in Fig. 17.1, the implementation in Fig. 17.2 requires only one kernel invocation. In the meanwhile, the number of off-chip memory accesses has been reduced by 90 % due to data exchange in shared memory. With this implementation, each TB has 64 threads and uses 8,736-Byte shared memory (8,192 B for data, additional bytes for padding to avoid bank conflicts and a few bytes reserved by CUDA). In other words, this optimized FFT implementation requires a large amount of shared memory. For the purpose of clarity of our discussion in remaining sections, we simplify the pseudo code in Fig. 17.2 to the code shown in Fig. 17.3. As shown in Fig. 17.3, we reduce a code section into a function call, and replace the parameters of function calls with the pass sequence numbers.

```
__global__ void kfft(float *greal, float *gimag) {
    __shared__ float lds[68*4*4*2]; // This is 8704 bytes
    uint gid = threadIdx.x+blockIdx.x*blockDim.x;
    uint me = gid & 0x3fU;
    uint dg = (gid >> 6) * VSTRIDE;
    float *gr = greal + dg;
    float *gi = gimag + dg;

    // section 1: load from global memory
    float4 *gp = (float4 *)(gr + (me << 2));
    float4 zr[4], zi[4];
    zr[0] = gp[0*64]; zr[1] = gp[1*64]; zr[2] = gp[2*64]; zr[3] = gp[3*64];
    gp = (float4 *)(gi + (me << 2));
    zi[0] = gp[0*64]; zi[1] = gp[1*64]; zi[2] = gp[2*64]; zi[3] = gp[3*64];

    // section 2: first FFT4 pass
    FFT_P1();
    savetoSM(66*4, 1, ((me << 2) + (me >> 3)), 66*4*4); __syncthreads();
    loadfromSM(66, 66*4, (me + (me >> 5)), 66*4*4);

    // section 3: second FFT4 pass
    FFT_P(me << 2); __syncthreads();
    savetoSM(1, 66*4, (((me << 2) + (me >> 3))), 66*4*4); __syncthreads();
    loadfromSM(66, 66*4, (me + (me >> 5)), 66*4*4);

    // section 4: third FFT4 pass
    FFT_P((me >> 2) << 4); __syncthreads();
    savetoSM(66*4, 66, me, 66*4*4); __syncthreads();
    loadfromSM(16, 66, ((me & 0x3) + ((me >> 2) & 0x3)*(66*4) + ((me >> 4) <<
2)), 66*4*4);

    // section 5: fourth FFT4 pass
    FFT_P((me >> 4) << 6); __syncthreads();
    savetoSM(68*4, 68, me, 68*4*4); __syncthreads();
    loadfromSM(16, 68, ((me & 0xf) + (me >> 4)*(68*4)), 68*4*4);

    // section 6: fifth FFT4 pass
    FFT4();

    // section 7: write back results to global memory
    gp = (float4 *)(gr + (me << 2));
    for (int i=0; i<4; i++) gp[i*64] = zr[i];
    gp = (float4 *)(gi + (me << 2));
    for (int i=0; i<4; i++) gp[i*64] = zi[i];
}
```

Fig. 17.2 The kernel of a 1 k-point FFT implementation, which uses 8,736-Byte shared memory per thread block and there are 64 threads per thread block

17.4 Characterization of Shared Memory Usage

As discussed in Sect. 17.3, the optimized FFT kernel uses 8,736 B shared memory per TB, and each thread block has 64 threads. Therefore, if each SM has only 16 kB shared memory, only one thread block can run in a SM at a time, resulting in relatively low TLP.

```
loadFromGlobal();
FFT4(0);
saveToSM(0);      //define by multiple threads in a TB
__syncthreads();
loadFromSM(0);    //use by multiple threads in a TB
FFT4(1);
__syncthreads();
saveToSM(1);      //(re)define by multiple threads in a TB
__syncthreads();
loadFromSM(1);    //use by multiple threads in a TB
....
FFT4(4);
writeToGlobal();
```

Fig. 17.3 The simplified pseudo code of a 1 k-point FFT implementation, which uses 8,736-Byte shared memory per thread block and there are 64 threads per thread block

Next, we resort to microarchitectural simulation to analyze how shared memory is actually utilized (see Sect. 17.6 for the detailed experimental methodology) in a TB. Since the compiler may schedule shared memory accesses to interleave with other types of instructions to improve instruction-level parallelism (ILP), if we simply consider the lifetime of shared memory usage as between the first instruction writing to shared memory and the last instruction reading from shared memory, we may find that shared memory is used for almost the entire lifetime of kernel execution. To isolate the usage of shared memory from other parts of the kernel code, we insert '__syncthreads()' instructions before the first *write/define* to and after the last *read/use* from shared memory. We denote a code region surrounded by our inserted '__syncthreads()' as a shared memory access region. A redefine of the shared memory variables will start a new shared memory access region. Here, any define or use of shared memory variables is based on all the threads in a TB. Then, we use the accumulated execution time of all shared-memory-access regions as the duration for shared memory usage. Our simulation result shows that shared memory is only used in 28.2 % of the execution time for FFT, which means in 71.8 % of the execution time, shared memory is occupied but not used.

17.5   Shared Memory Multiplexing: Software Approaches

As discussed in Sect. 17.3, FFT suffers from insufficient TLP due to the limited shared memory capacity. In the meanwhile, the allocated shared memory is only utilized for a small fraction of the overall execution time. In this section, we propose two software approaches to time-multiplex shared memory so as to boost TLP. The key idea is to combine original TBs to a larger one and introduce control flow to manage how shared memory is accessed by original TBs. The difference between the two approaches lies in how to overlap shared memory accesses with other parts of the code.

17.5.1   Virtual Thread Blocks (VTB)

In this approach, we first isolate the part(s) of a kernel function that accesses shared memory variables. Second, we combine two original TBs into a new one. Here, we refer to an original TB as a virtual TB. In other words, after TB combination, one TB contains two virtual TBs. Third, we introduce the control flow "*if(v_tb_id==0)*" and "*if(v_tb_id ==1)*" to manage which virtual TB will access shared memory at a time. The amount of the required shared memory of the combined TB remains the same as either of the virtual TBs.

For the FFT code shown in Fig. 17.3, the code after we apply VTB is shown in Fig. 17.4. The '*if-statements*' on lines 4, 6, 8, and 10 are introduced to ensure that only one virtual-TB is accessing the allocated shared memory at a time. The '*syncthreads()*' function on line 7 implicitly marks the last use of shared memory of virtual TB 0 so that virtual TB 1 can use shared memory immediately afterwards.

```
1.  int v_tb_id = threadIdx.x/64;  //virtual thread block id
2.  loadFromGlobal();
3.  FFT4(0);
4.  if (v_tb_id==0) saveToSM(0); //def. from threads in v_tb_0
5.  __syncthreads();
6.  if (v_tb_id==0) loadFromSM(0);//use. from threads in v_tb_0
7.  __syncthreads();
8.  if (v_tb_id==1) saveToSM(0); //def. from threads in v_tb_1
9.  __syncthreads();
10. if (v_tb_id==1) loadFromSM(0);//use. from threads in v_tb_1
11. FFT4(1);
12.    ....
13. FFT4(4);
14. writeToGlobal();
```

Fig. 17.4 The pseudo code of a 1 k-point FFT implementation using VTB. Each thread block uses 8,736-Byte shared memory and there are *128* threads in each thread block

Next, we illustrate the reason why our proposed VTB can improve the GPU throughput and also highlight its overhead. Assuming a GPU with 16 kB shared memory in each SM, since each TB requires more than 8 kB shared memory, two TB dispatched to the same SM have to execute back to back with the code in Fig. 17.3. This execution process is shown in Fig. 17.5a. For the purpose of clarity, in Fig. 17.4 we only show the execution time corresponding to the global memory access, the first 4-point FFT and the data exchange via shared memory. The remaining code in the kernel function simply repeats 4-point FFT and data exchange multiple times. With the code in Fig. 17.4, the combined TB is equivalent to the two original TBs. Due to the increased TLP, the execution time of the function *loadFromGlobal()* and *FFT4()* of 128 threads is significantly less than the back-to-back execution of the same functions of 64 threads, as shown in Fig. 17.5b. However, to control the accesses to shared memory between the two virtual TBs, additional synchronization functions are added to ensure correctness. Besides the latency to perform such '__syncthread()' functions, the barrier also limits the compiler's capability to schedule instructions across the barriers, which may result in reduced ILP and additional register usage. The added control flow instruction "$if(v\_tb\_id==0)$" has minimal overhead as it does not generate any control divergence within a warp since all 64 threads in the same virtual TB will follow the same direction and each warp has 32 threads on NVIDIA GPUs. The global memory access functions 'loadFromGlobal' and 'writeToGlobal' benefit from VTB as the increased TLP translate to increased memory-level parallelism (MLP).

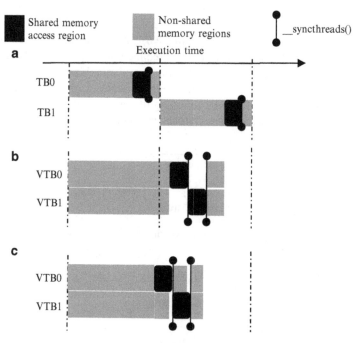

Fig. 17.5 A comparison of execution time for (**a**) Two thread blocks (TB0 and TB1) use one copy of shared memory (baseline), (**b**) Two virtual thread blocks (VTB0 and VTB1) multiplex one copy of shared memory (VTB), and (**c**) Two virtual thread blocks (VTB0 and VTB1) multiplex one copy of shared memory with a pipelined schedule (VTB_pipe)

From Fig. 17.5, we can also see that when a virtual TB accesses shared memory, the other virtual TB is forced to be idle due to the control flow and the '__syncthread()' functions. Our proposed second approach addresses this limitation and we include the execution time information of these approaches in Fig. 17.5 for comparison. We discuss this approach in detail in Sect. 17.5.2.

Note that although Figs. 17.4 and 17.5 show the case of combining two original TBs into one, we can apply the same principle to combine more than two TBs.

The optimal number of TBs to combine is dependent on how many concurrent threads can run on an SM. Typically, combining two TBs is sufficient to reap most performance benefits.

17.5.2 Pipelined Virtual Thread Blocks (VTB_PIPE)

As discussed in Sect. 17.5.1, VTB combines two virtual TBs into a larger one and it ensures that only one virtual TB is accessing shared memory by forcing the other virtual TB to be idle. To reduce such idle cycles, we propose to overlap computation with shared memory accesses. To do so, we make the first virtual TB to run faster than the second one using an '*if(v_tb_id==0)*' statement. Then, when the first virtual block reaches the code section of shared memory access, the second virtual TB continues its computation instead of being forced idle. When the second virtual block reaches the code section of shared memory accesses, the first will continue to run ahead. This process is similar to letting the two virtual TBs to go through a pipeline. Therefore, we refer to this approach as pipelined VTB (VTB_pipe).

```
1.   int v_tb_id = threadIdx.x/64;
2.   loadFromGlobal();
3.   if (v_tb_id==0)  FFT4(0);
4.   if (v_tb_id==0) saveToSM(0);
5.   __syncthreads();
6.   if (v_tb_id==0) loadFromSM(0);
7.   else FFT4(0);
8.   __syncthreads();
9.    if (v_tb_id==1) saveToSM(0);
10.  __syncthreads();
11.  if (v_tb_id==1) loadFromSM(0);
12.  else FFT4(1);
13.  __syncthreads();
14.  if (v_tb_id==0) saveToSM(1);
15.  __syncthreads();
16.  if (v_tb_id==0) loadFromSM(1);
17.  else FFT4(1);
18.   ....
19.  FFT4(4);
20.  writeToGlobal ();
```

Fig. 17.6 The pseudo code of a 1 k-point FFT implementation using VTB_pipe. Each thread block uses 8,736-Byte shared memory and there are *128* threads in each thread block. (The complete code of our proposed VTB_pipe approach for FFT is available at: http://people.engr.ncsu.edu/hzhou/fft_pipe.cu)

For the FFT, the code after we apply our proposed VTB_pipe is shown in Fig. 17.6. From Fig. 17.6, we can see that initially the two virtual TBs will both execute the '*loadFromGlobal()*' function. Then, the '*if(v_td_id==0)*' statements on lines 3 and 4 as well as the '*__syncthreads()*' on line 5 enable virtual TB 0 to execute the '*FFT4()*' and '*saveToSM*' functions, making it running ahead of virtual TB1. The code on line 6 and line 7 shows the overlapping between the function '*loadFromSM()*' of virtual TB0 and the '*FFT4()*' function of virtual TB1. Since virtual TB1 is lagging behind, when it reads from shared memory via '*loadFromSM()*' on line 11, the virtual TB0 proceeds to compute its next 4-point FFT, the '*FFT4()*' on line 12. The execution process is shown in Fig. 17.5c. Due to the overlapping between shared memory accesses and computation, we can reduce the idle cycles experienced by virtual TBs.

The complexity of VTB_pipe, however, is that we may need to partition the non-shared memory access code section to create small computational/global memory access tasks so that they can overlap with shared memory accesses. The ideal case is that the small computational tasks have similar execution latency to the shared memory accesses and can completely utilize the otherwise idle cycles. In the FFT case, the *FFT4()* function is a convenient choice and does not require such partition.

17.6   Experimental Methodology

To evaluate our proposed software approaches, we use both an NVIDIA GTX 480 GPU with CUDA SDK 4.0 and an NVIDIA Tesla K20c GPU with CUDA SDK 5.5. Because the shared memory size is configurable on these GPUs, we present two sets of results: one with 48 kB shared memory and the other with 16 kB shared memory.

For our experiment in Sect. 17.4, we use the GPGPUsim V3.0 simulator [5]. The simulator models an NVIDIA GTX285 GPU, which has 16 kB shared memory and a 64 kB register file on each SM. The off-chip memory frequency is set to 1,100 MHz.

17.7   Experimental Results

For 1 k-point FFT, we use batch execution [4] to evaluate the throughput and vary the batch size from 128 to 2,048. The throughput results of GTX480 are reported in Fig. 17.7. From the figure, we can see that our baseline implementation running on GTX480 with 16 kB shared memory outperforms CUFFT [6] for small batch sizes and not as good for large batch sizes. With the 48 kB shared configuration, our baseline implementation consistently outperforms CUFFT. The average throughput of '48K_BL' is 168.3 GFLOPS compared to the average of 72.4 GFLOPS throughput of CUFFT. Our VTB_pipe further improves the throughput by up to 33 % and achieves an average throughput of 205.9 GFLOS (a 2.84× speedup over CUFFT). The results on Tesla K20c using CUDA SDK 5.5 are shown in

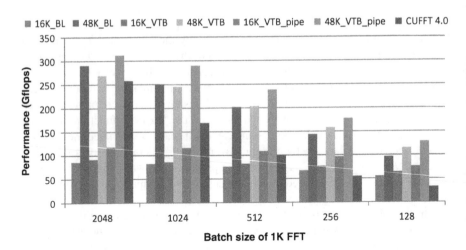

Fig. 17.7 Performance comparison of FFT among the baseline (xK_BL), VTB_pipe (xK_VTB_pipe) and CUFFT 4.0 on GTX 480. 'xK' denotes the size of shared memory

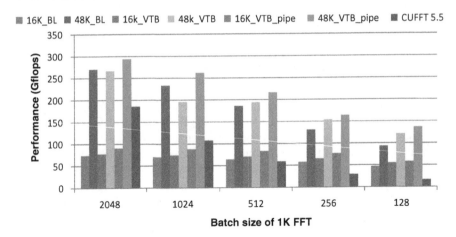

Fig. 17.8 Performance comparison of FFT among the baseline (xK_BL), VTB_pipe (xK_VTB_pipe) and CUFFT 5.5 on Tesla K20c. 'xK' denotes the size of shared memory

Fig. 17.8. As we can see from Fig. 17.8, the performance trend is similar to the one on GTX 480. On average, our VTB_pipe achieves 1.21× speedup over the baseline and 3.67× speedup over CUFFT 5.5 for the 48 KB shared memory configuration. We also observe that K20c does not yield higher performance than GTX 480 for this FFT implementation.

In the code shown in Fig. 17.5, we overlap the *loadFromSM()* function of a virtual TB with the *FFT4()* function of the other virtual TB. Therefore, the function *saveToSM()* is actually not utilized in this pipeline approach. Alternatively, a different pipeline choice is to overlap the *saveToSM()* function of a virtual TB

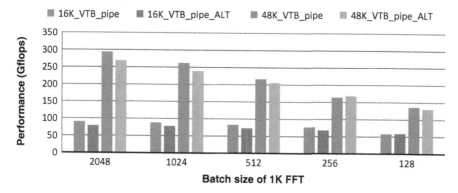

Fig. 17.9 Performance comparison of FFT among two different pipeline implementations. xK_VTB_pipe is the implementation shown in Fig. 17.5 and xK_VTB_pipe_ALT is the implementation using saveToSM to overlap with FFT4. 'xK' denotes the size of shared memory

with the *FFT4()* function of the other virtual TB. We show the results of such alternative pipeline approach in Fig. 17.9. As shown in figure, the alternative approach has similar but slightly worse performance than the previous pipeline approach. Since our approach tries to schedule the workload in a software way which is interfered by the hardware schedule, it is difficult to predict the best pipeline implementation. Therefore, a hardware approach proposed in [7], which supports dynamic allocation and de-allocation of shared memory, reduces the software overhead and avoids such interference between the software and hardware schedules.

17.8 Related Work

On-chip shared memory is a critical resource for GPGPU applications. Previous works mainly focus on utilizing shared memory to achieve coalesced memory accesses [8–16], to provide data exchange among threads [4], to use shared memory as software managed cache [14], etc. Although it is well known that heavy usage of shared memory may limit TLP [14, 17], it is common that the benefits of using shared memory overweigh the shortcomings of reduced TLP. As a result, many GPGPU workloads have exhibited high shared memory usage. This is also a reason why the NVIDIA Fermi and Kepler architecture provides larger shared memory and an L1 cache. On the other hand, the high number of SPs in an SM in the NVIDIA Fermi and Kepler architecture makes TLP more important to hide instruction execution latencies.

17.9 Conclusions

In this chapter, we propose novel software approaches to time-multiplex shared memory. Our approaches are based on our observation that for many GPGPU applications with heavy use of shared memory, the duration of time, when shared memory is utilized, is actually low. Our experimental results confirm that shared memory is utilized for only 28.2 % of the execution time of a TB for an optimized FFT implementation. Therefore, there exist significant opportunities to time multiplex shared memory. Among our software approaches, VTB is simplest and it combines two TBs into a new one and adds control flow to ensure only one original TB accesses shared memory at a time. VTB_pipe reduces the performance overhead of VTB by overlapping non-shared memory access regions (e.g., computation or global memory accesses) with shared memory accesses. Our experimental results show that our proposed software schemes improve the performance significantly on current GPUs.

Acknowledgments This work is supported by an NSF project 1216569, an NSF CAREER award CCF-0968667, a grant from DARPA PERFECT program, and a gift fund from AMD Inc.

References

1. AMD Accelerated Parallel Processing SDK V2.3 (2011)
2. NVIDIA CUDA C Programming Guide 4.0 (2011)
3. OpenCL. http://www.khronos.org/opencl/
4. Govindaraju, N., Lloyd, B., Dotsenko, Y., Smith, B., Manferdelli, J.: High performance discrete Fourier transforms on graphics processors. In: Proc. Supercomputing (2008)
5. Bakhoda, A., Yuan, G., Fung, W.W.L., Wong, H., Aamodt, T.M.: Analyzing CUDA workloads using a detailed GPU simulator. In: IEEE International Symposium on Performance Analysis of Systems and Software (2009)
6. NVIDIA CUDA Toolkit 4.0 CUFFT Library (2011)
7. Yang, Y., Xiang, P., Mantor, M., Rubin, N., Zhou, H.: Shared memory multiplexing: a novel way to improve GPGPU throughput. In: Proc. International Conference on Parallel Architecture and Compiler Techniques (2012)
8. Jang, B., Schaa, D., Mistry, P., Kaeli, D.: Exploiting memory access patterns to improve memory performance in data-parallel architectures. In: IEEE Transactions on Parallel and Distributed Systems (2010)
9. Zhang, E.Z., Jiang, Y., Guo, Z., Tian, K., Shen, X.: On-the-fly elimination of dynamic irregularities for GPU computing. In: International Conference on Architectural Support for Programming Languages and Operating Systems (2011)
10. Ruetsch, G., Micikevicius, P.: Optimize matrix transpose in CUDA. NVIDIA (2009)
11. Ryoo, S., Rodrigues, C.I., Stone, S.S., Baghsorkhi, S.S., Ueng, S., Stratton, J.A., Hwu, W.W.: Optimization space pruning for a multi-threaded GPU. In: Proc. International Symposium on Code Generation and Optimization (2008)
12. Ryoo, S., Rodrigues, C.I., Baghsorkhi, S.S., Stone, S.S., Kirk, D.B., Hwu, W.W.: Optimization principles and application performance evaluation of a multithreaded GPU using CUDA. In: Proc. ACM SIGPLAN Symposium on Principles and Practice of Parallel Programming (2008)

13. Volkov, V., Demmel, J.W.: Benchmarking GPUs to tune dense linear algebra. In: Proc. Supercomputing (2008)
14. Yang, Y., Xiang, P., Kong, J., Zhou, H.: A GPGPU compiler for memory optimization and parallelism management. In: ACM SIGPLAN Conference on Programming Language Design and Implementation (2010)
15. Yang, Y., Xiang, P., Mantor, M., Zhou, H.: Fixing performance bugs: an empirical study of open-source GPGPU programs. In: International Conference on Parallel Processing (2012)
16. Zhang, Y., Cohen, J., Owens, J.D.: Fast tridiagonal solvers on the GPU. In: Proc. ACM SIGPLAN Symposium on Principles and Practice of Parallel Programming (2010)
17. Sim, J., Dasgupta, A., Kim, H., Vuduc, R.: A performance analysis framework for identifying performance benefits in GPGPU applications. In: Proc. ACM SIGPLAN Symposium on Principles and Practice of Parallel Programming (2012)

Chapter 18
Increasing Parallelism and Reducing Thread Contentions in Mapping Localized N-Body Simulations to GPUs

Bharat Sukhwani and Martin C. Herbordt

18.1 Introduction

N-body simulations are applied to a variety of scientific problems, from space simulations to modeling molecular interactions. Due to their computational complexity, various parallelization techniques have been applied to reduce their execution time. In particular, the use of graphics processors to accelerate these computations has been widely studied [1, 2]. Depending on the actual geometry of the underlying system being simulated, however, the data structures and the approach for parallelization vary significantly. For example, widely studied molecular dynamics simulations that evaluate the interaction of the entire system of particles often employ cell-lists, a data structure that may not be best-suited for other problems that aim to simulate a localized region of the system involving only a small subset of particles. An example of one such application is energy minimization as employed in drug discovery. Here, the aim is to simulate a small probe molecule in a potential drug binding site (pocket). A commonly-used representation of particle-pairs in such applications is neighbor lists (Fig. 18.1), wherein for each particle in the system, a list of its neighboring particles that affect its energy is maintained. As the particles move around during subsequent simulation iterations, the neighbor lists are updated. Computing the energy of a particle then involves iterating through its neighbor list, computing the partial pair-wise energy value due to each of its neighbor and accumulating the partial values into a total sum. The total energy of the system is the sum of the energies of all the particles.

B. Sukhwani (✉)
IBM T. J. Watson Research Center, Yorktown Heights, NY, USA
e-mail: bharats@us.ibm.com

M.C. Herbordt
Boston University, Boston, MA, USA
e-mail: herbordt@bu.edu

V. Kindratenko (ed.), *Numerical Computations with GPUs*,
DOI 10.1007/978-3-319-06548-9_18, © Springer International Publishing Switzerland 2014

Fig. 18.1 Neighbor list for
an atom

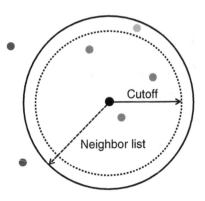

Note that neighbor lists are also often employed in molecular dynamics (MD) simulations but the geometry of the problem and thus the characteristics of the resulting neighbor lists are significantly different from those in energy minimization. In MD, the geometry is dense, with each particle having a large number of particles in its neighborhood and the neighbor lists for different particles are similar sized for the bulk of the computation. In energy minimization, on the other hand, the neighbor lists are relatively small and different particles have widely varying number of particles in their neighborhood. Moreover, unlike MD where the neighbor lists are updated every few iterations, the neighbor lists in energy minimization are seldom updated, once every few hundred iterations. Though these differences appear subtle, they lead to the need for significantly different approaches for effective parallelization.

Effective parallelization of computations using the neighbor lists structure, particularly for localized N-body problems, poses some challenges. Although there is parallelism while computing the partial energies of the different particles, due to the local nature of the problem, the runtime for a single iteration is usually relatively small and much of the time per iteration is spent in the serial accumulation step, making the parallelization, particularly on GPU threads, difficult due to inter-thread communication overheads. The computation is typically repeated for a large number of iterations, resulting in large total execution times. The loop-carried dependence among the iterations, however, does not allow concurrent execution of multiple iterations.

This chapter discusses efficient mapping of such class of localized N-body problems to GPUs. Though modern GPUs have hundreds of processor cores, their communication and synchronization pattern is very restrictive. This, combined with the nature of the underlying computation, makes it difficult to achieve efficient utilization of these available processors. This chapter discusses how this limitation can be addressed by performing significant restructuring of the original data structures and introduces modified data structures to achieve increased parallelism and reduced serialization of, and communication among, GPU threads. In this chapter, we assume an NVIDIA Tesla GPU with CUDA programming model, though the techniques discussed in this chapter should be more broadly applicable to newer generation GPUs and the OpenCL programming model.

18.2   Overview of Energy Minimization

Throughout this chapter, energy minimization is used as the candidate N-body application for mapping to the GPU, though the techniques discussed can be applied to other localized N-body problems employing neighbor lists or similar data structures.

Energy minimization is a widely applied routine in many molecular modeling algorithms. The purpose of energy minimization is to compute the minimum-energy conformation of a system of particles that interact with each other under various bonded and non-bonded forces. It is an iterative process, requiring many hundreds to a thousand or more iterations to converge. Each iteration aims to compute the total potential energy of the system by computing the different bonded and non-bonded energies for all the atoms in the system. A move to a neighboring position is then made using an optimization technique such as Newton–Raphson or quasi-Newtonian (L-BFGS). This iterative process of energy evaluation followed by position update is repeated until the energy of the system converges within a threshold.

Though the underlying computation of energy minimization is superficially similar to widely studied molecular dynamics (MD), it differs from MD simulations in several ways. First, unlike MD, where the movement of the atoms is based on Newtonian dynamic laws and produces a trajectory based on kinetic energy, minimization simply adjusts the atom coordinates so as to lower the total energy of the system [3, 4]. Minimization does not include the effect of temperature, and the final state of the system corresponds to the atom configurations when the temperature is approximately zero [4]. For these reasons, the final state achieved after minimization does not depend on the initial state. Second, unlike MD, where the system typically consists of millions of particles, energy minimization is often performed on a local region of the complex, resulting in only a few thousand atoms being simulated and requiring only up to a few tens of thousands of atom-pair evaluations per iteration. Finally, even though the energy terms computed in minimization are similar to those in MD, the actual energy expressions evaluated are quite different. For example, rather than evaluating the van der Waals term with a 12–6 Lennard-Jones function, a minimization routine often approximates it with a sum of two or four Gaussians [5].

18.3   Mapping Localized N-Body Simulations to GPUs

In this section, three different methods for mapping the energy computations onto the GPU processors are presented. The first uses the original neighbor-lists data structure and does not result in any performance improvements compared to the original serial software code. The difficulties in obtaining good performance from this scheme are discussed, followed by a modified data structure for improved performance. The second method uses this modified data structure, leading to improved distribution of work among GPU threads and better performance. This approach, however, still requires serialized accumulations in the global memory and thus results in modest overall speedup. In the third method, the data structure is further modified to enable multiple parallel accumulations from shared memory, resulting in significant performance improvements. Also, an additional data structure is used to statically assign the work to different GPU threads leading to better work distribution and reduced inter-thread communication and conflicts.

Figure 18.2a shows the neighbor-lists and the energy array used for computing the atom energies in serial CPU code. For each atom in the system, a neighbor-list is maintained containing a list of its neighboring atoms that affects its energy. To compute an atom's energy, the atoms in its neighbor-list are accessed and the partial energies of the two atoms are computed. These partial energies are accumulated, as they are computed, into the energy array that stores the total individual energy of each atom. There are several reasons why the neighbor-list structure is not suited for mapping to GPUs. First, since the individual total energy value of each atom needs to be computed and not just the total energy of the system, multiple accumulations are required, one for each entry of the energy array. This leads to serialization during energy accumulation. Second, due to the arbitrary, non-sequential occurrences of the *neighbor* atoms in the neighbor-lists, the energy array cannot simply be divided and distributed into the shared memories of different GPU multiprocessors. Rather, it must be present in the GPU global memory, accessible from all the multiprocessors. And third, having the energy array in the global memory can potentially lead to write conflicts, since a particular *neighbor* atom can be present in the neighbor-list of more than one atom (e.g. atom number 2 in Fig. 18.2).

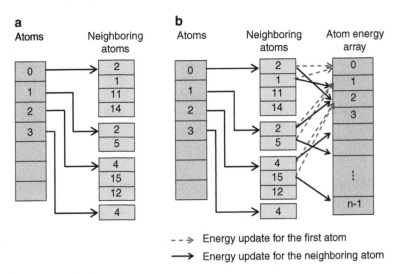

Fig. 18.2 (a) Neighbor-lists data structure (b) Arbitrary, non-sequential updates in the energy array cause write conflicts

18.3.1 Mapping Neighbor Lists to GPUs

The approach presented here attempts to map the unmodified neighbor-lists data structure onto the GPU threads. Though there are various ways to map this neighbor-list computation structure onto the GPU threads for parallel energy evaluations, most of them run into two problems: (1) memory conflicts during parallel updates from different threads and (2) serialization during the accumulation of partial energies into the energy array. To enable parallel updates and accumulations on different GPU multiprocessors, this first scheme maps one atom and its associated neighbor-list onto a single multiprocessor at a time. In other words, each multiprocessor computes the partial energies for exactly one atom due to all of its neighbors, plus the partial energies of all its neighbors due to that atom. For ease of understanding, we call the current mapped atom the *first* atom and atoms in its neighbor-list the *second* atoms. On each multiprocessor, two different energy arrays are created in shared memory. The first array stores the partial energies of the currently mapped *first* atom, with one entry for its partial energy due to each *second* atom in its neighbor-list. The second array stores the partial energies of the *second* atoms, with one entry for each atom in the system (see Fig. 18.3).

Each thread in a thread block computes two partial energy values: the partial energy of the current *first* atom (assigned to the thread block) due to one of the *second* atoms in its neighbor-list and the partial energy of the *second* atom due to the *first*. As the energies are computed by different threads, they are updated in these shared memory arrays. Note that since a *second* atom will appear in the neighbor list of a particular atom only once, no two threads will update the same shared memory location at the same time. This enables parallel, conflict-free updates.

Once the entire neighbor-list of the current *first* atom is processed, a barrier is reached. A master thread (thread 0 from each block) then computes the total energy of the *first* atom by adding all the partial values in the *first* atom energy array. The energies in the *second* atom array, however, are for different *second* atoms and are only partial. As shown in Fig. 18.3, analogous partial arrays are present on the shared memories of all the other multiprocessors. These must be combined to obtain the total energy of each of the *second* atom. This is done by copying the *second* atom energy arrays from the shared memories of the different multiprocessors to the global memory. The corresponding values from these arrays are then summed to obtain a single array with the total energies. This can be done in parallel by employing multiple threads.

Though this method allows parallel execution and updates, it has three drawbacks. First, since only one *first* atom is processed by a multiprocessor, the GPU threads are heavily underutilized and the distribution of work on different multiprocessors is uneven. This is because the distribution of the atoms in the neighbor-lists is non-uniform with different *first* atoms having widely varying number of atoms in their neighbor-lists, ranging from a few to a few hundred. Second, transferring multiple large *second* atom energy arrays from the shared to the global memory incurs high data transfer cost per iteration. Finally, accumulation

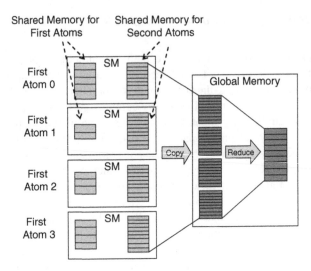

Fig. 18.3 Mapping the neighbor-lists onto GPU threads. Replicating the energy array enables parallel updates

in global memory is slow due to the inherently slow global memory access. Overall this method results in poor performance and is not preferred. It also underlines the unsuitability of the original neighbor-lists data structure for mapping to the GPUs.

18.3.2 Improved Data Structures for Efficient Mapping

Due to the small amount of computation per iteration—each requires only a few milliseconds on a serial computer—obtaining performance improvements from a GPU implementation requires efficient distribution of work to maximize parallelism and reduce communication overhead. Two methods are now presented; both use a modified data structure.

18.3.2.1 Pairs List

Neighbor lists are pointers to lists of atoms. To better map this structure to GPUs, it can be *flattened* into a table called the pairs list [6, 7, 8] (Fig. 18.4a).

A pairs list, as the name suggests, is a list of atom pairs that need to be processed for energy computations. For each pair, the list contains the indices of the two atoms, along with fields for storing the partial energies of the two atoms. For each atom in the system, every atom in its neighbor list forms a pair with that atom and is entered into the pairs list. Different atom pairs in this list are independent of each

a

Pair #	Atom 1	Atom 2
0	0	2
1	0	1
2	0	11
3	0	14
4	1	2
5	1	5
6	2	4
7	2	15
8	2	12
9	3	4

b

	Atom Index		Atom Energy	
Pair #	Atom 1	Atom 2	Atom 1	Atom 2
0	0	2		
1	0	1		
2	0	11		
3	0	14		
4	1	2		
5	1	5		
6	2	4		
7	2	15		
8	2	12		
9	3	4		

- - → Energy update for the first atom

——→ Energy update for the neighboring atom

Fig. 18.4 (a) Pairs list data structure (b) Sequential updates in the energy array

other and can be processed in parallel. Note that flattening the neighbor lists into a pairs list increases the storage requirements. This can potentially be a problem if the neighbor lists are dense, e.g., with thousands of particles as is common for MD computations. For localized N-body problems such as energy minimization, however, the neighbor lists are usually small and flattening does not increase the storage requirement significantly.

To compute the atom energies, the pairs list is stored in global memory. Atom pairs are distributed equally among threads. Each thread processes the pairs assigned to it and stores the partial energies of the two atoms at the corresponding index in the pairs list. Once all the pairs have been processed, these partial energies are accumulated to compute the total energy of each individual atom. Note that the accumulation step must be done serially due to the unordered occurrences of the second atoms in the pairs list.

There are two alternatives for performing the serial accumulation: on the GPU using a single thread or on the host. On the GPU, since the energy values are stored in the global memory, accumulation requires multiple slow accesses. Since the accumulation is done serially using a single thread, depending on the size of the lists, the accumulation on the host may outperform that on the GPU. Accumulation on the host, however, incurs the extra overhead of transferring the two arrays of atom energies from the GPU to the host memory in each iteration. In the case for energy minimization example, accumulation on the host outperform that on the GPU, thus the transfer overhead is lower than the gain from performing the accumulation using the faster CPU core.

The limitation of this method is the serial accumulations. Though it enables uniform distribution of work, parallel energy computations, and conflict-free parallel updates, serialization during the accumulation of the partial energies limits the overall performance. With the accumulations performed on the host, this method results in an overall speedup of around 3× over the original single-threaded CPU code running on a contemporaneous 3 GHz quad-core Intel Xeon Harpertown processor.

18.3.2.2 Split Pairs Lists and Static Mapping

To enable faster and parallel accumulations from the GPU shared memory, the third approach further modifies the data structure used in the previous section. This approach still utilizes the pairs list structure, but with two changes in how the pairs get mapped to the GPU threads.

The first change is to split the pairs list into two separate pairs lists. Recall that the serialization during the accumulation in the previous section is caused mainly due to the arbitrary occurrences of *second* atoms (atom 2) in the neighbor lists (now the pairs list). The *first* atoms (atom 1) still appear in an ordered fashion. Thus, splitting the list into two separate lists and processing each one separately will add determinism to how the atoms appear in the list (see Fig. 18.5).

	Atom Index		Energy
Pair #	Atom 1	Atom 2	Atom 1
0	0	2	
1	0	1	
2	0	11	
3	0	14	
4	1	2	
5	1	5	
6	2	4	
7	2	15	
8	2	12	
9	3	4	

	Atom Index		Energy
Pair #	Atom 1	Atom 2	Atom 1
0	1	0	
1	2	0	
2	2	1	
3	4	2	
4	4	3	
5	5	1	
6	11	0	
7	12	2	
8	14	0	
9	15	2	

Fig. 18.5 Split pairs lists: (*left*) forward list, (*right*) reverse list

The first pairs list is based on the original neighbor lists and is called the forward list. The second list, called the reverse list, is generated by reversing the original neighbor lists, i.e., by treating each *second* atom of the original neighbor list as a *first* atom for the reverse neighbor list. While processing each of these lists, only the energy of atom 1 in each pair must be computed and updated. This way, the energies of the *first* atoms (in the original list) get updated while processing the forward list and those of the *second* atoms (in the original list) while processing the reverse list. This is shown in Fig. 18.5. Note that there is no column for storing the energies of the second atom in the pair.

The second modification involves statically mapping the pairs from the new pairs lists onto the GPU threads. This comes from the observation that the pairs in the new lists can be grouped by the first atom in each list. This can be done since now only the energies of the first atoms in the pair need to be computed and not of the second atoms. These two changes allow better and more uniform distribution of atom-pairs on the GPU and enable parallel and much faster accumulations in GPU shared memory.

18.3.2.3 Assignment Tables for Static Mapping

Once the forward and reverse pairs lists have been generated, these can be statically distributed to the GPU threads running on different multiprocessors. The central idea is to have all the partial energies that need to be combined (accumulated) be computed on the same multiprocessor so as to perform the accumulations using the shared memory. The static mapping scheme does this by grouping together all pairs in a list having the same first atom and mapping the entire group onto the threads in the same thread block. Having all the pairs of a group mapped to the same thread block allows accumulations using shared memory since all the partial energies are present within the same multiprocessor. Moreover, multiple parallel accumulations can be performed, both within each multiprocessor, one for each of the group mapped to the thread block, as well as on different multiprocessors.

To determine the static assignment of work among the GPU threads, another data structure called the assignment table is generated (Fig. 18.6). Table contains one (or multiple) row(s) per thread id. Each row contains five fields: pair id, the two atom indices, a *master* field indicating if this thread is the first thread for this pairs group, and the number of pairs in the group. The master field and the number of pairs in the group are used when the energies of the atoms are accumulated in the shared memory.

Thread Id	Pair Id	Atom 1	Atom 2	Master	Pairs	
0	0	0	2	1	4	Group 0
1	1	0	1	0	4	Group 0
2	2	0	11	0	4	Group 0
3	3	0	14	0	4	Group 0
4	9	3	4	1	1	Group 3
5	4	1	2	1	2	Group 1
6	5	1	5	0	2	Group 1
7	6	2	4	1	3	Group 2
8	7	2	15	0	3	Group 2
9	8	2	12	0	3	Group 2

Thread Block 0 spans Thread Ids 0–4; Thread Block 1 spans Thread Ids 5–9.

Fig. 18.6 Assignment table for the GPU; atom-groups are mapped to the GPU thread-blocks in their entirety

The assignment table indicates which thread must work on exactly which atom pair and which threads are responsible for the accumulations. Two assignment tables need to be generated, one for the forward pairs list and the other for the reverse.

The generation of the assignment tables is a preprocessing step that is performed on the host computer. Once generated, these assignment tables can be transferred to the GPU and stored in the GPU global memory. Note that this is required only once, at the beginning of the minimization process. There is no further data transfer per iteration, unless the neighbor lists are updated, in which case the assignment tables must be regenerated and transferred again to the GPU. As stated earlier, unlike MD, the neighbor lists in energy minimization typically get updated only a few times per 1,000 minimization iterations; thus the preprocessing and transfer time is negligible.

As shown in Fig. 18.6, the groups of pairs are mapped to the thread blocks in their entirety. More than one group can be mapped onto a particular thread block, provided there are enough threads to accommodate all the pairs of the group. If the current thread block does not have enough threads left to accommodate the entire next group, the group is mapped onto the next available thread block (e.g., group 1). Unused spaces in different thread blocks are claimed by other smaller pair-groups (e.g., group 3). For the cases where the number of groups is larger than what can be mapped onto the available thread blocks, the computation is performed in multiple passes, with the threads being responsible for more than one group of pairs. This is shown in Fig. 18.7. This leads to more than one row per thread in the assignment table.

	Thread Id	Pair Id	Atom 1	Atom 2	Master	Pairs	
Thread Block 0	0	0	0	2	1	4	Group 0
	1	1	0	1	0	4	
	2	2	0	11	0	4	
	3	3	0	14	0	4	
	4	9	3	4	1	1	Group 3
Thread Block 1	0	4	1	2	1	2	Group 1
	1	5	1	5	0	2	
	2	6	2	4	1	3	Group 2
	3	7	2	15	0	3	
	4	8	2	12	0	3	
Thread Block 0	0	10	4	12	1	3	Group 4
	1	11	4	7	0	3	
	2	12	4	5	0	3	
	3	-1	-1	-1	-1	-1	Unused
	4	-1	-1	-1	-1	-1	
Thread Block 1	0	13	5	13	1	4	Group 5
	1	14	5	8	0	4	
	2	15	5	9	0	4	
	3	16	5	14	0	4	
	4	-1	-1	-1	-1	-1	Unused

Work assignment for the first pass (Thread Block 0, Thread Block 1); Work assignment for the second pass (Thread Block 0, Thread Block 1)

Fig. 18.7 Multi-pass assignment table; threads process pairs from different groups in different passes. Some of the threads may remain idle during some of the passes (shown with −1 in the corresponding rows)

18.3.2.4   Computation Using the Assignment Table

The GPU threads work in parallel on the rows of the assignment table with each thread computing the energy of only the first atom. Energies computed by the different threads are stored in an array in the GPU shared memory. The length of this array is equal to the number of threads in the thread block, with each thread storing the computed energy at the index equal to its local thread id (within the block). Note that the threads write to deterministic, sequential locations in the memory irrespective of the atoms being processed by them.

Once all the threads finish processing their assigned pairs, the master threads execute the accumulation round (see Fig. 18.8). Each master thread reads the number of atoms for the group associated with it and accumulates that many values from the shared memory, starting at its local thread id. This way many threads perform accumulations both in parallel and from shared memory, resulting in significantly better performance compared with the previous schemes. The master threads then store the accumulated values in GPU global memory. If multiple passes are required, the above process is repeated for each pass. Note that this method of computation is enabled by splitting the pairs list into forward and reverse lists, with each requiring the energies of just the first atoms to be computed and updated. For

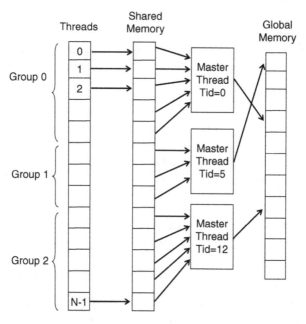

Fig. 18.8 Threads compute partial energies in parallel and store in the shared memory. Master threads then perform the accumulations from the shared memory and store the accumulated value in the global memory

the second atoms, this process is repeated with the assignment table corresponding to the reverse pairs list.

It is worth noting that processing the forward and reverse lists separately may lead to repeating some of the computations in pair-wise evaluations. This can be avoided by storing those values in the GPU global memory during the kernel for the forward list and reusing them in the kernel for the reverse list. Depending on the complexity of the underlying computations being reused this may result in a slowdown due to the slower global memory access. For the current example of energy minimization, performing redundant computations on the GPU is preferred as opposed to accessing the slow global memory for reuse.

18.3.2.5 Support for Large Neighbor Lists: Modified Assignment Table

The main purpose of the static assignment scheme is to ensure that all the pairs from the pairs list that have the same first atom (i.e., the entire group) are mapped to the same GPU thread block. This allows for efficient accumulations, but also puts a limit on the size of the neighbor lists. Since each *second* atom in the neighbor list of a particular atom creates one pair in the group and since the above method requires one thread for processing each pair in the group, the maximum number of atoms in the neighbor list of any atom is equal to the number of GPU threads per block. The actual number of threads per block is limited by the resources used by each thread. Though the neighbor lists used in energy minimization are usually sparse, nevertheless there can be cases where the neighbor lists are larger than the number of threads per block. This could be especially true for complex computations where the kernel resource requirements (registers and shared memory) are large, limiting the number of threads per block.

To overcome the limitation on the size of the neighbor lists, a modified assignment table with support for group partitioning is presented. Here each group having more pairs than the number of threads per block is divided into multiple partitions, each with its own *pseudo* first atom. These partitions can now be mapped to separate multiprocessors as separate groups. The groups with pairs fewer than or equal to the number of threads per block remain unchanged and can be mapped as before. The computation of the partial energies and the accumulation within each group or partitioned group is also done as before. A second round of accumulation is required to gather the partially accumulated energies from the partitions and finish computing the total energy.

The modified assignment table to support group partitioning is shown in Fig. 18.9. It includes three new fields: *Global Id*, *Number of partitions* and *Partition Id*. *Global Id* indicates the position in the global memory to store the accumulated energy for this group (or partition). For the atoms which have not been partitioned, the *Global Id* field takes the same values as *Atom 1* field in the original table. When a particular *first* atom is divided into multiple partitions, then for the first (primary) partition, *Global Id* = id of *Atom 1* and for the subsequent partitions, *Global Id* = next sequential unassigned atom id. Since the atom ids 0 to $n_{atoms} - 1$

	Thread Id	Pair Id	Atom 1	Atom 2	Master	Size	Global Id	# of Partitions	Partition Id	
Thread Block 0	0	0	0	1	1	13	0	3	31	Partition 0
	1	1	0	5	-1	13	0	3	-1	
	2	2	0	12	-1	13	0	3	-1	
	3	3	0	15	-1	13	0	3	-1	
	4		0	18	-1	13	0	3	-1	
Thread Block 1	5		0		0	13	31	3	-1	Partition 1
	6	:	0		-1	13	31	3	-1	
	7		0		-1	13	31	3	-1	
	8		0		-1	13	31	3	-1	
	9		0		1	13	31	3	-1	
Thread Block 2	10		0		0	13	32	3	-1	Partition 2
	11		0		-1	13	32	3	-1	
	12		0		-1	13	32	3	-1	
	13	1		0		2	1	-1	-1	
	14	1		-1		2	1	-1	-1	

(Partition 0 / Partition 1 / Partition 2 span Group 0; rows 13–14 form Group 1.)

Fig. 18.9 Assignment table to support group-partitioning; Group 0 has 13 pairs and there are only five threads per block—it is thus divided into three partitions. With the number of atoms in this example = 30, the Global Ids start from 31. The second-level master (with Master = 1) has the corresponding Partition Id = 31. It reads three partially accumulated energies from the global memory array, starting from position 31 and adds these values to its own accumulated partial energy. Other master threads (Master = 0) do not perform second level accumulations. (Table shown is just an example; actual assignment tables may contain tens of thousands of rows)

are used for the primary partitions or un-partitioned atoms, the unassigned atom id starts from n_{atoms} and is incremented to $n_{atoms} + 1$, $n_{atoms} + 2$ and so on as they are assigned to different partitions (n_{atom} is the total number of atoms in the system). For all the partitions the value in the *Atom 1* column remains the same as before partitioning since that is the actual atom id of the first atom whose energy is being computed.

The other two new columns are used by the master thread for the second level accumulation. These are used for gathering the partially accumulated energies of the different partitions from the global memory array and storing their sum back in the global memory at the appropriate first atom id. The thread responsible for the accumulation of the partial energies in the first partition is also responsible for this second level accumulation. *Number of partitions* indicates how many partitions the current first atom group has been split into. This depends on the size of its neighbor list and the number of threads per block. This is how many entries of the global memory array that a second-level master thread must add together. For the groups that have not been partitioned, this field takes a value of −1. The *Partition Id* field stores the *Global Id* for the second partition for the current group. This is used to determine where in the global memory array should the second-level accumulation start from. A second-level master thread accesses *Num Partitions−1* partial energies from the global memory array, starting from the index equal to the *Partition Id*, adds them to its own accumulated energy and stores the result back in the global memory at an index equal to the id of its *Atom 1* (or the *Global Id* since for the second-level masters, these two are the same). The *Partition Id* field is used only by the second-level master threads. For all the other rows, this field takes a value of −1.

In addition to the new fields in the assignment table, the previous entries also take slightly different values. The *Size* column now indicates the number of second atoms in the partitioned group, not the total. Total is not stored anywhere and is not needed. In the new assignment table, the *Master* field can take one of three values: -1 (not a master) 0 (master) and 1 (second-level master). If a thread is a master (or a second-level master), it is responsible for accumulating the partial energies from the shared memory and storing in the global memory at an index equal to the *Global Id*. The threads which are marked as second-level masters additionally perform the second-level accumulation, as described above. This second level accumulation is done in a separate kernel that is called after the main energy computation kernel.

The modified assignment table described above provides the flexibility to support large neighbor lists with no requirement on the minimum number of threads per block. Another benefit of this scheme is better utilization of the available multiprocessors. In the original assignment table, since the groups are required to be mapped on the multiprocessors in their entirety, each thread-block is required to have a large number of threads but only a few thread blocks are used. As a result, some of the multiprocessors remain idle. Partitioning the groups enables the distribution of work across multiple multiprocessors. With this, different combinations of thread block and grid sizes can be applied, allowing for improved processor utilizations and hence improved performance.

18.4 Performance Measurements

This section presents the performance improvements obtained from mapping the energy minimization computations on a GPU using the split-pairs-list method with assignment tables. The GPU-accelerated code runs on an NVIDIA Tesla C1060 GPU, containing 240 processor cores and running at 1.3 GHz. The GPU is housed in a Dell Precision workstation with a 3GHz quad-core Intel Xeon Harpertown processor running Windows XP. The GPU code was written using NVIDIA CUDA and compiled using Microsoft Visual Studio 8 with standard optimizations and the NVIDIA nvcc compiler. The original unaccelerated energy minimization code runs on a single core of the 3 GHz quad-core Intel Xeon Harpertown processor (2008 era 45 nm contemporaneous processor). The speedup reported is over a single core; see the next section for a discussion about extrapolation to a full CPU.

The runtimes shown in Table 18.1 are per iteration times, averaged across minimizing five different protein complexes, with 1,000 iterations per complex. Table 18.2 shows the overall end-to-end speedup for minimizing these five complexes. Each complex contains around 2,260 atoms and requires evaluations of 9,780 atom–atom pairs per iteration. The energy expressions evaluated during the minimization iterations are shown in Appendix A. The computation is divided into three GPU kernels: (1) computing the self energies, (2) computing the pairwise interaction energies and the van der Waals energies, and (3) updating the forces acting on the atoms. All of these kernels use the same data structures and

Table 18.1 Speedups for different energy evaluation and force update steps of energy minimization

Computation	Serial runtime (ms)	GPU runtime (ms)	Speedup
Self energies	6.15	0.23	26.7×
Pair-wise energies	2.75	0.19	17×
van der Waals energies	0.5		
Force updates	0.95	0.14	6.7×

Table 18.2 Overall speedup for energy minimization for different complexes

Complex	Serial runtime (s)	GPU runtime (s)	Speedup
Complex 1	11.9	1.098	10.8×
Complex 2	11.87	1.078	11×
Complex 3	11.8	1.078	10.9×
Complex 4	10.74	0.906	11.8×
Complex 5	11.87	1.094	10.8×

the performance improvement is proportional to the computational complexity of the expressions being evaluated—kernels with more complex evaluations achieve higher speedups. In the case of the split pairs list, the GPU runtime reported is the total time for kernel execution for both forward and reverse list.

Compared with the method presented in Sect. 18.3.1 that results in no performance improvement and the one presented in Sect. 18.3.2.1 which achieves a modest 3× improvement, the split pairs lists method with assignment table achieves up to 26× per-iteration improvement and 11× overall speedup. Clearly, efficient mapping of these computations on the GPU, so as to achieve better processor utilizations and minimize data transfer overheads, requires significant restructuring of the original data structures used in the serial CPU code. Moreover, the assignment tables for static mapping of the particle-pairs on the GPU threads enable uniform distribution and improved parallelism during energy accumulation.

18.5 Discussion

In this chapter we have described methods for creating efficient GPU implementations of an important variation of N-body simulation that is used, e.g., in modeling interactions between molecules. Several characteristics distinguish these localized simulations: one of the molecules is often much smaller than the other, the larger molecule often moves little if at all, processing begins with the smaller molecule close to its final orientation and in the binding pocket of the larger molecule, and convergence is usually achieved after at most a few thousand iterations. These characteristics lead to substantial changes in data structures and thus the algorithm. In particular, neighbor lists are highly non-uniform but vary little once computed. We optimize by first unrolling the list into an array of particle pairs and then creating a second array consisting of the first array reversed and sorted. Doing so adds uniformity in the occurrence of particle-pairs, thus allowing for clustering of these pairs into different independent workgroups and efficient distribution of these workgroups to GPU thread blocks. This results in higher concurrency during evaluation and accumulation of the partial energy values and thus improved performance.

As with most performance studies, the results are for processors that are necessarily limited in variety and have already been superseded by the time the published work appears. The overall trend of the results, however, is likely to remain. Addressing GPU variety: while we tuned the code specifically to the NVIDIA Tesla C1060, the code fits into the OpenCL framework well supported by most GPU families. As to more recent GPU models, we have found that the code ports easily to more recent GPUs including the Kepler class from NVIDIA.

As noted in the results section, the speed-up factors achieved for the localized N-body algorithm range from roughly 7× to 27× with respect to a single core of a contemporaneous four core CPU. Assuming that perfect parallelism can be achieved on the CPU, these results reduce by a factor of 4 to 1.75× to 6.75×. When application specific overhead is included, the end-to-end speed-up is roughly 11×, reduced to 2.75× when assuming perfect scalability with the CPU. The speed-up of GPU over CPU, however, could well remain closer the original figure (11× rather than 2.75×). This is because coarse-grained parallelization of energy minimization across different CPU cores is non-trivial and may not yield any significant performance improvement. Considering next generation CPUs and GPUs, since both families of devices continue to ride Moore's Law, a reasonable assumption is that without substantial change in architecture, these proportions will remain roughly in line.

Appendix A: Energy Expressions

The total energy of a system of atoms is given as a sum of various bonded and non-bonded energies for all the atoms:

$$E^{total} = \underbrace{E^{vdw} + E^{elec}}_{non-bonded} + \underbrace{E^{bond} + E^{angle} + E^{torsion} + E^{improper}}_{bonded} \tag{18.1}$$

Energy minimization involves repeated evaluation of this expression, once during each minimization iteration. In the current discussion, only the non-bonded terms are mapped to the GPU and evaluated using the neighbor lists.

The non-bonded energy of each atom is the sum of the contributions due to the neighboring atoms within a cutoff distance. Non-bonded energy is a sum of the electrostatic and van der Waals energy terms. A variant of the Lennard-Jones 6–12 potential representing the van der Waals energy term is shown in Eq. (18.2)

$$E_{ik}^{vdw} = eps_{ik} \left(\frac{rm_{ik}^6}{r_{ik}^{12}} - \frac{8rm_{ik}^6 \Big/ r_c^6}{r_{ik}^6} + \frac{rm_{ik}^6}{r_{ik}^{12}} \left(1 + \frac{2r_{ik}^6}{r_c^6} \right) \right) \tag{18.2}$$

$$eps_{ik} = eps_i . eps_k$$

$$rm_{ik} = (rm_i + rm_k)^2$$

where eps_i and rm_i represent the van der Waals parameters of atom 'i', r_{ik} is the distance between the atoms 'i' and 'k' and r_c is the cut-off distance.

The electrostatic energy of a solute can be decomposed into two components; a sum of the self energies E_i^{self} of all the charges and a sum of pair-wise interaction energies E_{ij}^{int} [9] [Eq. (18.3)]

$$E^{elec} = \sum_i E_i^{self} + \sum_{i<j} E_{ij}^{int} \tag{18.3}$$

Using the Analytic Continuum Electrostatics (ACE) model [9], the self-energy of an atom can be represented as a sum of its Born self-energy in the solvent and the sum of effective pair-wise interactions, E_{ik}^{self}, due to all the other solute atoms [see Eqs. (18.4) and (18.5)].

$$E_i^{self} = \frac{q_i^2}{2\varepsilon_s R_i} + \sum_{k \neq i} E_{ik}^{self} \tag{18.4}$$

$$E_{ik}^{self} = \tau q_i^2 \left(\frac{e^{-\left(r_{ik}^2 / \sigma_{ik}^2\right)}}{\omega_{ik}} + \frac{\tilde{V}_k}{8\pi} \left(\frac{r_{ik}^3}{r_{ik}^4 + \mu_{ik}^4} \right)^4 \right) \tag{18.5}$$

where q_i represents the charge on atom 'i', r_{ik} is the distance between the two atoms, $\tilde{V}_k$ is the size of the solute volume associated with atom 'k', ω_{ik} and σ_{ik} determine the height and width of the Gaussian that approximates E_{ik}^{self}, and μ_{ik} is an atom-atom parameter.

The pair-wise interaction term of Eq. (18.3) is given by the generalized Born (GB) equation shown in Eq. (18.6), which is the sum of Coulomb's law in a dielectric and the Born equation [10]

$$E_{ij}^{int} = 332 \sum_{j \neq i} \frac{q_i q_j}{r_{ij}} - 166\tau \sum_{j \neq i} \frac{q_i q_j}{\sqrt{r_{ij}^2 + \alpha_i \alpha_j e^{-\left(r_{ij}^2 / 4\alpha_i \alpha_j\right)}}} \tag{18.6}$$

where α_i and α_j represent the Born radii for atoms 'i' and 'j', respectively. These in turn depend on the self energies of the atoms. The self energy of each individual atom thus needs to be computed before computing the pair-wise interactions.

Appendix B: Source Code

<u>Host CPU code for generating the assignment table from neighbors lists</u>
<u>(with atom splitting, if needed)</u>

```
/* Split (partition) the first atoms if needed */
int listPos = 0;     /* Position in the atom pairs list */
int next_global_id = numAtoms;
for(i = 0; i < numAtoms; i++) {
        numSecondAtoms = nbList->numSecondAtoms[i];
        if(numSecondAtoms > 0) {    /* If the neighbor list of
                                       this atom is not empty */
                num_per_split = numSecondAtoms;
                num_splits = 1;

                /* See if there is a need to split */
                if (numSecondAtoms > NUM_THREADS_PER_BLOCK) {
                        num_splits = (((numSecondAtoms-1) /
                        NUM_THREADS_PER_BLOCK) + 1);

                        num_per_split = ((numSecondAtoms-1) /
                        num_splits) + 1;        /* To divide evenly
                        among different splits */
                }

                /* Splitting into multiple first atoms (parti-
                tions) */
                for (j = 0; j < num_splits; j++) {
                        nbList->numSecondAtoms[listPos] =
                        min(num_per_split, (nbList-> numSecond-
                        Atoms[i] - j*num_per_split));
                        /*num_per_split or the whatever is re-
                        maining */

                        nbList->firstAtomId[listPos] = i;

                        if (j == 0) {/* If first partition */
                                nbList->global_id[listPos] = i;

                                if (num_splits > 1) {
                                        nbList->split_count[listPos]
                                        = num_splits;
                                        nbList->partitionId[listPos]
                                        = next_global_id;
                                        nbList->real_master[listPos]
                                        = 1;
                                }
                        }
                        else {
```

```
                                    nbList->global_id[listPos] =
                                    next_global_id;
                                    next_global_id++;
                            }
                            listPos++;
                    }
            }
}


/* --- subroutine to create the assignment table --- */

/* Determine the required size of the assignment table */
numPasses = (nbList->numPairs / (NUM_THREADS_PER_BLOCK *
NUM_BLOCKS)) + 1;
if(numPasses == 0) {
    numPasses = 1;
}
tableSize = (NUM_THREADS_PER_BLOCK * NUM_BLOCKS) * numPasses;

/* -- Malloc the arrays for the assignment table -- */
firstAtomIndex_assign  = (int *) myMalloc (tableSize *
sizeof(int));
........
........
/* --------------- */

/* Clear the assignment table arrays */
for(i = 0; i < tableSize; i++) {
    firstAtomIndex_assign[i] = -1;
    secondAtomCount_assign[i] = -1;
    ij_assign[i] = -1;
    secondAtomIndex_assign[i] = -1;
    numSecondAtoms_assign[i]  = -1;
    global_id_assign[i] = -1;
    split_count[i] = -1;
    split_first_global_id[i] = -1;
}

/* Iterate through the atom partitions and map them
on thread blocks*/
for(i = 0; i < nbList->numFirstAtoms; i++) {

    /* Look for a thread block with enough unallocated
        threads to cover the entire current partition */
numSecondAtoms_temp = nbList->numSecondAtoms[i];
/* See if the number of remaining threads in this block is >=
num. of second atoms */
rep_id = 0;    /* This indicates how many rows of the assign-
ment table are computed by each thread */
block_index = 0;
```

```
while(threadStartId[rep_id][block_index] + numSecond-
Atoms_temp > NUM_THREADS_PER_BLOCK) {
       block_index++;

       /* If all the blocks in the rep have been checked,
       move to next rep */
       if(block_index == NUM_BLOCKS) {
              block_index = 0;
              rep_id++;
       }
}


/* Found the block with enough threads - now assigning values
to the arrays */
/* Get the array of second atoms for the current partition */
secondAtomIds = nbList->secondAtomIds[i]
for(j = 0; j < numSecondAtoms_temp; j++) {

       firstAtomIndex_assign [threadStartId [rep_id]
       [block_index] + j + block_index*NUM_THREADS_PER_BLOCK
       + NUM_THREADS_PER_BLOCK * NUM_BLOCKS * rep_id]
              = nbList->firstAtomId[i];

       global_id_assign[threadStartId[rep_id][block_index] +
       j + block_index*NUM_THREADS_PER_BLOCK +
       NUM_THREADS_PER_BLOCK * NUM_BLOCKS * rep_id]
              = nbList->global_id[i];

       split_count[threadStartId[rep_id][block_index] + j +
       block_index*NUM_THREADS_PER_BLOCK +
       NUM_THREADS_PER_BLOCK * NUM_BLOCKS * rep_id]
              = nbList->split_count[i];

       split_first_global_id[threadStartId[rep_id][block_inde
       x] + j + block_index*NUM_THREADS_PER_BLOCK +
       NUM_THREADS_PER_BLOCK * NUM_BLOCKS * rep_id]
              = nbList->partitionId[i];

       if (j == 0) {
              if (nbList->real_master[i] > 0) {
                     secondAtomCount_assign [threadStartId
                     [rep_id][block_index] + j +
                     block_index*NUM_THREADS_PER_BLOCK +
                     NUM_THREADS_PER_BLOCK * NUM_BLOCKS *
                     rep_id] = 1;
              }
              else {
               secondAtomCount_assign [threadStartId
               [rep_id][block_index] + j +
               block_index*NUM_THREADS_PER_BLOCK +
               NUM_THREADS_PER_BLOCK * NUM_BLOCKS * rep_id]
```

```
                              = 0;
                 }
            }

        numSecondAtoms_assign [threadStartId
        [rep_id][block_index] + j +
        block_index*NUM_THREADS_PER_BLOCK +
        NUM_THREADS_PER_BLOCK * NUM_BLOCKS * rep_id]
                = nbList->numSecondAtoms[i];

        secondAtomIndex_assign [threadStartId
        [rep_id][block_index] + j +
        block_index*NUM_THREADS_PER_BLOCK +
        NUM_THREADS_PER_BLOCK * NUM_BLOCKS * rep_id]
                = secondAtomIds[j];
    }

    threadStartId[rep_id][block_index] += numSecondAtoms_temp;

        running_ij = 0;
        for(i = 0; i < tableSize; i++) {
            if(firstAtomIndex_assign[i] >= 0) {
                ij_assign[i] = running_ij;
                running_ij++;
            }
        }
}
```

GPU kernel for computing the energies using the assignment table (without atom splitting)

```
----------------------------------------------------------------------------------
/* assign_table_size represents the size of the assignment
table. It represents the number of entries of the assignment
table that need to be processed. This can be larger than the
actual number of atom-pairs since there are gaps in the as-
signment table (places where there were not enough threads to
process all the second atoms of a first atom.
*/
__global__ void compute_energy_assign (int assign_table_size,
float* atom_X, float* atom_Y, float* atom_Z, float* at-
om_ftypen, int* firstAtomIndex_assign, int* master_assign,
int* secondAtomIndex_assign, int* numSecondAtoms_assign, int*
ij_assign, float* energy_arr) {

const int tid = IMUL(blockDim.x, blockIdx.x) + threadIdx.x;
const int threadN = IMUL(blockDim.x, gridDim.x);

/* Shared memory to store the values computed by different
threads. These are then accumulated by a master thread and
stored on the global memory
*/
__shared__ float energy_sh_first[NUM_THREADS_PER_BLOCK];

int i, j, i1, i2, my_id, master, my_ij, ij;
int atom1_param, atom2_param;    /* some atom parameters */
float dx, dy, dz;
int n2; /* Number of second atoms for this first atom */
float energy_total;

int threadId;
threadId = threadIdx.x;
int shared_memory_index = threadId;

for(i = tid; i < assign_table_size; i += threadN) {

        /* Reading the assignment table entries for the cur-
        rent thread */
        my_id = firstAtomIndex_assign[i];        /* Id of the
        first atom assigned to this thread */
        master = master_assign[i];
        energy_sh_first[shared_memory_index] = 0;

        if(my_id >= 0) {
        /* Whether the current thread is assigned an atom-pair
        in the assignment table */
                n2 = numSecondAtoms_assign[my_id];
                my_ij = ij_assign[i];         /* This is the index
                into the array of atom pairs */
```

```
        i1 = my_id;                /* Index of the first atom
        in the array of atoms - read from the array of
        pairs */

        i2 = secondAtomIndex_assign[i];    /* Index of
        the second atom in the array of atoms - read
        from the array of pairs */

        /* Read the atom co-ordinates and other atom-
        specific paramters based on the atom Ids (used
        for energy calculation) */
        dx = atom_X[i1] - atom_X[i2];
        dy = atom_Y[i1] - atom_Y[i2];
        dz = atom_Z[i1] - atom_Z[i2];

        atom1_param = (int)atom_ftypen[i1];
        atom2_param = (int)atom_ftypen[i2];

        /* PERFORM ENERGY COMPUTATIONS HERE */
        ...
        energy_val = ....
        ...
        //

        /* Store the computed energy in shared memory */
        energy_sh_first[shared_memory_index]
                = energy_val;
    }

    __syncthreads();

    /* First thread from the set of threads for this
    first-atom is responsible for accumulation */
    if(my_id >= 0 && master == 0) {
        energy_total = 0;

        /* Accumulate the partial energy values */
        for(j = threadId; j < threadId+n2; j++) {
            energy_total += energy_sh_first[j];
        }

        /* Update in global memory */
        energy_arr[my_id] += energy_total;
    }
    __syncthreads();
}
}
```

GPU kernels for computing the energies using the assignment table
(with atom splitting)

```
------------------------------------------------------------------------------
__global__ void compute_energy_assign_split (int as-
sign_table_size, float* atom_X, float* atom_Y, float* atom_Z,
float* atom_ftypen, int* firstAtomIndex_assign, int* mas-
ter_assign, int* secondAtomIndex_assign, int* numSecond-
Atoms_assign, int* ij_assign, int *global_id_assign, float*
energy_arr) {

const int tid = IMUL(blockDim.x, blockIdx.x) + threadIdx.x;
const int threadN = IMUL(blockDim.x, gridDim.x);

__shared__ float energy_sh_first[NUM_THREADS_PER_BLOCK];

int i, j, i1, i2, ij;
int atom1_param, atom2_param;
int my_id, master, my_ij;

int my_global_i1;        /* Used to support splitting This
gives the location in the global memory where the current
master thread must update the accumulated energy. This will
be the same as my_id if no splitting is done or for the first
master after splitting */

float dx, dy, dz;

int n2;
float energy_total;

int threadId;
threadId = threadIdx.x;

int shared_memory_index = threadId;

for(i = tid; i < assign_table_size; i += threadN) {
        my_id = firstAtomIndex_assign[i];
        master = master_assign[i];
        energy_sh_first[shared_memory_index] = 0;

        if(my_id >= 0) {
                n2 = numSecondAtoms_assign[i];

        /* This is the change to support splitting */

                my_ij = ij_assign[i];
                i1 = my_id;
                i2 = secondAtomIndex_assign[i];

                dx = atom_X[i1] - atom_X[i2];
```

```
        dy = atom_Y[i1] - atom_Y[i2];
        dz = atom_Z[i1] - atom_Z[i2];

        atom1_param = (int)atom_ftypen[i1];
        atom2_param = (int)atom_ftypen[i2];

        /* PERFORM ENERGY COMPUTATIONS HERE */
        ...
        energy_val = ....
        ...
        //

            energy_sh_first[shared_memory_index] = temp;
    }
    __syncthreads();

    /* More than one thread will be executing this block -
    - each updates different part of the array in global
    memory */
    /* First thread from the set of threads for this
    first-atom is responsible for accumulation */
    if(my_id >= 0 && master >= 0) {
            energy_total = 0;
            my_global_i1 = global_id_assign[i];
            /* This is used to support splitting */

            /* First level of accumulations */
            for(j = threadId; j < threadId+n2; j++) {
                    energy_total += energy_sh_first[j];
            }

            energy_arr[my_global_i1] -= energy_total;
    }
    __syncthreads();
}
}


/* Kernel for performing the second level accumulation (from
global memory to global memory) */

__global__ void accum_energy_assign_split (int as-
sign_table_size, int* firstAtomIndex_assign, int* mas-
ter_assign, int *split_count, int *split_first_global_id,
float* energy_arr) {

const int tid = IMUL(blockDim.x, blockIdx.x) + threadIdx.x;
const int threadN = IMUL(blockDim.x, gridDim.x);

int i, j, i1, i2;
```

```
int my_id, master;
int my_split_count, my_split_start_id;
float energy_total;

/* This loop goes through the assignment table */
for(i = tid; i < assign_table_size; i += threadN) {
    my_id = firstAtomIndex_assign[i];
    master = master_assign[i];

        /* If the thread is a second-level master, then it ac-
        cumulates from different split first Ids and stores at
        the real first atom id */
        if((my_id >= 0) && (master == 1)) {
                energy_total = 0;
                my_split_count  = split_count[i];
                my_split_start_id = split_first_global_id[i];

                /* Accumulate all the values from split first
                ids (from global memory) */
                for(j = my_split_start_id; j <
                my_split_start_id+my_split_count-1; j++) {
                        energy_total += energy_arr[j];
                }

                /* Update in global memory -- this time at the
                real first atom id */
                energy_arr[my_id] += energy_total;
        }
        __syncthreads();
}
}
```

References

1. Stone, J.E., Hardy, D.J., Ufimtsev, I.S., Schulten, K.: GPU-accelerated molecular modeling coming of age. J. Mol. Graph. Model. **29**(2), 116–125 (2010)
2. OpenMM: GPU-accelerated toolkit for molecular simulations. https://simtk.org/home/openmm/
3. Brooks, B.R., et al.: CHARMM: a program for macromolecular energy, minimization, and dynamics calculations. J. Comput. Chem. **4**, 187–217 (1983)
4. http://cmm.cit.nih.gov/intro_simulation/node22.html
5. Pappu, R.V., Hart, R.K., Ponder, J.W.: Analysis and application of potential energy smoothing and search methods for global optimization. J. Phys. Chem. B **102**, 9725–9742 (1998)
6. Sukhwani, B., Herbordt, M.C.: Fast binding site mapping using GPUs and CUDA. In: Proceedings of the Ninth International Workshop on High Performance Computational Biology (HiCOMB'10) (2010)

7. Sukhwani, B., Herbordt, M.C.: FPGA-based acceleration of CHARMM-potential minimization. In: Proceedings of the Third International Workshop on High-Performance Reconfigurable Computing Technology and Applications (HPRCTA'09) (2009)
8. Sukhwani, B., Herbordt, M.C.: Accelerating energy minimization using graphics processors. In: Proceedings of the 2009 Symposium on Application Accelerators in High Performance Computing (SAAHPC'09) (2009)
9. Schaefer, M., Karplus, M.: A comprehensive analytical treatment of continuum electrostatics. J. Phys. Chem. **100**(5), 1578–1599 (1996)
10. Still, W.C., et al.: Semianalytical treatment of solvation for molecular mechanics and dynamics. J. Am. Chem. Soc. **112**(16), 6127–6129 (1990)

Printed in the United States
By Bookmasters